The Toxic Museum

The Toxic Museum examines the use of pesticides in German museum collections at the end of the 19th and early 20th centuries.

It reconstructs the research of substances against harmful insects in museum collections within the historical context of the formation of nation-states, colonialism, a strengthening chemical industry, the First World War, and the resulting broad-based hygiene movement through the lens of the Ethnologisches Museum (Ethnological Museum) in Berlin. Because of their persistence, the consequences of the use of pesticides in museum collections are now unmistakable and well documented in many places. Numerous objects are highly contaminated and are only accessible under difficult conditions regarding occupational health and safety. This creates obstacles for conservation and scientific processing, as well as for mediation in the context of exhibitions and external loans. The most precarious and difficult situations arise when contaminated museum objects are repatriated to their countries of origin. This monograph examines contemporary challenges in the 21st century museum landscape and contextualises the history of pesticide use at the turn of the 20th century.

The Toxic Museum will be of great interest to students and scholars working in conservation, museology, monument preservation, art and cultural studies, ethnology, history, and economics.

Helene Tello has worked as a freelance senior conservator since 2020. After starting her career in 1980, she founded her own conservation studio in 1983. Later, she worked at the Vonderau Museum in Fulda, Germany, and was in charge of the Indian collections at the Ethnologisches Museum of the Staatliche Museen zu Berlin (Ethnological Museum of the National Museum in Berlin), Germany, from 1998 to mid-2020. There, she came across the topic of pesticides used previously on objects. She researches methods for decontaminating cultural assets that have undergone treatment, as well as safe handling practices for all who must handle them. Her expertise has been shared through numerous journal articles, teaching engagements, and lectures given at home and abroad.

The Toxic Museum

Berlin and Beyond

Helene Tello

LONDON AND NEW YORK

First published 2024
by Routledge
4 Park Square, Milton Park, Abingdon, Oxon OX14 4RN

and by Routledge
605 Third Avenue, New York, NY 10158

Routledge is an imprint of the Taylor & Francis Group, an informa business

British Library Cataloguing-in-Publication Data
A catalogue record for this book is available from the British Library

ISBN: 978-1-032-52634-8 (hbk)
ISBN: 978-1-032-52635-5 (pbk)
ISBN: 978-1-003-40760-7 (ebk)

DOI: 10.4324/9781003407607

Typeset in Times New Roman
by Apex CoVantage, LLC

Contents

Figures

x *Figures*

Tables

Preface

To understand the use of pesticides in museum collections during the late 19th and early 20th centuries, German museum history is placed within the context of current debates about neocolonialism and colonialism. German conservation practices during the period under study and the sociopolitical aspect are compelling to an international readership because Berlin was the center of culture, arts, and industry at the time. I look at the formation of nation-states, the situation around the German colonies at that time, the First World War, industrialization, and the subsequent hygiene movement. Within this context, museums combatted the deterioration of their artifacts caused by harmful insects by using various active ingredients and agents. Focusing on this broad context, museums are like small, self-contained institutions that are heavily dependent on external support. The current sociopolitical significance and geographical location of Ethnologisches Museum (EM) der Staatlichen Museen zu Berlin (National Museums in Berlin) – formerly known as the Königliches Museum für Völkerkunde/Museum für Völkerkunde (Royal Museum/National Museum of Ethnology in Berlin) – make it an ideal case study. Because of the far-reaching sociopolitical influence of the Prussian state, the museum has already played an important role within European museums in its heyday.

The founding director of the museum, Adolf Bastian, was granted an impressively generous budget. At the end of the colonial era of Germany, he was inspired by his nine world trips to send expeditions, researchers, and traders to non-European countries. From these, materials from Indigenous populations were transported to Berlin. This policy for collecting artifacts resulted in severely overcrowded exhibition halls and storage facilities, where harmful insects found ample food sources in the artifacts. These pests also caused significant problems in other areas. Because of the First World War and the immense population growth in industrial conurbations, many people were left exhausted and impoverished. To preserve both the military defenses and the urgently needed manpower, the authorities waged a battle against lice, scabies mites, bedbugs, and fleas. Effective measures to address these problems were implemented when the state's scientific and cultural institutions started to collaborate with industry. Active ingredients and pest control agents were developed, manufactured, distributed, and made available to the museum. Within the museum, scientists from other disciplines or administrative staff were often responsible for deciding which active ingredients and agents for pest control needed to be applied. They issued instructions to Eduard Krause, the conservator, and to the Chemisches Laboratorium (chemical laboratory) of the Königliche Museen/Staatliche Museen (Royal/National Museums) in Berlin. The research conducted on pest control at the museum played a relatively minor role.

The staff at the time also sought support for pest control issues outside the museum through specialist conferences and journals, which marked the beginning of their network both nationally and internationally. A prominent figure emerged in the form of Johann Bolle, a botanist and the founding director of the kaiserlich königlichen landwirtschaftlich – chemischen Versuchsstation (imperial royal agricultural – chemical experimental station) in Görz (now Gorizia in Italy). He sporadically conducted experiments on pest control at major ethnological museums. Thanks to his initiative, a new technology for mass fumigation of insect pests in a custom-built facility with vacuum pumps, specifically designed for museums, spread from Sweden to Germany and across Europe.

Building on this theoretical and empirical study of the almost 150-year history of pest control at the EM, it is unavoidable that further reflections on the use of active ingredients and agents during the same period have arisen on a national and international level. As a result, the EM created a blueprint that provides methodological starting points for museums worldwide with collections from colonial contexts to investigate their own exposure to previously used pesticides. This contribution to the field of conservation complements the knowledge possessed by current conservators regarding the preservation of museum artifacts and the assessment of hazards posed by the active ingredients and agents previously used in collections. In addition, the interdisciplinary investigation and historical contextualization of pesticides represents a new and innovative research approach. Aspects from the faculties of conservation, history, social history, and cultural studies were considered during the period under review. These links provide a valuable resource for both the scientific community and interested individuals. This book examines museums with a colonial history, primarily in Europe, North America, and Canada. Additionally, museums in the Global South, such as the Natural History Museum of Zimbabwe in Africa, can benefit from the research conducted on the use of pesticides in their own collections. Given the widespread repatriations, the safety and health hazards posed by toxic agents used in cultural assets have become a growing concern for representatives of societies.

The current historical evaluation was researched, structured, evaluated, and contextualised using documents from primary and secondary sources, and was based on epistemology, from which decisive impulses for science emerged. The historical-critical method was necessarily employed in this methodology and was used to investigate historical texts (sources). The reconstruction of the presumed prehistory and origin of a text, as well as its integration into the events of that time, plays a special role. Based on the historical-critical method, for example, relevant letters and advertising brochures related to pesticide distribution were subjected to a critical analysis. These materials were also analyzed based on the specific question of temporal context and interpreted linguistically. This holistic, interdisciplinary view provides additional knowledge for all the disciplines addressed. The described methodological procedures include letters, notes, drawings, files, and even photographs. The results of these modules are available in scientifically verifiable content. Based on epistemology, different methods were applied and presented. This is the basis of the findings of the current investigations, which were obtained from the analysis of individual disciplines and sources. In this way, a uniform theoretical foundation could be established. This leads to a unique selling point, which makes this book a well-regarded resource.

Acknowledgments

I would like to thank the following people who supported my work in archives and libraries: Kristen Mable from the Archive of the American Museum of Natural History, Division of Anthropology Archives; Hans-Hermann Pogarell from the Archive Bayer AG, Corporate History & Archives; Harri Ahlgrén from the Archives of the Finnish Heritage Agency; Chiara Maran from the Archivi della Galleria Nazionale di Roma, Fondo Ugo Ojetti (Archives of the National Gallery of Rome, Ugo Ojetti Fund); Mark Gill from the Archive of the Industrie- und Filmmuseum Wolfen, Dirk Handorf in the Archiv des Landesamtes für Kultur und Denkmalpflege (Archive of the State Office for the Preservation of Culture and Monuments) Mecklenburg-Vorpommern; Bernd Hoffmann from the Archives of the Max-Planck-Gesellschaft Berlin; Susanne Walther, Jantje Bruns, Catharina Winzer, and Heidelies Wittig from the Archive of the Museum am Rothenbaum – Kulturen und Künste der Welt; Svetlana Korepanova from the Archive of the Regional Museum Yekaterinburg; Carola Okrug from the Bundesarchiv (Federal Archives) Berlin; Jutta Billig, Barbara Hille, Sabine Pöggl, Cordula Treimer, Birgit Wichmann, and Anja Zenner from the archive and library of the Ethnologisches Museum der Staatlichen Museen zu Berlin (Ethnological Museum at the National Museums in Berlin); Anja Wagner from the Geheimes Staatsarchiv Preußischer Kulturbesitz (Secret National Archives Prussian Cultural Heritage); Iris Kausemann from the Historisches Archiv der Stadt Köln (Historical Archive of the City of Cologne); Kerstin Bötticher from the Landesarchiv Berlin (Berlin State Archive); Marion Schatz from the Landesarchiv Sachsen-Anhalt, Abteilung Merseburg (Saxony-Anhalt State Archives, Merseburg Department); Taina Partanen from the National Archives of Finland; Christel Kraft from the Nordiska museet Arkiv Stockholm; Romy Hartmann from the Sächsisches Staatsarchiv-Hauptstaatsarchiv Dresden (Saxon State Archives-Dresden Main State Archives); and Beate Ebelt-Borchert, Detlef Botschek, and Carolin Pilgermann from the Zentralarchiv of the Staatliche Museen zu Berlin (Central Archive of the National Museums in Berlin). They all have provided me with invaluable services over long periods of time, often in unbureaucratic ways.

I would like to thank my colleagues at home and abroad: Karin Björing-Olufsson, Belinda Blum, Matthias Farke, Nancy Fonicello, Diana Gabler, Thomas Gütebier, Richard Haas, Greta Hansen, Alan McCright, Boaz Paz, Andrea Scholz, and Jörg Weber. I have received special support from my family: Lieselotte Kupferstein, with whom I spent wonderful hours transcribing, and Thomas Kupferstein, Maria Tello, Anton and Emil Meencke, and Johannes Paetzold. My friends, Adele Boiger, Andreas Eis, Sabine Bröck, Angelika Hüffell, Dorothea Ihme, Christine Petersen, and Thomas Petersen, Jens Dietrich, Karl-Heinz Tönges, and Anna Zielke, encouraged me never to give up. Jochen Sprenger helped me to overcome all technical obstacles while writing this book. Daniela Liebscher and Sven Arnold inspired me with their infectious enthusiasm and patience for writing. Through them, I met Carolin Vogel, with whom I have been spending wonderful writing retreats ever since. Ludolf Kuchenbuch's insights dispelled my uncertainties, thereby compelling me to finally embark on my doctoral thesis. I would like to express my gratitude to my doctoral supervisors, Przemyslaw Paul Zalewski and Jirina Lehmann, for their numerous valuable hints, constructive questions, and thorough review of my manuscript. Finally, I am grateful to Achim Unger, who directed my attention to the past use of pesticides at the Ethnologisches Museum and shared his expertise with me. He inspired me to publish my diploma thesis in English and consistently encouraged me to concentrate on technical questions, even in difficult situations.

Credits

Figures

Figure 3.1 From: Karl Heinz Roth (2009): Die Geschichte der I.G. Farbenindustrie AG von der Gründung bis zum Ende der Weimarer Republik. Norbert Wollheim Memorial, J.W. Goethe-Universität/Fritz Bauer Institut, Frankfurt am Main, 2009. www.wollheim-memorial.de. Copyright: Stiftung für Sozialgeschichte des 20. Jahrhunderts, Fritz-Gansberg-Str. 14, 28213 Bremen.

Figure 6.1 Copyright: ERSA, Agenzia regionale per lo sviluppo rurale, Bibliothek Luigi Chiozza.

Figure 6.2 Copyright: Staatliche Museen zu Berlin, Ethnologisches Museum, photographer unknown.

Figure 8.1 Copyright: Staatliche Museen zu Berlin, Ethnologisches Museum, photographer unknown.

Figure 8.2 Copyright: Staatliche Museen zu Berlin, Museum für Vor- und Frühgeschichte.

Figure 9.1 Copyright: Ethnographische Sammlung der Marburger Philipps-Universität, Nachlass Koch-Grünberg.

Figure 11.1. Staatliche Museen zu Berlin, Zentralarchiv, photographer unknown.

Figure 11.2. Copyright: Mopanepool Berlin, Beate Reußner.

Figure 11.3. Copyright: National Archives of Finland, RakH II Iba 37: 1 b, 56 M 1/11.

Figure 11.4. Source: BArch-Filmarchiv. Begasungsanlage. Heilbehandlung von Kunstwerken. Copyright: Filmproduktion Ruth Cürlis, Ehrenbergstr. 3, 14195 Berlin.

Figure 11.5. Source: BArch-Filmarchiv. Begasungsanlage. Heilbehandlung von Kunstwerken. Copyright: Filmproduktion Ruth Cürlis, Ehrenbergstr. 3, 14195 Berlin.

Figure 11.6. From: Meyer, Adolph Bernhard (1902/1903):
Abhandlungen und Berichte des Königlichen Zoologischen und
Anthropologisch-Ethnographischen Museums zu Dresden. 3. Bericht
über einige Neue Einrichtungen des Königlichen Zoologischen
und Anthropologisch-Ethnographischen Museums in Dresden. XI.
Desinfektionsapparat. Bd. X, 1902/1903 (5), Tafel XIX, Desinfektionsapparat, 23.
Figure 11.7. Fotosammlung/Historische Diasammlung, Kat. Nr. F
2016–3/1562, photographer unknown.
Copyright: Staatliche Kunstsammlungen Dresden, Museum für
Völkerkunde Dresden.
Figure 11.8. Copyright: Nordiska museet Arkiv. NMAR.0000272.
Architect: Clason, Isak Gustaf, Digitalisat. https://digitaltmuseum.se/
search/?q=NMAR.0000272, (last visited June 16, 2018).
Figure 11.9. Copyright: Nordiska museet Arkiv. Foto 291.az. and Foto 291.ay.
Figure 11.10. Copyright: Staatliche Museen zu Berlin, Ethnologisches
Museum, photographer unknown.
Figure 13.1. Source: SMB-PK, EM, I/MV 730, Vol. 30, Pars I. B.,
E. Nr. 578/04. Acta betreffend die Restauration von Alterthümern,
photographer: Helene Tello.
Figure 15.1. Copyright: Norsk Folkemuseum Oslo. NF.01272–203.

Tables

Table 7.1 Copyright: Helene Tello.
Table 7.2 Copyright: Helene Tello.
Table 7.3 Copyright: Helene Tello.
Table 7.4 Copyright: Helene Tello.
Table 7.5 Copyright: Helene Tello.
Table 11.1 Copyright: Helene Tello
Table 12.1 Copyright: Helene Tello.
Appendix 1. Table 1. Copyright: Helene Tello.
Appendix 2. Table 1. Copyright: Helene Tello.
Appendix 3. Table 1: Copyright: Achim Unger.
Appendix 4. Table 1: Copyright: Helene Tello.
Appendix 5. Table 1: Copyright: Helene Tello.

Abbreviations

AAM	American Association of Museums
BM	British Museum
BRA	Biologische Reichsanstalt für Land- und Forstwirtschaft (Biological Reich Institute for Agriculture and Forestry)
DLG	Deutsche Landwirtschafts-Gesellschaft (German Agriculture-Society)
CCI	Canadian Conservation Institute
E. Nr.	Eingangs Nummer (Accession number)
EM	Ethnologisches Museum (Ethnological Museum)
IPM	Integrated Pest Management
KBA	Kaiserliche-Biologische-Anstalt für Land- und Forstwirtschaft (Imperial Biological Institute for Agriculture and Forestry)
KWG	Kaiser-Wilhelm-Gesellschaft (Kaiser-Wilhelm-Society)
KGA	Kaiserliches Gesundheitsamt (Imperial Health Office)
KWI	Kaiser-Wilhelm-Institut (Kaiser-Wilhelm-Institute)
KWIB	Kaiser-Wilhelm-Institut für Biologie (Kaiser-Wilhelm-Institute for Biology)
KWIpCh	Kaiser-Wilhelm-Institut für physikalische Chemie und Elektrochemie (Kaiser-Wilhelm Institute for Physical Chemistry and Electrochemistry)
k.k. l-c-V	kaiserlich königliche landwirtschaftlich-chemische Versuchsstation (Imperial Royal agricultural-chemical experimental station)
KM	Königliche Museen zu Berlin (Royal Museums in Berlin)
KMfV	Königliches Museum für Völkerkunde zu Berlin (Royal Museum of Ethnology in Berlin)
MfV	Staatliches Museum für Völkerkunde zu Berlin (National Museum of Ethnology in Berlin)
NPS	National Park Service
PTR	Physikalisch-Technische Reichsanstalt (Physical-Technical Reich Institute)

RF	Rathgen-Forschungslabor (Rathgen-Research Laboratory)
RJM	Rautenstrauch-Joest-Museum – Kulturen der Welt (Rautenstrauch-Joest-Museum – Cultures of the World)
rF	relative humidity
RGA	Reichsgesundheitsamt Berlin (Reich Health Office Berlin)
RK-I	Robert Koch-Institut (Robert Koch-Institute)
SAT	Save America's Treasures
SMB	Staatliche Museen zu Berlin (National Museums in Berlin)
VDLUFA	Verband Deutscher Landwirtschaftlicher Untersuchungs- und Forschungsanstalten
V & A	Victoria & Albert Museum
WaBoLu	Landesanstalt für Wasser-, Boden- und Lufthygiene (State Institute for Water-, Soil- and Air Hygiene)

Abbreviations of the archives

AAMNH-DAA	Archive of the American Museum of Natural History, Division of Anthropolgy Archives, New York City
AIFM Wolfen	Archiv des Industrie- und Filmmuseums Wolfen (Archive of the Industry- and Filmmuseums Wolfen)
AMPG	Archiv der Max-Planck-Gesellschaft Berlin (Archive of the Max-Planck-Society Berlin)
BAL	Archive of the Bayer AG, Corporate History & Archives
BArch	Bundesarchiv (Federal Archives)
FHA	Archives of the Finnish Heritage Acency
GStA PK	Geheimes Staatsarchiv Preußischer Kulturbesitz (Secret National Archives Prussian Cultural Heritage)
HAStK	Historisches Archiv der Stadt Köln (Historical Archive of the City of Cologne)
LAB	Landesarchiv Berlin (Berlin State Archive)
LASA	Landesarchiv Sachsen-Anhalt, Abteilung Merseburg (Saxony-Anhalt State Archives, Merseburg department)
LAKD	Archiv des Landesamtes für Kultur und Denkmalpflege Mecklenburg-Vorpommern (Archive of the state office for the preservation of culture and monuments Mecklenburg-Vorpommern)
NMA	Nordiska museet Arkiv Stockholm
SMB-PK, EM	Archiv des Ethnologischen Museums der Staatlichen Museen zu Berlin (Archive of the Ethnological Museum of the National Museums in Berlin)
SMB-ZA	Zentralarchiv der Staatlichen Museen zu Berlin (Central Archive of the National Museums in Berlin)

Without abbreviations

Archiv des Museums am Rothenbaum – Kulturen und Künste der Welt
(Archive of the Museum am Rothenbaum – Cultures and Arts of the World)
Archive of the Regional Museum Yekaterinburg
Archivi della Galleria Nazionale di Roma, Fondo Ugo Ojetti (Archives
of the National Gallery of Rome, Ugo Ojetti Fund)
National Archives of Finland
Sächsisches Staatsarchiv – Hauptsaatsarchiv Dresden (Saxon State
Archives – Dresden Main State Archives)

Introduction

My interest in the topic of pesticides in museum collections began with my work at the Ethnologisches Museum (EM) at the Staatliche Museen zu Berlin (Ethnological Museum at the National Museums in Berlin) at the end of the 1990s. As a conservator in the study collection of American Ethnology, I was responsible for the care of approximately 70,000 objects until June 2020. The items were collected from Alaska to Patagonia and brought to Berlin in various ways. Stored in many cabinets, some of which were completely overcrowded, these objects from North and South America exhibited a high diversity of organic materials, such as wood, plant fibers, feathers, skins, hide, leather, wool, and cotton, as well as works on paper and historical archival materials. Because of their material properties, all objects made of organic materials were exposed to constant danger of being attacked by plant and animal pests, as well as from molds. As the cabinets were opened, an unfamiliar smell unfolded in the supervised collection, which at once seemed pungent and sharp, yet musty sweet. In addition to the objects in the cabinets, yogurt cups and tin cans of pipe tobacco were filled from time to time with synthetic camphor by the then-collection manager to prevent against harmful insects.

It was a fortunate circumstance that during this time, the chemist, Dr. Achim Unger, worked at the Rathgen-Forschungslabor (RF) of the Staatliche Museen zu Berlin (Rathgen-Research Laboratory at the National Museums in Berlin). Through him, I learned that my predecessor in the office repeatedly complained of diffuse health problems. She personally attributed these health problems exclusively to the fumigation plant operated by the EM, where ethnological objects were fumigated to combat and prevent pest infestation. Initial investigations into the material samples of individual objects revealed that collection items were contaminated with toxic substances, such as synthetic camphor or the fumigant "Illo-Spezial-T", which had been used in the in-house fumigation plant. Scientific investigations conducted within the RF uncovered the first signs of heavy metals and organochlorine compounds. After conducting measurements of air, dust, and materials from the collections from 2001 to 2006, it was discovered that many collections had been treated with different active ingredients over an extended period and thus were contaminated. Dr. Unger's team had already successfully decontaminated objects made of wood by the end of the 1990s. It has been proven that it is possible to gently remove certain wood preservatives from wooden objects using supercritical carbon dioxide.

DOI: 10.4324/9781003407607-1

In a follow-up project, the EM was involved in investigations conducted in 2003. In my diploma thesis, I evaluated the test results (Tello 2006). Studies of museum artifacts contaminated with pesticides inevitably led me to the question of where the former employees of the EM and the RF obtained their active ingredients and agents for the prevention and control of harmful insects during the heyday of the Königliches Museum für Völkerkunde (Royal Museum of Ethnology in Berlin) (KMfV).

The effects of pesticide use on museum collections are evident and well documented. Many objects are contaminated by these active ingredients and agents, making them difficult to access for conservation and scientific processing, mediation in the context of exhibitions, external loans, and repatriation processes. Because of the importance of occupational safety during the handling of pesticide-contaminated objects, it is mandatory for both internal and external personnel to wear personal protective clothing. In recent years, many museums have started to implement chemical-free methods, such as integrated pest management, for the pest control of their collection items. Efforts have been made to minimise the risks posed by the objects and to research and develop effective solutions for the remediation and decontamination of contaminated collection materials. In the context of ethnological collections, there has been widespread research from the USA and Canada about the effects of contaminated collections on Indigenous communities who are working to repatriate and manage their cultural assets. The detoxification efforts collections in these countries have prioritised the needs of the First Nations and their ceremonial uses, rather than focusing on the safety of museum personnel, which can be more easily managed in a controlled environment (Charola and Koestler 2010; Preserving Aboriginal Heritage 2007).

Discussions regarding the historical use of pesticides in museum collections have been taking place in conservation literature nationally and internationally since the mid-1990s. Notable studies included those conducted for the Smithsonian Institution in Washington, D.C. (Goldberg 1996), the history of conservation and restoration in Russia and the Soviet Union (Lehmann 2005), as well as the investigations for the EM (Tello 2006), Museum Jardin des Sciences de Dijon (Pfister 2008), National Museum of Wales Herbarium (Purewal 2012), and Muséum d'Histoire Naturelle de Neuchâtel (Dangeon 2013/2014). We owe all six studies valuable information on the use of active substances and agents used both preventively and combatively against insect pests in anthropological, ethnological, and natural history collections during the period under investigation. They were compiled based on empirical research in museum archives and the sources located there. Relevant references in acquisition, inventory, and travel books, as well as occasionally in index cards, were supplemented by topic-related secondary literature. Finally, oral traditions of employees contribute to supplementing the knowledge that may be lost in the institutions mentioned and the studies conducted therein. Within this theoretical framework, in Jansen's doctoral thesis "Schädlinge" Geschichte eines wissenschaftlichen und politischen Konstrukts 1840–1920" (Jansen 2003), the conceptual transformation from insect to pest and the linguistic declaration of insects as the enemy of a more rationalised forestry and agriculture during the examined period are comprehensively discussed and proven.

For research into pesticides developed for civil and military purposes and their adaptation to cultural institutions, information bases are obtained from available sources. The *Acta betreffend die Restauration von Alterthümern*[1] serves as the foundation for the reconstruction of the active components and agents used at the EM in Berlin. Here, we will find initial evidence of the purchase of funds, exchange with colleagues at both the national and international levels, and hints for the museum's own experiments to combat insect pests. The museum's archive also contains numerous sources in the form of acquisition files, old files from individual collections, travel reports, and diary entries. They led to people who shaped the development of the collections in the former KMfV and provided insight into the hierarchies within the institution. The impact of these structures on the museum's daily operations was significant and had a considerable effect on the use of active ingredients and agents to combat insect pests. The relationship between the use of pesticides in museum collections and their sources of supply is a hitherto unnoticed area in the field of conservation science. Based on the source situation, fundamental insights are gained for the first time, which are then analyzed and evaluated in terms of content. After an initial review of the source material and its nature, there was a shift in the perspective toward considering the KMfV in Berlin as a single institution. This change takes place in such a way as to perceive this institution as an integral part of society. Thus, the theoretical framework is unfolded against the background of an expanding context. This led to the assumption that museums were integrated into external social and political currents in the late 19th and early 20th centuries. Through this historical classification, the scope of action expands to multiple perspectives, in which the topic of pesticides in museum collections and their specific problems can be classified more precisely. It was inevitable that a question arose about where the staff obtained the pesticides at the time. The interfaces for the present interdisciplinary work are seen in the historical, cultural, natural, and conservation sciences. Based on this, methodological procedures are used, which are commonly used in the mentioned disciplines.

Likewise, the present studies contribute to the maintenance of collections for current and future generations of curators, conservators, collection managers, and source communities. Individuals within these occupational groups are often exposed to stressors that pose a risk to their health, which can be caused by hazardous substances that were used previously in museum collections for pest control. Improving one's own knowledge allows for more careful handling of entrusted objects. The use of pesticides on ethnological materials and objects has been linked to the perpetuation of colonialism when these cultural assets are repatriated by or passed on to Indigenous community members. As a result of contamination with pesticides, further "objectification" has been introduced into the cultural assets from indigenous people, which must now be accepted as largely irreversible.

Note

1 Acta concerning the restoration of antiquities (translation by the author).

References

Other Sources

Charola, Elena; Koestler, Robert (eds.) (2010): Pesticide Mitigation in Museum Collections: Science in Conservation, Proceedings from the MCI Workshop Series, Smithsonian Contributions to Museum Conservation, No. 1, 2010.

Dangeon, Marion (2013/2014): Conservation des collections naturalisées traitées aux biocides: étude de la collection Mammifères et Oiseaux du Muséum d'Histoire Naturelle de Neuchâtel. Bachelor of Arts. Haute École Arts, Appliqués-La Chaux-de-Fonds, Filière Conservation-Restauration, Neuchâtel.

Dignard, Carole; Helwig, Kate; Masory, Janet; Nanowin, Kathy; Stone, Thomas (eds.) (2008): Preserving Aboriginal Heritage: Technical and Traditional Approaches. Proceedings of a Conference Symposium 2007, September 24–28. Canadian Conservation Institute, Ottawa, 2008.

Goldberg, Lisa (1996): A History of Pest Control Measures in the Anthropology Collections, National Museum of Natural History, Smithsonian Institution. *Journal of the American Institute for Conservation/American Institute for Conservation of Historic and Artistic Works*, 35(1), 23–43.

Jansen, Sarah (2003): "Schädlinge". Geschichte eines wissenschaftlichen und politischen Konstrukts 1840–1920. Techn. Univ., Diss. Braunschweig, 1997. Campus Verlag GmbH, Frankfurt/Main (Campus historische Studien, 25). Available online at http://www.gbv.de/dms/faz-rez/FD1200307281954595.pdf.

Lehmann, Jirina (2005): Geschichte der Konservierung und Restaurierung in Russland und in der Sowjetunion. im Buch von Professor M.W. Farmakowskij. *VDR Beiträge zur Erhaltung von Kunst- und Kulturgut*, (2), 47–62.

Pfister, Aude-Laurence (2008): L'Influence des Biocides sur la Conservation des Naturalis. Diplomarbeit. Haute École Arts, Appliqués-La Chaux-de-Fonds, Filière Conservation-Restauration, La Chaux-De-Fonds. Filière Conservation-Restauration.

Purewal, Victoria Jane (2012): Novel Detection and Removal of Hazardous Biocide Residues Historically Applied to Herbaria. Dissertation, University of Lincoln, Lincoln.

Tello, Helene (2006): Investigations on Super Fluid Extraction (SFE) with Carbon Dioxide on Ethnological Materials and Objects Contaminated with Pesticides. Diplomarbeit. Fachhochschule für Technik und Wirtschaft. Berlin. Fachbereich 5, Gestaltung, Studiengang Restaurierung/Grabungstechnik.

Social and political currents from the end of the 19th to the beginning of the 20th century

Part I

1 The nation-state of Prussia, colonialism, and the age of industrialization

The formation of Prussia into a nation-state and its consequences for cultural policy

The transformation of the Prussian state from a feudalistic to a nationalist system will be examined in greater detail with regard to the cultural–political implications for the Ethnologisches Museum Berlin (EM) (Ethnological Museum) in Berlin. The rise of European nation-states and their subsequent development under persistent economic and ultimately knowledge-based competition led to the development of defining features of modernity, such as efficiency orientation, application of scientific knowledge, and expertise. Compared to other European countries, German states demonstrated a higher level of professionalization in several academic professions. The reason for this was a state-standardised university education, as Prussian doctors could refer to it from 1851 to 1852 by state regulations. For the founding director of the Königliches Museum für Völkerkunde (KMfV) Berlin, Adolf Bastian, the process of selecting young scientists for his museum may have been influenced by this aspect (see Part III, chapter 8.6). Following the joint victory of the German states in the Franco–Prussian War, the founding of the German Empire and the proclamation of Wilhelm I as German Emperor occurred in 1871, with the decisive participation of Otto von Bismarck. During the imperial era and under Prussian rule, Germany manifested itself as a common nation. Berlin became increasingly important as the capital of the German Reich for Central Europe and maintained its international standing until the end of World War II (see Kocka 1995, 72; Cf. Werner 1997, 74).

Numerous activities reflect how national identity was created. These include the erection of numerous memorials and monuments adorned with national symbols. In the fields of education and science, libraries were established nationwide, and in 1884, the Technische Universität (Technical University) in Berlin-Charlottenburg was extablished by merging the Berliner Bauakademie (Academy of Architecture) and the Königliche Gewerbeakademie (Royal Trade Academy). It united the Prussian state ideals of power, military might, art, culture, and science. In terms of public image, it now centered around how one's nation compared to others in terms of greatness. Prussia evolved into a highly cultured state and thus fostered a sense of intellectual superiority. The main reasons for this were undoubtedly the economic

DOI: 10.4324/9781003407607-3

and cultural competition with other European states. Therefore, it is not surprising that numerous museums were established at the same time as nation-states were formed. A museum differs from the Wunderkammern or Kunstkammern (Savoy 2015) of the 16th century by some important criteria and is defined by the International Council of Museums

> "as a non-profit, permanent institution open to the public at the service of society and its developments, which procures, preserves, researches, makes known and exhibits material testimonies of people and their environment for study, educational and subsistence purposes".
>
> (Savoy 2015, 17, translation by the author)

It should be noted that the public character of museums developed in the 18th century, following Savoy. The architecture of a museum building and its strategic location were carefully chosen to make it a prominent and accessible institution in the inner city. This is still a notable feature of museums today, emphasizing their significance as an institution (Savoy 2015, 19). The use of material artifacts from one's own and foreign cultures was employed symbolically to showcase one's own strength and power. By exhibiting ethnological objects in public spaces, a large group was portrayed as superior to others. The establishment of the Kaiser Friedrich Museum in Berlin in 1904, which is now known as the Bode Museum and is part of the Staatliche Museen (National Museums) zu Berlin (SMB), marked a significant milestone in the pursuit of national and cultural identity. This museum was established according to the ideas of Bode (Cf. Neugebauer 2007, 10–12).

The Prussian state, known for its cultural diversity and strong integrative power, acted as a model for other European states in the realms of science and culture (Neugebauer 2007, 2–3; Schlenke 1981, 136–161). Since the 15th century, there has been intense competition between European colonial powers on an international level, with individual countries seeking to assert their influence overseas. To achieve this, large expeditions were organised to bring exotic goods from distant lands back to their home countries (Beßler 2012). When examining strange and foreign objects, an individual experienced a collective confirmation of what had been achieved. At the same time, a sense of superiority and one's own power was formed, in which one could reflect oneself collectively and culturally in the objects of "the savages" as part of a nation (Zimmermann 2013, 247–258). This development was expressed through the linguistic distinction between civilised peoples and primitive peoples. At the end of the 19th and beginning of the 20th century, the young discipline of ethnology referred to statements, such as those of the German historian Leopold von Ranke. He referred to non-European peoples

> ". . . as without history and not worthy of scientific investigation".

This led ethnologists to establish their right to study "material sources", i.e., people and their artifacts, in a scientific manner (Zimmermann 2013, 248–249). The emergence of nation-states and their search for their own identity are also reflected in a proliferation of publications (Funk 2010; Konrád und Paetzke 2013; Löhr und Wenzlhuemer 2013; Malešević 2013; Weisband und Thomas 2015). With the self-portrayal of one's own history, an important prerequisite was created for using "the savages" and the testimonies of their intellectual and material cultures as a blue-print for the formation of their own identity. The distinction between civilised and primitive peoples (Hermannstädter 2002, 9, 46) eventually led to a pan-national consciousness in the field of art and culture.

Colonialism and its consequences

In this chapter, after a brief introduction, the German colonial policy from 1880 onward, including its heyday known as imperialism for all of Europe, is presented. It is noteworthy that the vocabulary already illustrates a Eurocentric view in some places. The history of colonization is global, spanning five centuries and starting in the late 15th century with the great discovery of America, India, Central Asia, Australia, and Africa by Europeans. From there on, the first colonies were estab-lished by the Spanish and Portuguese kingdoms, which eventually led to colonial-ism. A serious consequence of the discovery of these non-European continents is their dependence on European powers. Objects now considered as rarities and curiosities found their way into royal and princely courts as well as the homes of wealthy citizens. Exotic items made from unknown materials were imported in small quantities from overseas and kept for a select audience in the Wunderkam-mern or Kunstkammern. With this form of preservation and presentation, the first forms of museum emerged, which were then replaced during the 19th century by today's museums with their specializations. Modern colonialism peaked at the end of the 19th century and the beginning of the 20th century. The end of Prus-sian colonial power was sealed by the end of the First World War (Authaler 2019, 4–10). This was followed by a new division between the German colonies and the Ottoman Empire, today's Turkey. The colonies in England and France experi-enced the greatest expansion. As colonialism came to an end, the colonies began to aspire for independence. However, the breakthrough of the freedom movements in Africa and Asia did not succeed until the end of World War II. In 1932, Moritz Julius Bonn introduced the concept of decolonization for the detachment of colo-nies from colonial powers (Cf. Reinhard 2008, 310; Cf. Pelizaeus 2008, 231–240; Ibid. 310–311).

Colonies are territories conquered by foreign rulers and brought into perma-nent dependence. The reasons for this appropriation are manifold, which is why colonies are divided into different categories. Settlement colonies are described as archetypes of the colonies. In newly settled countries, hunters, gatherers, and nomads are often displaced by sedentary farmers, thus securing private owner-ship of land. The term "domination colonies" does not refer to individual bases, but rather to the complete takeover of a country. Subjugation does not take place with the help of a new settlement, but rather through the exercise of dominion over the existing population by immigrant settlers. A model case in history is the colonization of India by the British. Without the collaboration of locals, this would never have been successful. The earliest colonial settlements were established in integration colonies, which had the aim of establishing European settlements far from the mother country after a conquest. Pure base colonies served both the commercial development of a country and the expansion of its sales markets, and the maintenance of military presence ensured dominance in a region. Examples of economic and military connections are Great Britain, with its worldwide network of naval bases, and Portugal, with its colonies on the Indian Ocean (Pelizaeus 2008, 5, 20–21; Reinhard 2008, 4). The term "colonialism" is associated with various phenomena. Reinhard defines colonialism as

> ". . . the control of a people over a foreign one, exploiting economically, politically, and ideologically the development efficiency between the two"
> (Ibid. 1, translation by the author).

From a political point of view, this refers to colonization and decolonization; historically, one speaks more of colonialism and post-colonialism. Whether colonised people accepted their situation with equanimity, undermined it with their own cleverness, or whether they collaborated with the colonial masters or even resisted requires detailed examination elsewhere. Colonialism and post-colonialism have sometimes led to victims actively shaping events, making it difficult to clearly distinguish between perpetrators and victims in some cases, as Reinhard has pointed out. The historical consequences of colonialism on the history of science are also closely intertwined with individuals who, through a mission of exploration or motivated by a sense of destiny beyond Europe, sought to understand the world in a new way at that time (Reinhard 2008, 7, 2; Habermas und Przyrembel 2013, 9–10). At the end of the 19th century, for example, Amalie Dietrich, the daughter of a Saxon herbalist and glove maker, collected over 20,000 botanical specimens in Australia, thus making important contributions to botany. Ludwig Adzakko, a Togolese, may be another example of how knowledge was transferred from the other side of the world to Europe. Together with an unspecified person without scientific education, he translated the Lutheran Bible into his language in Tübingen in 1907, and thus created the written form of a language that had previously only been handed down orally. In contrast to notable figures, such as Alexander von Humboldt and Robert Koch, many individuals have often fallen into oblivion or underappreciated by the history of science. Nevertheless, their contributions are still relevant today. This little digression into the lives and works of individuals illustrates the unusual exchange of knowledge and information gathering practices during the colonial period. The specific implications for the KMfV are discussed extensively in Part III (Ibid. 9, 14–15).

The German colonies from 1884 to 1918 included German Southwest Africa, Togo, Cameroon, German East Africa, Wituland, German New Guinea, the Marshall Islands, Nauru, the Caroline Islands, the Palau Islands, the Mariana Islands, Samoa, and Kiautschou. It should be noted that all the listed colonies had other colonial successors and usually changed their names after their independence. An exceptional and unusual event in this context occurred in Berlin in 1896, which was a trade exhibition. In addition to industrial achievements, the German colonies were displayed to a hitherto unknown extent. Many citizens were able to experience and view German colonial policy through a geopolitical presentation, which showcased its actors. In contrast to ethnological museums, visitors to this exhibition were provided with a detailed understanding of the social and economic conditions in the colonies through the exhibition's meticulous staging. More than 300 colonial companies showcased their business activities during the exhibition, which was visited by more than two million visitors (Gottschalk et al. 2016; Gründer und Hiery 2017, 326–327). In a purpose-built tropical house, the reconstructed offices of German colonial officials in East Africa, Southwest Africa, and Oceania enabled them to gain a detailed insight into their work. An urban colony with clear spatial divisions between Europeans and Indigenous people was also on display. The Deutsche Frauenverband für das Gesundheitswesen (German Women's Association for Health Care) in the colonies presented a tropical hygiene exhibition with a laboratory, a pharmacy, an operating room, and an exact replica of a tropical hospital. These realistic, recreated, and staged scenarios from the colonies offered visitors a direct experience, allowing them to become colonial officials, doctors, nurses, businessmen, or even settlers. The example of this exhibition reveals the significance of the German Empire's colonial possessions and subjugation of other peoples not only for its own interests but also for the purpose of showcasing its power and dominance over other states. The international reputation of a state is also often influenced by the size and number of its colonies (Steinmetz 2017, 50–56; Ibid. 56–57; Hobsbawn 2008, 181–182). Prussia's economic, political, and cultural–political influence grew. While a Western mindset of superiority over the "schrift- und metalllosen Kulturen"[1] had emerged due to their own industrial achievements, concern began to rise as the rapidly advancing progress of civilization put many of these primitive peoples and their possessions at risk of annihilation in existing colonies. This way of thinking fueled a race among the existing European colonial centers and had drastic effects on the collection policies of ethnological and natural history museums. Extensive expeditions and research trips have led to unprecedented expansion in museum collections within a very short time. Individuals, such as professional traders and adventurers, were complementary actors in this "Sammelwut".[2] The infestation of insect pests aggravated the already challenging problem in the collections with their abundance of organic materials. One of the most important protagonists in this context was Adolf Bastian, who in 1876 was appointed director of the then-KMfV, known today as EM. Bastian has undertaken nine major world trips and thus had a detailed view of the momentous changes for non-European peoples. His collecting policy resulted in one of his most striking sentences (Cf. Westphal-Hellbusch 1973, 3–4):

"Der letzte Augenblick ist gekommen, die zwölfte Stunde ist da! Dokumente von unermesslichem Wert für die Menschheitsgeschichte gehen zugrunde. Rettet! rettet! Ehe es zu spät ist".[3]

(Cf. Bolz 2005, 5; Fischer et al. 2007, 185)

The central issue concerning the collections was to safeguard the vast number of objects from deterioration, which had been generated by the frenzied collection, purchase, and exchange of objects. The transport of objects from a tropical to a temperate Central European climate often resulted in damage. The lack of storage capacity and the limited human resources often led to an overabundance of food for insect pests that feed on objects made of cellulose, keratin, or collagen materials.

The age of industrialization

The active ingredients and agents used to protect the collections from insects in the Königliches Museum für Völkerkunde/Museum für Völkerkunde (KMfV/MfV) were determined based on the context of the age of industrialization, the hygiene movement, storage, and plant protection, as well as the widespread impact of the First World War. Numerous inventions contributed to a rapid change in the world of work. Thanks to steamships and steam trains, distances became shorter, and goods could be traded more quickly. In the field of communications, the invention of the telegraph revolutionised the concept of space and time. This was demonstrated with the opening of a modern railway ferry between Trelleborg and Sassnitz on July 6, 1909, which was a major event. Swedish companies moved closer to the expanding German Empire and sought to benefit from Germany's status as an outstanding industrial nation within European states. At that time, German was the language of science, and about a third of all Nobel Prizes in science were awarded to German researchers. The close connection between politics, industry, and science in Germany acted as the driving force for progress, leading to large investments in laboratories, libraries, test facilities, and institutions. Thanks to Friedrich Althoff's initiative, a new science location was established in Berlin-Dahlem from 1900 onward. Althoff had been responsible for the entire education and medical system in Prussia since 1882 as ministerial director of the Preußisches Ministerium der geistlichen, Unterrichts- und Medizinalangelegenheiten (Prussian Ministry of religious, educational, and medical affairs) (Klemm 1989, 141–144; Runeby 1997, 389–396; Ibid. 390–392). Wilhelm von Humboldt's idea of a "großen Wissenschaftsplan"[4] manifested itself with the establishment of the Königliche Domäne Dahlem (Royal Domain Dahlem) as a unique research environment (Max-Planck-Gesellschaft 2011). The structural narrowness in the Berlin city area and the "Vorbild englischer Campus-Universitäten"[5] were important impulses for Althoff's successors Friedrich Schmitt-Ott and Adolf von Harnack to pursue his dream

" . . . von einer durch hervorragende Wissenschaftsstätten bestimmten vornehmen Kolonie, eines Deutschen Oxfords".[6]

Kaiser Wilhelm II founded the Kaiser-Wilhelm-Gesellschaft für Wissenschaft und Forschung (Kaiser-Wilhelm-Society for science and research) for the common interests of industry and the state. From 1911, the Kaiser-Wilhelm-Institut für Chemie (Kaiser-Wilhelm-Institute for chemistry), the Kaiser-Wilhelm-Institut für physikalische Chemie und Elektrochemie (Kaiser-Wilhelm-Institute for Physical

Chemistry and Electrochemistry), the Kaiser-Wilhelm-Institut für Biologie (Kaiser-Wilhelm-Institute for Biology), the Kaiser-Wilhelm-Institut für Kohlenforschung (Kaiser-Wilhelm-Institute for Coal research), and the Kaiser-Wilhelm-Institut für Arbeitsphysiologie (Kaiser-Wilhelm-Institute for occupational Physiology) were established. Over the years, a number of Nobel Prize winners have emerged from the institutions of this society since 1914, earning the science location the epithet "German Oxford" (Kaltenbach 2011, 33–34; Brocke vom und Vierhaus 1990; Brocke vom 1996).

In the first half of the 19th century, the production of chemical preparations was exclusively the domain of pharmacists. This led to the establishment of companies that initially produced pharmaceutical–chemical preparations, such as Merck KgaA, which emerged from a pharmacy in Darmstadt. The transformation of pharmacies into pharmaceutical–chemical production sites gave the German Reich a unique competitive advantage in industrial production before the First World War. Large chemical companies, such as the Actien-Gesellschaft für Anilin-Fabrication (AGFA), the Badische Anilin- & Soda-Fabrik Ludwigshafen (BASF), the Farbwerke Hoechst AG, vormals Meister, Lucius & Brüning, and the Wacker Chemie AG emerged from small, innovative family businesses. As a result, Germany became the clear number two on the world market, ahead of England and behind the USA (Cf. März 2014, 27–29, 112). Fundamental innovations are mainly attributable to the chemical industry. The invention of large-scale chemical syntheses in inorganic chemistry, played a significant role in feeding an ever-growing population. Factories were built, in which soda, sulfuric acid, potash salts, and nitrogen and phosphorus fertilizers were produced. These agents, used for artificial fertilization, greatly increased yields in agriculture. The organic-chemical industry in Germany was characterised by the work of notable chemists, such as Justus Liebig, August Wilhelm von Hofmann, and later Johann Friedrich Wilhelm Adolf von Baeyer. The chemical syntheses allowed for the production of high-quality products from simple raw materials for the first time, including the invention of ammonia synthesis by Fritz Haber and the production of fuels, lubricating oils, and numerous other products (Cf. Klemm 1989, 182–183).[7] Catalytic syntheses also enabled the production of organic–synthetic dyes. One possible source of raw materials was the waste product coal tar, which was obtained by dry distillation of hard coal (Schreiber 1923). Coal tar produced colored substances that were marketed as synthetic dyes. In 1863, the Farbwerke Hoechst AG, vormals Meister, Lucius & Brüning, and in 1865, the BASF arose (Bäumler 1963). Early organic–synthetic pesticides also contained coal tar, but this branch of production was much smaller than that of the paint industry. Through a close connection between theoretical chemistry and the technical use of substances and agents developed in chemical factories, Germany quickly became the market leader in Europe in the field of tar dye production. With regard to the lighting of private households and industrial sites, the domestic oil production of Germany at the end of the 19th century was insufficient, necessitating imports from abroad. During this time, the USA and England were the leading producers of both oil and petroleum, which was enough to power the lamps (Cf. Jansen 2003, 66–67; Karlsch und Stokes 2003, 15–19).

However, Germany's outstanding position in the late 19th and early 20th centuries as an emerging industrial nation and economic power also resulted in negative consequences. The disruptions caused by the age of industrialization led to widespread economic and social changes that were not limited to Germany but affected Western Europe as a whole. Explosive urbanization and metropolitanization processes contributed to social unrest and long-lasting poverty among large segments of the population. Low wages of the working class, high levels of unemployment, poor housing, and inadequate living conditions led to impoverishment of these groups. Poor hygienic conditions also had a negative impact on the health of many people in the conurbations (Lueger 1904; Geist und Kürvers 1989; Kieß 1991; Zalewski 2007, 28–36). As a result, the well-being of workers was at risk. A growing demand for food led to extensive research in Berlin and other European cities to improve agricultural and forestry yields. The extent to which crops had to be protected to ensure harvests was previously unknown. This led to the large-scale research and development of numerous agents for fighting harmful microorganisms, insects, and rodents. The aspect of nationalism and colonialism is embedded in the colonial wars, with the aim to wipe out not just religions but also entire cultures. From this point of view, museum collections from colonial contexts are a legacy that should not be underestimated from today's point of view in working together with representatives of communities of origin and their cultural assets.

Notes

1 Writing and metal-free cultures (translation by the author).
2 Collecting mania (translation by the author).
3 The last moment has come, the 12th hour has arrived! Documents of immeasurable value for human history are perishing. Salvages! salvages! Before it's too late" (translation by the author).
4 Great science plan (translation by the author).
5 Model of English campus universities (translation by the author).
6 . . . of a noble colony determined by outstanding scientific institutions, a German Oxford (translation by the author).
7 Large-scale synthesis or chemical synthesis is the chemical production of high-quality products from simple raw materials. The first and best-known industrially produced synthesis was achieved by Carl Bosch and Fritz Haber in 1913 at the Badische Anilin- & Soda-Fabrik Ludwigshafen. Under high pressure and high temperatures, they succeeded in synthesizing ammonia with atmospheric nitrogen and hydrogen. Bosch and Haber had thus discovered an important raw material to produce chemical fertilizers. As a result, fertilizers could be produced industrially. Ammonia synthesis went down in history with the name Haber–Bosch process, for which Fritz Haber received the Nobel Prize in chemistry in 1919 (see Klemm 1989, 182–185).

References

Other sources

Authaler, Caroline (2019): Das völkerrechtliche Ende des deutschen Kolonialreichs. Globale Neuordnung und transnationale Debatten in den 1920er Jahren und ihre Nachwirkungen. *Aus Politik und Zeitgeschichte*, 69, 40–42.

Bäumler, Ernst (1963): Ein Jahrhundert Chemie. Unter Mitarbeit von Gustav Ehrhart and Volkmar Muthesius. Düsseldorf: Econ Verl.

Beßler, Gabriele (2012): Wunderkammern. Weltmodelle von der Renaissance bis zur Kunst der Gegenwart. 2. erweiterte Auflage. Berlin: Reimer.

Bolz, Peter (2005): Ethnologische Sammlungen in Berlin bis zur Eröffnung des "Königlichen Museums für Völkerkunde". Bastian-Symposium im Ethnologischen Museum Berlin. Unveröffentlichter Beitrag.

Brocke vom, Bernhard (Hrsg.) (1996): Die Kaiser-Wilhelm-, Max-Planck-Gesellschaft und ihre Institute. Studien zu ihrer Geschichte. Kaiser-Wilhelm-Gesellschaft zur Förderung der Wissenschaften, Max-Planck-Gesellschaft zur Förderung der Wissenschaften. Berlin: de Gruyter. Available online at www.gbv.de/dms/faz-rez/F19970326NOTKE-100.pdf.

Brocke vom, Bernhard; Vierhaus, Rudolf (Hrsg.) (1990): Forschung im Spannungsfeld von Politik und Gesellschaft. Geschichte und Struktur der Kaiser-Wilhelm-, Max-Planck-Gesellschaft. Kaiser-Wilhelm-Gesellschaft zur Förderung der Wissenschaften, Max-Planck-Gesellschaft zur Förderung der Wissenschaften. Stuttgart: DVA. Available online at www.gbv.de/dms/faz-rez/900926_FAZ_0037_37_0001.pdf.

Fischer, Manuela; Bolz, Peter; Kamel, Susan (eds.) (2007): Adolf Bastian and His Universal Archive of Humanity. The Origins of German Anthropology. Ethnological Museum Berlin. Hildesheim: Georg Olms.

Funk, Albert (2010): Kleine Geschichte des Föderalismus. Vom Fürstenbund zur Bundesrepublik. Paderborn: Schöningh.

Geist, Johann Friedrich; Kürvers, Klaus (1989): Das Berliner Mietshaus. München: Prestel.

Gottschalk, Sebastian; Hartmann, Heike; Hilden, Irene (2016): Deutscher Kolonialismus. Fragmente seiner Geschichte und Gegenwart. Theiss Verlag, Stiftung Deutsches Historisches Museum, Berlin, Darmstadt.

Gründer, Horst; Hiery, Hermann (Hrsg.) (2017): Die Deutschen und ihre Kolonien. Ein Überblick. Berlin: be.bra Verlag.

Habermas, Rebekka; Przyrembel, Alexandra (Hrsg.) (2013): Von Käfern, Märkten und Menschen. Kolonialismus und Wissen in der Moderne. 1. Auflage, Göttingen: Vandenhoeck & Ruprecht.

Hermannstädter, Anita (2002): Symbole kollektiven Denkens. Adolf Bastians Theorie der Dinge. In: Hermannstädter, Anita (Hrsg.). Deutsche am Amazonas – Forscher oder Abenteurer? Expeditionen in Brasilien 1800 bis 1914. Begleitbuch zur Ausstellung im Ethnologischen Museum, Berlin-Dahlem in Kooperation mit dem Brasilianischen Kulturinstitut in Deutschland. Ethnologisches Museum Berlin, Ausstellung. 2., unveränderte Auflage Münster: LIT. Publikationen des Ethnologischen Museums Berlin, Fachreferat Amerikanische Ethnologie, Neue Folge, (71).

Hobsbawm, Eric Jonas (2008): Das imperiale Zeitalter 1875–1914. Flörsheim a. M.: Campus Verlag GmbH.

Jansen, Sarah (2003): "Schädlinge". Geschichte eines wissenschaftlichen und politischen Konstrukts 1840–1920. Techn. Univ., Diss. Braunschweig, 1997. Frankfurt/Main: Campus-Verl. (Campus historische Studien, 25), 29–32. Available online at www.gbv.de/dms/faz-rez/FD1200307281954595.pdf.

Kaltenbach, Angelika (Hrsg.) (2011): Bezirk Steglitz-Zehlendorf, Ortsteil Dahlem. Bearbeitete Fassung: Januar 2011. Petersberg: Imhof. Denkmale in Berlin, hrsg. Senatsverwaltung für Stadtentwicklung und Umweltschutz, Bezirk Steglitz-Zehlendorf.

Karlsch, Rainer; Stokes, Raymond (2003): Faktor Öl. Die Mineralölwirtschaft in Deutschland 1859–1974. München: Beck. Available online at www.gbv.de/dms/faz-rez/FD1200304101799547.pdf.

Kieß, Walter (1991): Urbanismus im Industriezeitalter. Von der klassizistischen Stadt zur Garden City. Berlin: Ernst.

Klemm, Friedrich (1989): Geschichte der Technik. Der Mensch und seine Erfindungen im Bereich des Abendlandes. Orig.-Ausg. Reinbek bei Hamburg: Rowohlt (rororo, 7714).

Kocka, Jürgen (Hrsg.) (1995): Bürgertum im 19. Jahrhundert. Deutschland im europäischen Vergleich, eine Auswahl. Göttingen: Vandenhoeck & Ruprecht (Kleine Vandenhoeck-Reihe).

Konrád, György; Paetzke, Hans-Henning (2013): Europa und die Nationalstaaten. Essay. Deutsche Ausgabe, 1. Auflage. Berlin: Suhrkamp.

Löhr, Isabella; Wenzlhuemer, Roland (2013): The Nation State and Beyond. Governing Globalization Processes in the Nineteenth and Early Twentieth Centuries. Berlin, Heidelberg: Springer (Transcultural Research – Heidelberg Studies on Asia and Europe in a Global Context). Available online at http://dx.doi.org/10.1007/978-3-642-32934-0.

Lueger, Otto (Hrsg.) (1904): Lexikon der gesamten Technik und ihrer Hilfswissenschaften. Im Verein herausgegeben mit Fachgenossen von Otto Lueger. Zweite, vollständig überarbeitete Auflage. 8 Bde. Stuttgart, Leipzig: Deutsche Verlags-Anstalt.

Malešević, Siniša (2013): Nation-States and Nationalisms. Organization, Ideology and Solidarity. Cambridge: Polity Press (Political Sociology Series).

März, Peter (2014): Nach der Urkatastrophe. Deutschland, Europa und der Erste Weltkrieg. Böhlau Verlag, Köln, Berlin: De Gruyter. Available online at http://www.degruyter.com/search?f_0=isbnissn&q_0=9783412216658&searchTitles=true.

Max-Planck-Gesellschaft zur Förderung der Wissenschaften (2011): Wissenschaft im "Deutschen Oxford". Stadtrundgang durch das Wissenschaftsquartier. Berlin-Dahlem: Hrsg. von Max-Planck-Gesellschaft. Available online at www.mpiwgberlin.mpg.de/PDF/Flyer_MPG_Spaziergaenge2011.pdf.

Neugebauer, Wolfgang (2007): Acta borussia nf. Preußen als Kulturstaat. Unter Mitarbeit von Bärbel Holtz, Rainer Paetau, Christina Rathgeber, Hartwin Spenkuch, Reinhold Zilch, Gaby Huch. Berlin: Berlin-Brandenburgische Wissenschaften (Acta Borussia).

Pelizaeus, Ludolf (2008): Der Kolonialismus. Geschichte der europäischen Expansion. Wiesbaden: Marix-Verlag (Marixwissen).

Reinhard, Wolfgang (2008): Kleine Geschichte des Kolonialismus. 2., vollst. überarb. und erw. Aufl. Stuttgart: Kröner (Kröners Taschenausgabe, 475). Available online at http://deposit.d-nb.de/cgi-bin/dokserv?id=3040366&prov=M&dok_var=1&dok_ext=htm.

Runeby, Nils (1997): Deutschland als technisches Vorbild. Möten och vänskapsband; [Deutsches Historisches Museum, 24. Oktober, 1996–6. January, 1997, Nationalmuseum Stockholm, 26 February–24. Mai, 1998, Norsk Folkemuseum . . .]. Skandinavien och Tyskland 1800–1914.

Savoy, Bénédicte (Hrsg.) (2015): Tempel der Kunst. Die Geburt des öffentlichen Museums in Deutschland 1701–1815. Zweite Auflage. Köln, Weimar, Wien: Böhlau Verlag.

Schlenke, Manfred (Hrsg.) (1981): Preußen. Beiträge zu einer politischen Kultur. Reinbek bei Hamburg: Rowohlt (Preußen – Versuch einer Bilanz, eine Ausstellung der Berliner Festspiele GmbH. 15 August–15 November, Martin-Gropius-Bau Berlin, ehemaliges Kunstgewerbemuseum. Katalog in fünf Bänden. Verlag: Berliner Festspiele GmbH, Bd. 2).

Schreiber, Fritz (1923): Die Industrie der Steinkohlenveredelung. Zusammenfassende Darstellung der Aufbereitung, Brikettierung und Destillation der Steinkohle und des Teers. Wiesbaden, s.l.: Vieweg+Teubner Verlag. Available online at http://dx.doi.org/10.1007/978-3-663-05097-1.

Steinmetz, George (2017): Empire in Three Keys. Forging the Imperial Imaginary at the 1896 Berlin Trade Exhibition. *Thesis Eleven*, 139(1). DOI: 10.1177/0725513617701958.

Weisband, Edward; Thomas, Courtney Irene Powell (2015): Political Culture and the Making of Modern Nation-States. London, New York: Routledge Taylor & Francis Group.

Werner, Frank (1997): Berlin: Neue alte Hauptstadt. In: Der Bürger im Staat., Heft 2, 74–79.

Westphal-Hellbusch, Sigrid (1973): Zur Geschichte des Museums. Hundert Jahre Museum für Völkerkunde. *Baessler-Archiv*, XXI, 1–99.

Zalewski, Przemyslaw Paul (2007): Altstadtsanierungen in Deutschland und in Europa bis zum Zweiten Weltkrieg. Eine Erinnerung an Motive und Methoden. *Journal of Comparative Cultural Studies in Architecture*, (1), 28–36.

Zimmermann, Andrew (2013): Bewegliche Objekte und globales Wissen. Die Kolonialsammlungen des Königlichen Museums für Völkerkunde in Berlin. In: Habermas, Rebekka; Przyrembel, Alexandra (Hrsg.). Von Käfern, Märkten und Menschen. Kolonialismus und Wissen in der Moderne. 1. Auflage. Göttingen: Vandenhoeck & Ruprecht.

2 The First World War and the hygiene movement

The First World War

In the temporal context, it is methodologically indispensable to include the First World War. Between 1914 and 1918, around 10 million people were killed and 20 million were wounded over the course of four years. Extremely precarious living conditions led to major health problems for the military and civilian populations during the war years. On battlefields, soldiers had to deal with the challenges of poor personal hygiene. Army units were plagued by parasites, such as scabies mites (*Sarcoptes* sp.) and lice, which even infested their clothing and uniforms. Millions of survivors were either expelled from their homes or deported to prison camps and forced to work as prisoners of war or laborers. As a result of being confined to shantytowns, epidemics broke out uncontrollably due to poor sanitary and hygienic conditions. This was also a time of economic scarcity, during which a severe shortage of goods arose.

At the end of the First World War, a multitude of problems accumulated in Germany, which negatively affected the general health of society. As a result, political, economic, scientific, and social hygiene solutions were sought to combat the spreading of diseases. Among these efforts, the Kaiser-Wilhelm-Institut für physikalische Chemie and Elektrochemie (Kaiser-Wilhelm-Institute for Physical Chemistry and Electrochemistry) (KWIpCh) in Berlin-Dahlem played a crucial role. The institute was divided into nine departments, worked exclusively for the military administration, and was subordinated to the War Ministry from the end of 1916. During his tenure as the first director of this institute from 1911 to 1933, the chemist Fritz Haber oversaw the development and testing of chemical warfare agents (see also: Ebbinghaus 1998, 39; Stoltzenberg 1994).[1] He personally monitored the first gas attack at Ypres, which began on April 22, 1915, using chlorine-filled metal cylinders (AMPG 1958). In another attack in the eastern Alps of Italy, the 14th Army of the German Empire first used the respiratory poison phosgene, which penetrated the filters of gas masks. Subsequently, another poison gas was used, which corroded the lungs of unprotected soldiers (Cf. Rauchensteiner und Broukal 2015, 184–185). Haber succeeded in disseminating his own inventions by inviting representatives of the Farbwerke Hoechst, vormals Meister, Lucius & Brüning, the Badische Anilin- und Soda-Fabrik Ludwigshafen (BASF), the IG-Farbenfabriken,

DOI: 10.4324/9781003407607-4

vormals Friedrich Bayer & Co., Elberfeld & Leverkusen, and the Aktiengesells-chaft für Anilin-Fabrikation (AGFA) Berlin to joint meetings (AMPG 1916). The production of their paints produced vast amounts of chlorine as a waste product, which earned this industry the epithet "chlorine chemistry". In the correspondence between the KWIpCh and the respective companies, their representatives were directly requested to contact the respective specialists of the institute for specific questions (AMPG 1917b). The military administration took strict care to retain official sovereignty over Haber's inventions and developments (AMPG 1917a; AMPG 1917c).

Two examples serve to illustrate the close relationship between political inter-ests, scientific research, and industrial production. Farbwerke Hoechst AG, vormals Meister, Lucius & Brüning provided the KWIpCh with confidential formulations to produce the most important substances for war. These contained detailed infor-mation on the filling of smoke grenades (100 % solid sulfur trioxide) and fog appa-ratus (liquid mixture of 60 % sulfur trioxide and 40 % chlorosulfonic acid). Haber negotiated directly with Heinrich Dräger, the manufacturer of breathing apparatus, for the industrial production of gas masks. As a result, the Preußisches Kriegsmin-isterium (Prussian War Ministry) commissioned the company to develop special-ised respiratory protection equipment for German soldiers at the front (AMPG (ohne Jahresangabe)). The first gas companies to be equipped in this way were called "Desinfektionskompanien" for camouflage (AMPG 1955).[2]

The escalation of the war led to a significant reduction in the export of domes-tic products to global markets, which had a detrimental impact on the German economy. The procurement of essential raw materials from foreign sources, essen-tial to the war effort, further exacerbated the situation. Until 1913, the German Empire imported mineral oil mainly from the USA, with approximately 30 % coming from Galicia and Romania (Cf. Karlsch und Stokes 2003, 93–94, 131–132). On August 13, 1914, on the initiative of Walter Rathenau[3] and Wichard von Moellendorff,[4] a war raw materials department was founded in the Preußisches Kriegsministerium to procure all raw materials and substitutes important to the war effort. Rathenau headed this department until March 1915 and employed Haber as a member of the team. He was given the task of finding and developing substitutes for nutrition and increasing crop yields. Haber had already achieved a major break-through with the ammonia synthesis process he had invented (see chapter 1: The Age of Industrialization). Based on this invention, he developed a process for the industrial production of artificial fertilizer with Carl Bosch and received the Nobel Prize in Chemistry for 1918 in 1919. Haber's work on developing combat gases for the German army command later extended into the development of gases for controlling insect pests in the civilian sector (Karlsch und Stokes 2003, 19; see also AMPG 1985; Schmaltz 2005, 18; Cf. Kaiser 2002, 1–5; Schulte von Drach 2015). He saw himself as a link between science, industry, and the military, as well as a savior of humanity, expressing this in his paradigmatic statement

"Im Frieden der Menschheit, im Krieg dem Vaterland".[5]

(Cf. Kaiser 2002, 218)

Although the Hague Conferences of July 29, 1899, and October 1, 1907, had prohibited the use of poisonous substances as weapons by a treaty, the First World War had transformed into a "gas war" and thus became the first industrially waged war. The classification of basic research in the field of pest control in terms of time is often challenging, making it difficult to place it historically. The consequences of industrialization and the commencement of the First World War are closely linked (Cf. März 2014, 263–264). However, what is indisputable is the historically proven fact that the first use of toxic gases to exterminate people took place during the First World War and was done by German soldiers (Kaiser 2002, 210–220). The use of gases as pesticides in museum collections is similar to a special use in the civilian sector and is discussed extensively in Part III.

The hygiene movement as an indirect consequence of industrialization and the First World War

Industrialization and the First World War were historical events of unimaginable dimensions that had serious consequences for civil society and the military. In confined spaces under dire hygienic conditions, people, whether in European cities, on battlefields, or in prison camps, were forced had to live in close proximity to one another. As a result, microorganisms, such as viruses, bacteria, and yeasts, as well as parasites, spread explosively. Insects and rodents transmitted infectious diseases and caused significant damage to food. These factors have made insects and microorganisms into (material or stock) "pests" and "vermin". These are not zoological terms; rather, they reflect the point of view of contemporaries. To address health hazards, the government has worked intensively to find solutions for better cleanliness and hygiene, collaborating with industry to develop, manufacture, and distribute active ingredients and agents. Pharmacists' own recipes were also used, including the arsenic soap developed by the French pharmacist Jean-Baptiste Bécouer (1718–1777). This soap was used for preservation of natural history collections until the mid-20th century (see glossary for explanation of arsenic soap). A turning point in the production of pesticides began at the end of the 19th century. The chemical industry has become increasingly important and specifically sought sales markets for products to combat insect pests. There was great interest in expanding the client base beyond the healthcare sector. Cultural institutions, such as the KMfV/MfV, were also potential buyers (see Part III, chapter 10).

During the late 19th and early 20th centuries, the emerging and developing hygiene and health authorities played a key role in the hygiene movement. The establishment of the Kaiserliches Gesundheitsamt (KGA) (Imperial Health Office) as the highest Reich authority for medical care on April 28, 1876, in Berlin, had widespread importance (Hüntelmann 2008). In the years 1884–1885, Robert Koch was appointed as its deputy leader (Ibid. 10). There was a bacteriological laboratory in the increasingly science-oriented authority where Koch conducted his first experiments. In 1891, the Königliche Preußische Anstalt für Infektionskrankheiten (Royal Prussian Institute for Infectious Diseases), which today is known as Robert Koch-Institut (RK-I) (Robert Koch-Institute), was founded where Koch devoted himself entirely to the study of bacteria (Hüntelmann 2008, 10, 172–173; Vasold 2002). The intervention of the KGA in Berlin and Koch's dispatch to Hamburg successfully combated the cholera epidemic there (Evans 1991, 372). During a tour of the city, where he witnessed the catastrophic circumstances, he made his famous statement on August 24, 1892:

"Meine Herren, ich vergesse, dass ich in Europa bin".[6]

(Ibid. 398)

From 1918, the authority was renamed the RGA and henceforth the highest health and veterinary authority in Germany, which played a significant role in shaping health policy. The RGA developed an extensive network across the realms of science, medicine, and politics, often serving as an intermediary for conflicting political and social interest groups. In competition with individual states and municipalities in Germany at that time, the RGA gained more significance in the fields of medicine, science, and agriculture. Hüntelmann attributes the founding of the Physikalisch-Technische Reichsanstalt (PTR) (Physical-Technical Reich Institute) in 1887 and the Kaiserliche-Biologische-Landesanstalt für Land- und Forstwirtschaft (KBA) (Imperial Biological Institute for Agriculture and Forestry) in 1905 as a result of this development. Industrialization led to poor housing, living, and working conditions in urban areas, driving a push toward the natural sciences (Hüntelmann 2008, 262). In the long term, it indirectly led to a "Verwissenschaftlichung des Sozialen",[7] i.e., in which scientific expertise became increasingly relevant to the daily lives of large sections of the population. Within five decades, the RGA developed into a "Hochburg der Hygiene" (see Hüntelmann 2008, 11–12)[8] and received the status of a Ministerium für Gesundheit (Ministry of Health) (Ibid. 76). In 1935, the authority was merged with the RK-I für Infektionskrankheiten and the Preußische Landesanstalt für Wasser-, Boden- und Lufthygiene (WaBoLu) (Prussian State Institute for Water, Soil and Air Hygiene). This institution developed into a ministry with a main task of combating prevalent diseases, such as typhoid, tuberculosis, cholera, and diphtheria, and ensuring adequate nutrition for the entire population (see https://wabolu.de).[9]

The results of basic and applied research have led to the classification of microorganisms, insects, and rodents into harmful and unharmful organisms. Knowledge about the spread and control of pests was disseminated in Berlin by the WaBoLu in Berlin-Dahlem. They offered courses aimed at professional groups of exterminators, disinfectors, and other interested people. Well-known professors, such as the zoologist Julius Franz Wilhelmi, introduced the basics of zoology and entomology. Furthermore, the chemical basics of pest control were taught, and different procedures and methods were applied. Even at that time, people were aware of the risks associated with occupational accidents that could result in harm to their health while executing pest control measures. This knowledge as well as first-aid measures were also incorporated into the lessons (Anonymous 1927, 49). As part of efforts to improve hygiene in homes, various aspects of air hygiene, drinking water supply, sewage, and waste disposal were examined. The knowledge of disease-transmitting pests in residential and public buildings has deepened and worked holistically in the field of urban hygiene in today's sense (Cf. Hüntelmann 2008, 193–208). The social hygiene activities of the RGA included, for example, "systematische Zahnpflege in den Schulen",[10] medical examinations, and school meals (Cf. Hüntelmann 2008, 209). The knowledge gained resulted in the Reich Epidemic Law, which was drafted primarily by the RGA. It was issued after long preparation on August 28, 1905, by the Preußische Minister der geistlichen, Unter- richts- und Medizinalangelegenheiten im Deutschen Reich (Prussian Minister of Religious, Educational, and Medical Affairs in the German Reich). The education and instruction on hygiene of large parts of the population was imparted through the distribution of numerous pamphlets, leaflets, and brochures, as well as a health booklet.

The common practice of "Schlafgänger"[11] at the end of the 19th century encour- aged the spread of insects and diseases when beds were used by several people through shifted sleeping. In garments made of coarse wool, textile moths (Cf. Münch 1995, 204) (*Tineola* sp.) nested and multiplied without problems. Sca- bies, caused by scabies mites as well as lice, bedbugs, and fleas, were a common part of everyday life. One of the greatest dangers was the plague transmitted by rats. Patients had to undergo targeted fumigation measures in clinics as well as in their living environments (Ibid. 203). For this purpose, new means of fumigation and sterilization have been developed, which have changed the entire spectrum of medical and agrotechnical pest and disease control. The spread of the disease through the importation of goods was regulated by export and import bans. For implementation, people were trained in eight-day courses and equipped with an ID card as state-certified disinfectors. Disinfection was carried out in special facilities using both steam and chemical agents. The level of concern about the spread of the plague was so high that attending a feast after a funeral was made a punishable offence. The extent to which urban clinics were contaminated with vermin is dem- onstrated, for example, by attempts to combat the pharaoh ant. In 1930, a report by the städtische Heil- und Pflegeanstalt in Berlin-Buch (municipal sanatorium and nursing home in Berlin-Buch) named the Rudolf-Virchow-Krankenhaus (Rudolf Virchow Hospital), the Kinderheilanstalt Buch (Children's Hospital Buch), the Hospital Buch-Ost (Buch-East Hospital), the Krankenhaus Britz (Britz Hospital), the Krankenhaus am Friedrichshain (Friedrichshain Hospital), the Kinderkrank- enhaus Reinickendorferstraße (Reinickendorferstraße Children's Hospital), the Heil- und Pflegeanstalt Herberge (Mental and Nursing Home Herberge), and the Heil- und Pflegeanstalt Buch (Mental and Nursing Home Buch) as infected (Ibid. 153–158, 166–167; LAB 1930). By the early 19th century, the negative effects of bacteria, microorganisms, and rodents led to the realization that bathing not only served medical purposes but also cleanliness and personal hygiene (Cf. Münch 1995, 205). In the second half of the 19th century, bathing establishments became part of hygiene campaigns to create awareness that cleanliness and health are inter- related (Ibid. 214). During the first cholera epidemic in 1866, sewage and cesspools were controlled using scientific methods and with the threat of police penalties. In

Berlin, Rudolf Virchow[12] (Ibid. 222–225), in collaboration with James Hobrecht,[13] provided a municipal sewage system and a central drinking water supply to contain the epidemics (Wagner et al. 1972, 280–281). In all changes to the health care system, the preservation of the workforce was a top priority.

This was the reaction of German society and its institutions to the rapidly increasing urban population numbers and the changing habits, needs, and requirements of their inhabitants. During this period, birth rates declined significantly. The birth rate dropped from 35.6 to 27.5 per 1000 of the population, while the death rate decreased from 22.1 to 15.0 per 1000 of the population. This "medical-hygienic revolution" led to an increase in life expectancy while also introducing improvements in working and production conditions. Important active substances and agents, such as "Autan", "Flit", and "Globol", were developed and distributed to combat microorganisms and rodents. These funds were also offered to the KMfV/ MfV and were used there. That also concerned the fumigant "T-Gas", but proof of its application could not be provided in the context of this book (see Part III, chapter 10; see März 2014, 118–119). The studies available in this book relate to factories in Germany that developed pesticides that were then disseminated around the world. An example of this is AGFA in Germany, which had its own laboratories and distributed "Globol". However, it is assumed that laboratories worldwide were producing and marketing toxic pesticides to museums.

Notes

1 Fritz Haber was born on December 9, 1868, and died on January 29, 1934.
2 Disinfection companies (translation by the author).
3 Walter Rathenau, born on September 29, 1867, and died on June 24, 1922, was a German industrialist, writer, and liberal politician. During the First World War, he participated in the organization of the war economy.
4 Wichard von Moellendorff, born on October 3, 1881, and died on May 4, 1937, was a German engineer and economic theorist. He became publicly known for his economic policy activities during and after the First World War.
5 In peace to mankind, in war to the fatherland (translation by the author).
6 Gentlemen, I forget that I am in Europe (translation by the author).
7 Scientification of the social (translation by the author).
8 Stronghold of hygiene (translation by the author).
9 In 1901, the Royal Research and Testing Institute for Water Supply and Sewage Disposal was founded in Berlin. In 1923, it was renamed the Prussian State Institute for Water, Soil and Air Hygiene. Klaus, Burkhard. Verein für Wasser-, Boden- und Lufthygiene e.V., Berlin. WaBoLu Homepage. Available online at https://wabolu.de, last accessed on February 24, 2021.
10 Systematic dental care in schools (translation by the author).
11 No term could be found for the German term "Schlafgänger" in the English language.
12 Rudolf Virchow, born on October 13, 1821, and died on September 5, 1902, was a German physician, pathologist, pathological anatomist, anthropologist, prehistorian, and politician. He gained an international reputation in Würzburg and Berlin.
13 James Hobrecht, born on December 31, 1825, and died on September 8, 1902, was a Prussian urban planner and responsible for Berlin's first perspective development plan, the Hobrecht Plan of 1862.

References

Archive material

AMPG (1916): Haber-Sammlung von Johann Jaedicke. Va ABT, Rep. 0005, Nr. 516. Loseblattsammlung. Haber, Fritz. Brief an die Direktion der Farbwerke Hoechst, vormals Meister, Lucius & Brüning. 30 Juni, Blatt 7–8, 2 Seiten.

AMPG (1917a): Haber-Sammlung von Johann Jaedicke. Va ABT, Rep. 0005, Nr. 516. Loseblattsammlung. Haber, Fritz. Brief an die Farbwerke Hoechst AG, vormals Meister, Lucius & Brüning. 30 Mai, Blatt 12–13, 2 Seiten.

AMPG (1917b): Haber-Sammlung von Johann Jaedicke. Va ABT, Rep. 0005, Nr. 516. Haber, Fritz. Brief vom. 12 Juni, Blatt 15, 1 Seite.

AMPG (1917c): Haber-Sammlung von Johann Jaedicke. Va ABT, Rep. 0005, Nr. 516. Anonymous. 19 Oktober, Blatt 16, 1 Seite.

AMPG (1955): Haber-Sammlung von Johann Jaedicke. Va ABT, Rep. 0005, Nr. 534. Lummitzsch, Otto. Brief an Heinrich Dräger vom. 10 Oktober, Blatt 1, 1 Seite.

AMPG (1958): Haber-Sammlung von Johann Jaedicke. Va ABT, Rep. 0005, No. 514. Loseblattsammlung. Niehaus, (first name unknown, author's note). Der allererste Anfang des Gaskrieges. Bericht vom März, Blatt 2, 2 Seiten.

AMPG (1985): Haber Sammlung von Johann Jaedicke. Va ABT, Rep. 5, Nr. 2295. Frucht, Adolf-Henning. Fritz Haber und die Schädlingsbekämpfung während des I. Weltkrieges und in der Inflationszeit. Vorläufige Bekanntmachung über die Einzeluntersuchung vom Dezember. Blatt 8–9, 2 Seiten.

AMPG (ohne Jahresangabe): Haber-Sammlung von Johann Jaedicke. Va ABT, Rep. 0005, Nr. 533. Aktenauszug der Farbwerke Hoechst AG, vormals Meister, Lucius & Brüning ohne Jahresangabe. Rezepturen zur Herstellung von Gaskampfstoffen sowie zum Befüllen von Rauchgranaten und Nebelgeräten. Blatt 1–10, 10 Seiten.

LAB (1930): A Rep. 003–04–01, No. 194. Akte Städtische Heil- und Pflegeanstalt Buch. Schädlingsbekämpfung. Loseblattsammlung. Pieper, (first name unknown, author's note). Bericht vom 8. Juli 1930, über die Bekämpfung der Pharaoameise in städtischen Kliniken, Blatt 36, 1 Seite.

Other sources

Anonymous (1927): Zweiter Lehrgang zur Bekämpfung der Gesundheitsschädlinge vom 14.-22. Februar 1927. *Zeitschrift für Desinfektions- und Gesundheitswesen*, (1), 49.

Ebbinghaus, Angelika (1998): Der Prozeß gegen Tesch & Stabenow. Von der Schädlingsbekämpfung zum Holocaust. *Zeitschrift für Sozialgeschichte des 20. und 21. Jahrhunderts*, 1999, 13(2), 16–71.

Evans, Richard J. (1991): Tod in Hamburg. Stadt, Gesellschaft und Politik in den Cholera-Jahren 1830–1910. 4.–5, 372. Tsd. Reinbek bei Hamburg: Rowohlt.

Hüntelmann, Axel Cäsar (2008): Hygiene im Namen des Staates. Das Reichsgesundheitsamt 1876–1933. Diss.-Bremen. Univ., 2005. Wallstein, Göttingen. Available online at http:// deposit.d-nb.de/cgi-bin/dokserv?id=3099685&prov=M&dok_var=1&dok_ext=htm.

Kaiser, Gerhard (2002): Wie die Kultur einbrach. Giftgas und Wissenschaftsethos im Ersten Weltkrieg In: Sonderdrucke aus der Albert-Ludwigs-Universität Freiburg, 1–5. Available online at URN: urn: nbn:de: bsz:25-freidok-5065. Originalbeitrag erschienen in: Merkur 56, 2002, Heft 635.

Karlsch, Rainer; Stokes, Raymond (2003): Faktor Öl. Die Mineralölwirtschaft in Deutschland 1859–1974. München: Beck. Available online at www.gbv.de/dms/faz-rez/ FD1200304101799547.pdf.

März, Peter (2014): Nach der Urkatastrophe. Deutschland, Europa und der Erste Weltkrieg. Köln, Berlin: Böhlau; de Gruyter. Available online at www.degruyter.com/ search?f_0=isbnissn&q_0=9783412216658&searchTitles=true.

Münch, Ragnhild (1995): Gesundheitswesen im 18. und 19. Jahrhundert. Das Berliner Beispiel. Zugl.: Berlin, Freie Univ., Diss., 1992 u.d.T.: Münch, Ragnhild: Öffentliches Gesundheitswesen und soziale Fürsorge in Berlin zwischen staatlicher Repression und Reformkonzepten (18. und 19. Jahrhundert). Berlin: Akademie Verlag (Publikationen der Historischen Kommission zu Berlin). Available online at www.gbv.de/dms/faz-rez/ F19950902ROST1–100.pdf.

Rauchensteiner, Manfried; Broukal, Josef (2015): Der Erste Weltkrieg und das Ende der Habsburgermonarchie 1914–1918. In aller Kürze. Wien, Köln, Weimar: Böhlau Verlag.

Schmaltz, Florian (2005): Kampfstoff-Forschung im Nationalsozialismus. Zur Kooperation von Kaiser-Wilhelm-Instituten, Militär und Industrie. Vollst. zugl.: Bremen, Univ., Dissertation, 2004. Göttingen: Wallstein (Geschichte der Kaiser-Wilhelm-Gesellschaft im Nationalsozialismus, 11). Available online at www.h-net.org/review/hrev-a0f1o1-aa.

Schulte von Drach, Markus C. (2015): Erster Giftgaseinsatz im Ersten Weltkrieg. Die schreckliche Erfindung des Patrioten Fritz Haber, Süddeutsche Zeitung vom 22.04.2015. Available online at www.sueddeutsche.de/politik/erster-giftgaseinsatz-im-ersten-weltkrieg-die-schreckliche-erfindung-des-patrioten-fritz-haber-1.2385082.

Stoltzenberg, Dietrich (1994): Fritz Haber. Chemiker, Nobelpreisträger, Deutscher, Jude, eine Biographie. Weinheim: VCH.

Vasold, Manfred (2002): Robert Koch, der Entdecker von Krankheitserregern. Heidelberg: Spektrum der Wissenschaft.

Wagner, Fritz; Rieckenberg, Hans Jürgen; Glaubrecht, Martin; Jaeger, Hans; Hentig, Hans Wolfram von; Körner, Hans (1972): Neue Deutsche Biographie. Hess-Hüttig. 1–9. Berlin: Duncker & Humblot (9).

3 The development of storage and plant protection

The development of storage protection during industrialization and the First World War

The Kaiserliches Gesundheitsamt/Reichsgesundheitsamt (Imperial Health Office/ Reich Health Office) was significantly involved in the shaping of food laws (Cf. Hüntelmann 2008, 263). On February 25, 1898, a separate biological department for agriculture and forestry was established within the authority, which was granted four laboratories within a year. On April, 1, 1905, the Kaiserliche Biologische Anstalt für Land- und Forstwirtschaft (KBA) (Imperial Biological Institute for Agriculture and Forestry) was created (Aderhold 1906, 2–6). Specialists dealt primarily with the preservation of food and the control of storage pests and other harmful insects (Ibid. 12–14). Friedrich Zacher, a German entomologist and head of the laboratory from 1911, coined the concept of storage protection. His standard work, a systematic description of storage and material pests as well as their control (Zacher 1927), received much recognition in science as well as in numerous journals at home and abroad (BArch 1927–1933). Along with his wife, he founded and directed the Gesellschaft für Vorratsschutz (Society for the Protection of Stores) in Berlin-Steglitz, whose central organ was the *Mitteilungen der Gesellschaft für Vorratsschutz e.V.* The society was a platform for industry, trade, agriculture, and consumers, where the latest scientific results and their practical applicability were published and discussed. In the immediate vicinity were the employees of the Königliches Museum für Völkerkunde/Museum für Völkerkunde (KMfV/MfV), who also read this journal (see Part III, chapter 10). Zacher regarded himself and his society as an important link between the state and industry, and between scientific research and the practical questions of trade and agriculture. As a pioneer of a nation that had to reconquer its role on the world market after the First World War was lost, he fought for Germany's competitiveness vis-á-vis other states.

The importation of potatoes from North America led to the introduction of the American Colorado potato beetle (*Leptinotarsa decemlineata*) into Germany. The authorities informed farmers about the ban on importing American potatoes (Anonymous 1875), but the ban was ineffective (GStA PK 1875). Despite the use

DOI: 10.4324/9781003407607-5

of crude benzene and arsenic-containing infusions to control the Colorado potato beetle, educational efforts and prohibitions were unsuccessful in eradicating the insect. During the term of office of Martin Schwartz, the senior government councilor and head of the economic department of the Biologische Reichsanstalt für Land- und Forstwirtschaft (Biological Reich Institute for Agriculture and Forestry) in Berlin-Dahlem, a decree was issued on March 7, 1923, aiming to prevent the import of living plants and fresh plant parts from America, Japan, Australia, China, and Hawaii in an attempt to eradicate the Colorado potato beetle, the phylloxera, the San José scale insect, and the potato crab (Schwartz 1925, 653–654). This early form of globalization even gave rise to fears of economic collapse and the ruin of farmers (GStA PK 1922c).

During the late 19th and early 20th centuries, horticulture for both commercial and private use became a permanent fixture for feeding broad sections of the population. In 1925, Theodor Landgraf, a head teacher of trades, issued an appeal to professional gardeners and allotment gardeners for the fundamental control of harmful insects. In the *Grundsätze der Schädlingsbekämpfung im Gartenbau*,[1] he staged a proxy war against microorganisms, using idioms to conjure up horror scenarios. It was a struggle against millions of people, a campaign in which the culprits and perpetrators were determined from the very beginning (Landgraf 1925, 37). In his pamphlet, he turned insects, also known as "plague pathogens", into active actors whose goal is to rage in a devastating way. Landgraf consequently saw it as a primary task to intervene in changing the population density of an animal species. From today's perspective, it is interesting to note that he also addressed aspects of the integrated pest management method commonly used in museums. He recommended cleaning manure beds, greenhouses, tools, and shoes to prevent contamination. Infested plant parts, wood, and unharvested fruits should also be burned. From his point of view, these measures serve only as preparation to burn out or burn "contaminated" earths with a mixture of "Formalin" and 0.25 % or 0.5 % copper sulfate or carbon disulfide (Ibid. 33–36). In terms of controlling stored pests, Zacher observed significant differences in terms of suitable methods compared to plant protection, and he justified this in the much more manageable range of species. Therefore, he advocated careful treatment of goods and used the concept of the "schädlingsbiologische Hygiene der Warenlagerung" (Zacher 1924a, 46, 87–88, 1924b, 45–48).[2] When it comes to protecting food from insect pests, he clearly preferred toxic gases. Its list of requirements includes equity, harmlessness to goods and furnishings, effectiveness, harmlessness to humans and pests, and non-explosive property. To date, carbon monoxide, carbon dioxide, carbon disulfide, carbon tetrachloride, dihydrogen sulfite (sulfurous acid), hydrogen cyanide, and "Zyklon B" have been discussed. Own tests revealed that each substance was not able to meet the requirements. Zacher's observations pursued a combination of volatile compounds with toxic effects and irritant gases.

Belief in progress and the achievements of technology are expressed in an article by Walter Heerdt. He saw in "Zyklon B" an improvement of the hydrocyanic acid process and directed his attention to the USA, where the liquid "Zyklon B" had already been converted on a large scale in the fumigation of orange and lemon trees. "Zyklon B" refers to liquid prussic acid (hydrogen cyanide), which is filled into tin cans and processed by diatomaceous earth (Kainer 1951). In addition, stabilizing and irritants (organic halogen compounds) are added (Heerdt 1924, 81). When the tin cans were opened, the prussic acid evaporated. Heerdt emphasised that "Zyklon B" requires lower amounts of hydrocyanic acid for fumigation compared to the hydrogen cyanide process. He estimated the consumption for a mill with a capacity of 5000 m³ to be a total of 850 kg of hydrogen cyanide but only 220 kg for "Zyklon B" (Ibid. 81–83). Not only are lower amounts of hydrogen cyanide advantageous, but "Zyklon B" is also less dangerous in its application than hydrogen cyanide in the hydrogen cyanide process. For this purpose, diluted sulfuric acid was poured into a vat. Next, the necessary amount of sodium cyanide was added, resulting in hydrogen cyanide (Jäckel 1927, 38). The evaporation process was much more uncontrolled.

By 1929, significant experience had been gained with pesticides for food preservation. At a scientific–technical meeting in Eltville am Rhein, the advantages of gases over solids were discussed with regard to minimizing harm for humans and animals. Carbon disulfide, which had been used since 1850 to combat grain pests, was classified as unsuitable due to its high toxicity, flammability, explosiveness, and long-lasting and unpleasant odor. Although the insecticidal effect of carbon tetrachloride was lower, it was considered a better option than carbon disulfide because it was nonflammable and had a lower toxicity rating. After the First World War, the use of hydrogen cyanide became more prevalent. However, it is hazardous and should only be handled by trained personnel. The dangers of the flammability of "Areginal" were also discussed. According to the knowledge at the time, it was considered to have low toxicity to the human body and no toxic effect on plants. Other tested chemicals, such as propylene, ethylene, and butylene oxide, were found to have good penetration and insecticidal properties but were also more flammable and highly damaging to germs compared to "Areginal". The insecticidal activity of butyl formate (*n*-butyl formate) and carbon tetrachloride in a ratio of 1:1 was found to be the same as "Areginal". There was also no impairment in the germination capacity of seeds. At this point, however, carbon tetrachloride was assigned an unspecified hazard (LASA 1929; Braßler 1925, 69–70).

If we look at the active substances and agents recommended for storage protection in the context of the KMfV/MfV, a particular picture emerges. "Areginal" and carbon disulfide were used in the museum (see Part III, chapter 8). In addition, Rathgen conducted a long-term study on the use of carbon tetrachloride on sensitive surfaces of art and cultural assets (see Part III, chapter 8). The term "Formalin" in this context refers to a questionnaire sent to other institutions by the KMfV (see Part III, chapter 13). However, no evidence of the use of the product during the investigation period could be provided in the present investigation. Conservation measures, including the use of copper sulfate, hydrogen cyanide, and "Zyklon B", were not applied, according to sources.

The development of plant protection during industrialization and the First World War

The protection of stocks and plants are closely related. While the former aims to control food pests, the latter aims to protect agricultural products in the fields, meadows, and forests from harmful insects. However, many active substances, agents, and methods used for plant and storage protection were not always designed with material protection in mind. They can also be used to protect museum collections made of organic materials when necessary. The KBA conducted research on plant protection by investigating animal and plant pests and developing methods for their control. A classification of beneficial and harmful microorganisms was performed. Observations and investigations of bee diseases also fell within the remit of this institution. The first field trials were performed on a test field at the Königliche Domäne (Royal Domain) in Berlin-Dahlem. In the field, the increase in crop yield was studied and the findings were presented to the public (BArch 2008). Another milestone was the "Deutsche Pflanzenschutz zur Beobachtung und Bekämpfung von Pflanzenkrankheiten".[3] It was established by representatives of the individual main offices of the countries and state institutions, which conducted experiments and trials in the corridors, resulting in the strengthening of Germany's economic position on the global market (see BArch 1921). There was an agreement that pioneering work be done in Germany on ways to control pests and the associated plant protection, which also required considerable financial resources. Thus, the Reichsausschuss für Öle und Fette (Reich Committee for oils and fats) provided 50,000 marks for joint experiments. The cooperation between non-official societies and state institutions intensified when it came to raising funds for scientific purposes.

For example, in 1922, the Deutsche Landwirtschafts-Gesellschaft (DLG) (German Agricultural-Society) wrote a letter to the Ministerium für Landwirtschaft, Domänen und Forsten (Ministry of Agriculture, Domains and Forestry) asking for long-term financial support from the Deutsche Pflanzenschutzdienst (German Plant Protection Service) (BArch 1922a). This request was not granted by the authorities because plant protection has not yet been subject to any legal provisions. References were made only to the funds already made available for the current year and the efforts to create a common basis for this with the Reichsernährungsministerium (Reich Ministry of Food) (BArch 1922b). Nevertheless, cooperation was established with non-official bodies, including the Gesellschaft für angewandte Entomologie (Society for applied Entomology) and its chairman Karl Escherich, whose aim was to elevate entomologists' significance and primacy over plant physiologists and plant pathologists. He justified this by citing the work of entomologists in controlling harmful insects and preventing great damage they may cause. As an advocate of specialism, he categorised the various plant damages and their treatments into different categories. Damages caused by external agents should be treated by plant physiologists, damages caused by fungal parasites by mycologists, and damages caused by animal pests primarily by zoologists, whose field was applied entomology (Escherich 1922, 17–21). The non-official institution was reluctant to be dictated how to cooperate with industry when it came to the publication of test results. Therefore, pride was not excluded here either, where representatives of the Reichsministerium (Reich Ministry) always affirmed their leading position in meetings and assemblies (GStA 1920b). However, at the beginning of the 20th century, there were still many unclear relationships and responsibilities. There was a period of experimentation, with numerous products being marketed by individuals, government agencies, and industry from different locations. Interestingly, a lack of nicotine-containing pesticides has led to the spread of arsenic-containing agents, although their toxicity was already known in 1920 (Ibid. Anonymous).

In joint consultations with the representatives of the Hauptämter für Pflanzenschutz im Deutschen Reich (Main Offices for Plant Protection in the German Reich), great attention has been paid to educating and teaching collectors, gardeners, commercial gardeners, horticultural teachers, farmers, agricultural teachers, and plant pathologists. A large-scale campaign was launched through the distribution of leaflets and brochures, and an offer of additional training to establish a permanent foothold for theoretical knowledge on pests and their control across various economic areas. The potential use of microorganisms as a means of controlling plant pests was also explored. For research in this field, the Kaiserlich Königliche Landwirtschaftlich-Bakteriologische Pflanzenschutzstation (Imperial Royal Agricultural and Bacteriological Plant Protection Institution) was maintained in Vienna at the beginning of the 20th century. In his basic article, "*Über die Bekämpfung tierischer landwirtschaftlicher Schädlinge mit Hilfe von Mikroorganismen*",[4] Kornauth discusses the difficulty of breeding such fungal cultures to be able to use them profitably (Kornauth 1904, 365–387). During a meeting with the special committee on plant health, the control of field mice was discussed in field trials. To deter birds of prey, poisoned grains and crutches were placed on the fields. For economic reasons, arsenic-containing agents were preferred over strychnine (GStA 1920a). The use of chemical agents in forests, fruit cultivation, and viticulture resulted in visible consequences. There were birds that must be protected, while starlings in wine-growing areas multiplied at an alarming rate and posed a significant threat to the wine harvests (GStA 1922a).

The industry established its own departments for crop protection and product sales. In May 1933, I.G. Farbenindustrie Aktiengesellschaft showcased its own stand at the traveling exhibition of the DLG in Berlin, presenting itself as an integral part of the overall social achievement of the Germans and emphasised the "nationalen Aufbauwillen des deutschen Volkes".[5] The company highlighted the importance of its own products, which were presented as serving the interests of agriculture through the union of theory and practice or "Geist und Arbeit" (LASA 1933).[6] Based on this background and from a certain distance, museums emerge as small, self-contained institutions that were heavily reliant on external support in their efforts to combat insect pests present in their collections during the period under investigation.

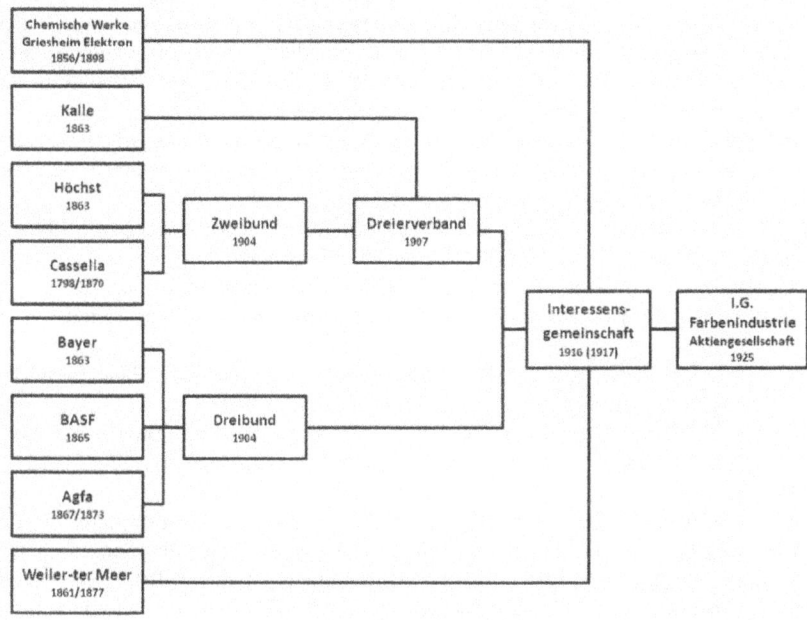

Figure 3.1 Graphic representation of the development of I.G. Farbenindustrie Aktiengesellschaft.

Notes

1 Principles of pest control in horticulture (translation by the author).
2 Biological hygiene of goods storage (translation by the author).
3 German plant protection for the monitoring and control of plant diseases (translation by the author).
4 On the control of animal agricultural pests with the help of microorganisms (translation by the author).
5 National will of the German people to rebuild (translation by the author).
6 Spirit and work (translation by the author).

References

Archive material

BArch (1921): R 3602. Anonymous. Gemeinsame Sitzung von Vertretern der Reichs- und Landesregierungen und des deutschen Pflanzenschutzdienstes und der landwirtschaftlichen Körperschaften am 21.05.1921. Blatt 35–37, 3 Seiten.

BArch (1922a): R 3602. Deutsche Landwirtschafts-Gesellschaft. Brief vom 08.05.1922 an das Ministerium für Landwirtschaft, Domänen und Forsten, Blatt 38, 2 Seiten.

BArch (1922b): R 3602. Eggert, first name unknown (author's note). Brief aus dem Ministerium für Landwirtschaft, Domänen und Forsten vom 31.05.1922 an die Deutsche Landwirtschafts-Gesellschaft, Blatt 45, 2 Seiten.

BArch (1927–1933): R 3602/2461. Akte der Biologischen Reichsanstalt für Land- und Forstwirtschaft. Berücksichtigung des Vorratsschutzes. Loseblattsammlung. Entwicklung und Tätigkeiten allgemein, historische Entwicklung, Rezensionen und Empfehlungen. Bundesarchiv Berlin-Lichterfelde 1927–1933, Blatt 305, 1 Seite.; Blatt 326–329, 4 Seiten.; Blatt 335–337, 3 Seiten.; Blatt 380–381, 2 Seiten.

BArch (2008): R 3602. Zusammenfassung der Bestandssignatur. Biologische Reichsanstalt für Land- und Forstwirtschaft. Schädlingsbekämpfung, Schutzmaßnahmen, Aufklärung von 1897–1920, 2 Seiten, ohne Paginierung.

GStA (1920a): I. HA Rep. 87, Nr. 19127. Anonymous. Die Überwachung von Pflanzenschutzmitteln. Bericht vom 15.10.1920, Blatt 15–16, 2 Seiten.

GStA (1920b): I. HA Rep. 87, Nr. 19127. Anonymous. Niederschrift der Beratung vom 15.10.1920 von Vertretern der Hauptstellen für Pflanzenschutz im Deutschen Reich, Blatt 7–8, 2 Seiten.

GStA (1922a): I. HA, Rep. 87, B, Nr. 19127. Berlepsch, Freiherr von, Hans. Bericht vom 17.02.1922 über den Vogelschutz in der Pfalz, Blatt 40, 1 Seite.

GStA (1922b): I. HA, Rep. 87, B, Nr. 19127. Loseblattsammlung. Anonymous. Niederschrift der 39. Sitzung des Sonderausschusses für Pflanzenschutz vom 17.02.1922, Blatt 40, 1 Seite.

GStA PK (1922c): I. HA, Rep. 87, B, Nr. 19127. Akte des Ministeriums für Landwirtschaft, Domänen und Forsten. Abteilung I A. Reichsausschuss für Pflanzenschutz und Schädlingsbekämpfung, 1920–1922. Loseblattsammlung. Behrens; Brick; Schultze, first names unknown (author's note). Sonderausschuss für Pflanzenschutz. Niederschrift der 39. Sitzung vom 17. Februar 1922. Protokoll, Blatt 39–42, 8 Seiten.

GStA PK (1875): I. HA Rep. 89, Nr. 30186, Acta betreffend die Maßregelung zur Vertilgung der den Feldfluren schädlichen Insecten. Loseblattsammlung. Ernährung der Bevölkerung; landwirtschaftliche Erträge. Wilmowski von, Karl. Brief an den deutschen Kaiser Wilhelm I. vom 24. Februar 1875, 1 Seite, ohne Paginierung.

LASA (1929): I 532, Nr. 599. Loseblattsammlung. Anonymous. Referat gehalten am 15. Mai 1929 für die 8. wissenschaftlich-technische Austauschsitzung in Eltville, 4 Seiten, ohne Paginierung.

LASA (1933): I 532, Nr. 600. Loseblattsammlung. Maier-Bode, first name unknown (author's note). Wanderausstellung der Deutschen Landwirtschafts-Gesellschaft vom 20.–28. Mai 1933. Standardbrief vom Mai 1933 der I.G. Farbenindustrie Aktiengesellschaft, Beratungsstelle für Pflanzenschutz, 1 Seite, ohne Paginierung.

Other sources

Aderhold, Rudolf (1906): Die Kaiserliche Biologische Anstalt für Land- und Forstwirtschaft in Dahlem. In: Mitteilungen aus der Kaiserlichen Biologischen Anstalt für Land- und Forstwirtschaft. Berlin: Verlagsbuchhandlung Paul Parey und Julius Springer, 1, 1–20.

Anonymous (1875): Der Kartoffelkäfer, Chrysomela (Doryphora) decemlineata. Herausgegeben im Auftrag des Königlich Preußischen Landwirtschaftsministeriums. Berlin: Ernst Schotte & Voigt.

Braßler, Karl (1925): Areginal, ein neues Mittel gegen Sammlungs- und Bücherschädlinge. *Anzeiger für Schädlingskunde*, 1(6), 69–70. DOI: 10.1007/BF02628433.

Escherich, Karl (1922): Die Stellung der angewandten Entomologie im Pflanzenschutz. Dritte Mitgliederversammlung zu Eisenach vom 28–30 September 1921. *Verhandlungen der Deutschen Gesellschaft für angewandte Entomologie*, 17–25.

Heerdt, Walter (1924): Zyklon B, ein verbessertes Blausäureverfahren. Vierte Mitglieder-versammlung zu Frankfurt a.m. vom 10. bis 13. Juli 1924. *Verhandlungen der Deutschen Gesellschaft für angewandte Entomologie*, 81–83.

Hüntelmann, Axel Cäsar (2008): Hygiene im Namen des Staates. Das Reichsgesundheitsamt 1876–1933. Diss. Bremen. Univ., 2005. Göttingen: Wallstein. Available online at http://deposit.d-nb.de/cgi-bin/dokserv?id=3099685&prov=M&dok_var=1&dok_ext=htm.

Jäckel, R., first name unknown (author's note) (1927): Schädlingsbekämpfung mit Zyklon B (Blausäure). *Zeitschrift für Desinfektion und Gesundheitswesen*, 19(1), 37–41.

Kainer, Franz (1951): Kieselgur, ihre Gewinnung, Veredlung und Anwendung. 2., umgearb. Aufl. Stuttgart: Enke (Sammlung chemischer und chemisch-technischer Vorträge, N.F. H. 32).

Kornauth, Karl (1904): Über die Bekämpfung tierischer landwirtschaftlicher Schädlinge mit Hilfe von Mikroorganismen. Mitteilung der kaiserlich königlichen landwirtschaftlich-bakteriologischen und Pflanzenschutzstation in Wien. Nach einem Vortrag gehalten am 12. Februar 1904 in den Kursen für praktische Landwirte unter Benutzung von Ver-suchsergebnissen der kaiserlich königlichen landwirtschaftlich-bakteriologischen und Pflanzenschutzstation in Wien. *Zeitschrift für das Landwirtschaftliche Versuchswesen in Österreich*, VII, 365–387.

Landgraf, Theodor (1925): Grundsätze zur Schädlingsbekämpfung im Gartenbau. Von Gew-erbe-Oberlehrer Theodor Landgraf, Wandsbek. In: Führer durch die Gartenbau-Ausstel-lung, Bergedorf, 33–37.

Schwartz, Martin (1925): Die reichsgesetzlichen Pflanzenschutzbestimmungen für die Ein-fuhr lebender Pflanzen und frischer Pflanzenteile nach Deutschland. *Der Deutsche Erw-erbsgartenbau*, (45), 653–654.

Zacher, Friedrich (1924a): Der Brotkäfer, ein schlimmer Haushaltsschädling. *Hof und Gar-ten*, 87–88.

Zacher, Friedrich (1924b): Methoden der Vorratsschädlingsbekämpfung. Vierte Mitglieder-versammlung zu Frankfurt a.M. vom 10–13. Juli. *Verhandlungen der Deutschen Gesells-chaft für angewandte Entomologie*, 45–48.

Zacher, Friedrich (1927): Die Vorrats-, Speicher- und Materialschädlinge und ihre Bekämp-fung. Berlin: P. Parey.

Conservation of cultural property from organic materials for the prevention and control of harmful insects in museum collections

Part II

Conservation of cultural
property from organic
materials for the prevention
and control of harmful
insects in museum collections

4   Definition of pesticides

The word "pesticides" has only become common in recent scientific literature, although an exact chronological classification for its introduction could not be proven. According to the current pesticidal/biocidal product regulation of the European Union, biocides and biocidal products are defined in Article 3, Paragraph 1 a) in

> "jeglichen Stoff oder jegliches Gemisch in der Form, in der er/es zum Verwender gelangt, und der/das aus einem oder mehreren Wirkstoffen besteht, diese enthält oder erzeugt, der/das dazu bestimmt ist, auf andere Art als durch bloße physikalische oder mechanische Einwirkung Schadorganismen zu zerstören, abzuschrecken, unschädlich zu machen, ihre Wirkung zu verhindern oder sie in anderer Weise zu bekämpfen"[1]

and in

> "jeglichen Stoff oder jegliches Gemisch, der/das aus Stoffen oder Gemischen erzeugt wird, die selbst nicht unter den ersten Gedankenstrich fallen und der/das dazu bestimmt ist, auf andere Art als durch bloße physikalische oder mechanische Einwirkung Schadorganismen zu zerstören, abzuschrecken, unschädlich zu machen, ihre Wirkung zu verhindern oder sie in anderer Weise zu bekämpfen".[2]
>
> (Verordnung (EU) Nr. 528/2012, 2012, 6)

Biocides and pesticides are often insoluble in water and can therefore cause long-lasting damage to the environment, humans, objects, and materials. Regarding their effects, a distinction is made between whether an active ingredient or a remedy can be used preventively or to combat the ailment. Furthermore, they are divided into respiratory toxins, which are absorbed from the air; feeding toxins, which are absorbed through food; nerve toxins, which are absorbed through the skin; and deterrents, which are mostly perceived through the sense of smell. The term "active ingredients" refers to substances that have a specific effect on an organism or cause a specific reaction. They are effective components of all plant protection products, pesticides, and deterrent agents, and are available for the prevention and

DOI: 10.4324/9781003407607-7

control of harmful insects in gaseous, liquid or solid state. The possibilities for their applications are manifold. Depending on the active ingredient or agent, it can be fumigated, nebulised, vaporised, diffused, brushed, rubbed in, and completely soaked. During the investigated period of the late 19th and early 20th centuries, hydrocyanic acid (which is now called hydrogen cyanide; see glossary), carbon disulfide (still known today as carbon disulfide; see glossary), and carbon tetrachloride (which is now called tetrachloromethane; see glossary) were important representatives of active substances for pest control. At room temperature, these active ingredients are liquids in their physical state. However, because they are very volatile, they can also be used in steam form, and thus for fumigation. Although these substances have no preventive effect because they normally escape from organic materials, it is quite possible that some of their active ingredients have remained locked in museum objects for years. There is a suspicion that residues of reactive fumigants, such as hydrogen cyanide, and volatile solids, such as 1,4-dichlorobenzene ("Globol"), camphor, and naphthalene, are found in organic materials due to chemical reactions and the slow sublimation of these substances into the environment.

Active substances and agents for pest control may also be present in the preparations. This refers to a recipe in which certain raw materials are combined into a product using a predetermined process. These preparations can consist of mixtures, blends, or solutions. Mixtures are usually based on granular materials, which cannot be homogeneously mixed; otherwise, they would lose their function. An example of a mixture is treated seed, which, when mixed with a fungicide, achieves a fungicidal effect. For blends, at least two pure substances are required, which are divided according to the degree of mixing into homogeneous, in which they are mixed at the molecular level, and heterogeneous, in which the blends are present in clearly defined phases. The arsenic soap, which is formulated using different recipes, represents a blend that has been processed in solid form. "Eulan" preparations and petroleum, on the other hand, are available as blends in liquid form. In chemistry, a solution is a homogeneous mixture of at least two substances that break down physically. Externally, solutions are not recognizable because they only form a homogeneous phase. In this process, molecules, atoms, and ions can be homogeneously and statistically distributed. Active substances containing heavy metals, such as potassium and sodium arsenate, arsenic (III) oxide, and mercury (II) chloride (sublimate), were used as solutions. After the solvents have evaporated, the active ingredients remain in the materials in solid form and pose a permanent danger to humans.

All pesticides share a common characteristic of acting physically, chemically, or mechanically, which enables them to deter harmful organisms and molds that may cause damage to plants and animals. Pesticides eliminate or reduce the negative impact of harmful organisms by making them harmless, preventing their effects, destroying them, or fighting them in various ways. Active substances are categorised based on their intended use, including disinfectants for controlling pathogenic microorganisms, pesticides, and other biocidal products. During the period under examination, at the end of the 19th and beginning of the 20th century, the term "pesticides" was already commonly used (Ibid. Anhang V, 142–144). The discussion of active substances and agents for the prevention and control of microorganisms, such as bacteria and molds, was limited in this context, as their growth was associated with increased material moisture. However, in museums, a constant climate is sought, which means that these microorganisms usually lack the necessary growth conditions. Undoubtedly, microorganisms are often a problem for today's museum objects in ethnological collections in tropical areas and for long-term ship transport from overseas to Europe. This situation differs from that of animal insect pests. Non-European insect pests have been introduced to Europe from overseas. Exotic and native species were often able to adapt to the climatic conditions in museum collections and multiply uncontrollably. Questions on the current significance of previously used pesticides and prevention agents in museum collections can only be answered if there is clarity about their distribution and application during the period under investigation.

Notes

1 Any substance or mixture, in the form in which it reaches the user, consisting of, containing, or producing one or more active substances intended to destroy, deter, render harmless, prevent, or otherwise control harmful organisms other than mere physical or mechanical action (translation by the author).
2 Any substance or mixture produced from substances or mixtures not themselves covered by the first indent and intended to destroy, deter, render harmless, prevent, or otherwise control harmful organisms other than mere physical or mechanical action (translation by the author).

Reference

Other source

Verordnung (EU) Nr. 528/2012 des Europäischen Parlaments und Rates der Europäischen Union 27.06.2012. (2012): Biozidproduktarten und ihre Beschreibung vom 22.05.2012 gemäss Artikel 2 Absatz 1; Anhang V der Biozid-Verordnung (EU) Nr. 528/2012, (EU) Nr. 528/2012. In: Amtsblatt der Europäischen Union, L 167/1 vom 27. Juni.

5   Control of wood-destroying insects, textile pests, and harmful insects on natural history objects

Control of wood-destroying insects

The preservation of wood with salt was mentioned as early as 50 BC by Vitruvius. Evidence of the complete impregnation of boards boiled in aqueously dissolved salts was recorded in 1445 for the construction of St. Mary's Church in Königsberg in Franconia (Cf. Schiessl 1984, 9). The best-known surviving example of the conservation of panel paintings against wood-destroying insects was probably Leonardo da Vinci in 1492. He applied the backs of his wooden panels with sublimate (mercury (II) chloride) dissolved in spirits or alternatively with arsenic (III) oxide (Cf. Eibner 1928, 97). An oak figure of St. Catherine was "gesotten und gebraten"[1] in 1494 for conservation (Cf. Schiessl 1984, 33). Salt impregnations were used in works of art made of wood, such as the Kefermarkt altar. According to their observations, the actual protection was attributed to the finely crystallised salt remaining in the wood fibers, which was intended to keep woodworms away (Ibid. 10). Substances of animal origin and plant extracts are known to combat wood-destroying insects. Balthasar Schnurr von Lendsiedel, author of the "Kunst-, Hauss- und Wunderbuches",[2] advised in 1657 to use ox bile as an insecticide. Furthermore, Schiessl mentioned extracts of nutshells; cooking broths made from tobacco leaves, blackthorn leaves, pepper, laurel, aloe, and myrrh decoctions; garlic juice; and even absinthe for impregnation of wood in the 17th century. In 1705, the German naturalist Wilhelm Homberg recommended a sublimate solution for the protection and control of wood-destroying insects on parquet floors. Arsenic trioxide, which is also poisonous, was mentioned around 1730 for the treatment of wood-destroying insects (Ibid. 11). Animal greases, such as beef, sheep tallow, and pork grease, were already applied in the 18th century and used in the preservation of wooden utensils. All types of wax were used as coatings on wooden surfaces to protect against insects (Ibid. 16). Distillates of wood and coal tar have been known for wood impregnation since 1750 and 1812 (Troschel 1916, 302). At the beginning of the 19th century, Schweinfurter Grün[3] (copper (II) arsenite acetate) was used on altars to protect against harmful insects (Schiessl 1984, 12). Smoking and fumigation were among the early methods of wood preservation. The smoking of timber in a smokebox was first mentioned in 1833 as a protective method. Furthermore, libraries and art cabinets were smoked with the sulfur

DOI: 10.4324/9781003407607-8

dioxide produced when sulfur burned, as well as strong-smelling herbs, resins, or arsenic trioxide (Troschel 1916, 306). Suardi expressed his view in a detailed manner toward the use of arsenic trioxide for the conservation of panel paintings in his key work on conservation in Italy, "Il manuale ragionato per la parte meccanica dell'arte del ristauratore dei dipinti", which was published in 1866.[4] He recommended using kerosene fumes instead of arsenic trioxide to treat paintings, as the latter is toxic and the former is effective after three to four months of use in an airtight box (Achsel 2012, 127–128). Around 1900, Rathgen himself criticised color changes, unpleasant odors, and an increased fire risk due to the use of kerosene on wooden objects. He also recommended the use of this mineral oil for full soaking of wooden objects until the 19th century (see Part III, chapter 8). The reduction in the shiny appearance on wood surfaces resulting from the application of mineral oils was achieved by the addition of solvents, such as carbon tetrachloride (Schiessl 1984, 14–15). Rare evidence of the preventive protection of ethnographica using repellent-looking wood can be found in Scharf & Kayser, an import and export company from Hamburg. In August 1884, they offered the Generalverwaltung der Königlichen Museen (General Administration of the Royal Museums) a list of various objects of the Rapanui from Easter Island (see Part III, chapter 8). Among these were feather hats and other headgears, which were transported in a chest made of camphor wood. Although no specific reference is made to the intended effect of the chest wood, it is conceivable that it was a preventive measure to keep harmful insects away from the valuable objects during their long-lasting journey (SMB-PK EM 1884). Further processing of coal tar produced carbolic acid, cresols, and naphthalene compounds. Their insecticidal effects have been used in numerous wood preservatives. A mixture of coal tar, phenols, and cresol was marketed in 1888 under the trade name "Carbolineum" by the German company Avenarius. Despite the strong browning of works of art made of wood, it has been applied for a long time in conservation practice. For example, at the beginning of the 20th century, the epitaph of the former mayor of Tönning, North Frisia, in the St. Laurentius Church was both coated and soaked with "Carbolineum", causing considerable damage (Weber und Unger 2018, 63–64). During development, carbon disulfide, hydrocyanic acid, bromomethane (methyl bromide), and ethylene oxide were used as fumigants at the end of the 19th and the beginning of the 20th century. However, because fumigations are not preventively effective, works of art treated in this way can be reinfested by harmful insects at any time (Schiessl 1984, 17–19). Fumigation with the aid of gas crates or gas tents was extended at the beginning of the 20th century by the invention of fumigation systems, which were specially developed for the mass treatment of art and cultural assets in museums (see Part III, chapter 8).

Control of textile pests

Because of their rapid volatility, essential oils do not offer permanent protection against harmful insects. However, as an insecticide against textile moths (*Tineola* sp.), essential vapors of lavender oil were discussed in the 17th century (Cf.

Schiessl 1984, 13). At the beginning of the 20th century, the curator Creassy Edward Cecil Tatershall was responsible for carpet collections at the Victoria and Albert Museum in London. His 1924 recommendations for the preservation of carpets and objects made of wool included storage in cool places as well as in sealed containers. He believed that camphor and insect powder were unsuitable as a means of prophylaxis, as they were ineffective according to his point of view. On the other hand, he warned against using naphthalene, as it should not come into direct contact with textile fabrics (Tatershall 1924, 199–200). Just one year later, a contribution was published by Böttcher (first name unknown; author's note) in the *Zeitschrift für Desinfektion und Gesundheitswesen*,[5] advising against home remedies, such as naphthalene and moth powders. Not only did Böttcher doubt their effectiveness, but he also pointed out that objects treated with these agents can sometimes make entire rooms uninhabitable due to the strong odor nuisance of these agents (Böttcher 1927, 143–144). The "Zyklon method" is the most effective method for combating textile moths, according to his perspective. He recommended "Eulan" (Ibid. 144–146) as a preventive measure. The discovery of the wool repellent "Eulan" will be discussed in more detail, as it triggered a true revolution in combating textile moths. The insecticide can be traced back to the chemist Ernst Meckbach, a chemist who worked in the laboratory of Friedrich Bayer & Co. in Leverkusen. While observing the effects of nitro dyes on textile moths, Meckbach discovered that the moths avoided the colorant known as Martius yellow (Homolka 2015a, 8–9). The demand for this product was so great that the company subsequently developed over 40 preparations with the name "Eulan" from 1920 onward, some with the same active ingredients and some with different ones (Homolka 2015b, 27–194; Unger 2012, 25–39). Initially, "Eulan" could only be dissolved in water, which caused hesitation among museums regarding its use on textiles as well as for zoological preparations. The risk of bleeding of textiles dyed with plant-based dyes and damage to unattended preparations was deemed too significant to ignore. The interdisciplinary cooperation between zoologist Erich Titschack, who worked at Friedrich Bayer & Co. in Leverkusen from 1919 to 1924, and Meckbach led to the development of a petrol-soluble moth repellent called "EULAN BL". The market launch of the product took place in 1933 (Cf. Titschak 1937, 19–20). A publication by Titschack four years later in the *Museumskunde, Zeitschrift für Verwaltung und Technik öffentlicher und privater Sammlungen*[6] was specifically designed to conquer the museum market. In the end, he resolutely noted that the development of this product has only led to progress in the context of manufacturers and consumers. In his view, this essential factor of competition is completely missing in museums (Titschak 1937, 24). The use of "Eulan" preparations was among the most frequently used methods in the preservation of both textiles and zoological preparations. In 1964, the conservator Detlev Lehmann still applied "Eulan" preparations on carpets at the Staatliche Museen zu Berlin (SMB) (National Museums Berlin) in the Islamic department, and this has been discussed extensively (Lehmann 1964, 69–72).

Control of insect pests on natural history objects

Natural history collections differ widely in how they are conserved in museums that are otherwise similar in nature. This is due to the peculiarity of plant and animal specimens, which tend to deteriorate quickly. To preserve these precarious materials, preparations made by pharmacists, doctors, biologists, zoologists, taxidermists, and naturalists were initially used. During the middle of the 17th century, a method involving the use of wine spirit (ethanol) and salmiak (ammonium chloride) for preserving bird eggs and small bird fetuses was described. In 1748, Ferchault de Réaumur recommended a method involving the use of brandy with and without sugar for zoological preparations, which had been popular about a hundred years later (Boyle 1665, 199, 201; Réaumur de 1748, 307–308, 319). At the same time, in 1770, Kuckhan became deeply involved in bird conservation. In addition to wine spirit, alum (aluminum potassium sulfate dodecahydrate), salt, and pepper have been mentioned (Kuckhan 1770, 303–304). He recommended coating the preparations with camphor dissolved in turpentine, and for the dry preparation, he suggested a mixture of sublimate, saltpeter (potassium nitrate), alum, flowers of sulfur, musk, black pepper, and ground tobacco (Ibid. 312). Finally, he recommended that the inside of bird heads be stuffed with well-dried tansy, wormwood, hops, and tobacco (Ibid. 315). The most widely used pesticide for zoological preparations is arsenic soap. It goes back to the invention of the pharmacist Jean-Baptiste Bécoeur that consisted of camphor, arsenic (III) oxide, potassium carbonate, soap, and shell lime flour. Throughout his lifetime, Bécoeur kept secret both the composition and recipe of his soap, which served as a unique selling point in the production and distribution of this agent. After his death, from the year 1800, the procedure in making arsenic soap was reproduced, and arsenic soap was used to preserve natural history collections until the 1980s (Marte et al. 2006, 144–145; Rookmaaker et al. 2006, 146–158). In 1827, Thon used a mixture of arsenic, sal tartari (tartar salt, mainly potassium hydrogen tartrate), camphor, white soap, and powdered, burnt lime (Thon 1827, 165–166). He was already aware of the toxicity of arsenic, which is why he recommended an alternative recipe from Boitard.[7] Thon issued an urgent warning about the damaging effects of the mercury compound on the human body before proceeding with the rubbing of the insides of preparations with a mixture of fat and sublimate, (Ibid. 168–169). Richter recommended that rooms containing large bird collections be aired out to eliminate the bad odors of sublimate or arsenic solutions (Richter 1829, 48). In 1877, to combat the museum mite *Thyreophagus entomophagus* in natural history insect collections, Murray recommended using alcohol with dissolved sublimate. He also mentioned solvents, such as ether, petrol, or concentrated naphtha, which can be applied to insects using an atomizer. To treat insects individually, he recommended exposing them to fumes from ammonia liquor act for ten to 15 minutes and preemptively equipping the insect boxes with naphthalene crystals (Murray 1877, 481–482). For the liquid preservation and preparation of animal skins and plants, the "Wickersheimersche solution"[8] was handed down from 1879, based on a recipe by Jean Wickersheimer (Keil 1879).[9]

The objects or preparations were immersed in the solution, dried, and impregnated in this way.[10] Wickersheimer's lack of knowledge regarding the immense health hazards posed by individual substances is reflected in the use of large quantities dissolved in boiling water. In the early 20th century, a clear professionalization in the production of preservatives can be seen at the Zoologisches Museum (Zoological Museum) of the Friedrich-Wilhelms-Universität zu Berlin (Vanhöffen 1918, 3–6)[11] (Giere et al. 2018, 89–102). In 1907, Brauer[12] published a detailed *Anleitung zum Sammeln, Konservieren und Verpacken von Tieren für das Zoologische Museum in Berlin*[13] (Brauer 1907). He differentiated between dry and wet preparations and recommended for baling mammals and birds to carry "Arsenikseife oder arsenigsaures Kali zum Vergiften der Häute"[14] (Ibid. 1). Dry hides were made supple with water and then coated from the inside with arsenic soap. For reptiles and amphibians, Brauer recommended naphthalene or insect powder against insect predation, or placing it in a saline solution with a small amount of alum (Ibid. 14–31). He recommended that abnormal cell formations on plants, such as galls, be immersed in a sublimate solution before pressing and labeled poisoned (Ibid. 77). For dry preserved animals, he advised filling cigar boxes with cotton plates and sprinkling naphthalene between the cotton wool layers (Ibid. 46). Brauer recommended using "Formol" as a preservative only when other agents are not available, as it is an innovative active ingredient for wet preparations. To preserve insect spawn, Brauer recommended a mixture.[15] Lastly, when catching plankton along with smaller organisms, Brauer recommended adding "Formol", sublimate, or picric acid (2,4,6-trinitrophenol) into a container (Ibid. 102–103). Although he had pointed out previously about the toxicity of sublimate, he proceeded to boil active ingredients in solvents, especially in highly concentrated sublimate, which is extremely dangerous for one's health.

During the transition from the 19th to the 20th century, the list of preparations for conservation of natural history collections illustrates a certain heterogeneity of active ingredients and agents with regard to their use and processing. It also became clear that mixtures of individual substances are increasingly recommended in natural history preparation techniques. This was probably done against the background of preventively achieving a strengthening effect on the inside of skins and prepared mammals against infestation by harmful insects. The different approaches of Thon, Richter, and Brauer showed that the health risks of preservatives were perceived and evaluated individually. The following chapter examines the extent to which chemical substances reserved for protection against diseases in the tropics were also used to preserve organic materials.

Notes

1 Simmered and fried (translation by the author).
2 Book of arts, houses, and miracles (translation by the author).
3 Schweinfurt green (translation by the author).
4 The annotated manual for the mechanical part of the art of the restorer of paintings (translation by the author).
5 *Journal of Disinfection and Health Care* (translation by the author).

6 *Museology, Journal for Administration and Technology of Public and Private Collections* (translation by the author).

7 The recipe consisted of white soap, potash (potassium carbonate), alum, well water, Steinöl ("rock oil") (presumably Tiroler Schieferöl), and camphor (Cf. Thon 1827, 167).

8 Wickersheimer's solution (translation by the author).

9 Jean Wickersheimer (1832–1896) received 5,000 RM from the Prussian Ministry of religious, educational, and medical affairs for his Wickersheimer solution and was then the first in Prussia to bear the title "taxidermist". See also: Behrhol, Hans; Henckendorrf, Evelyn; Schnalke, Thomas; Wilcke, Günther (2018): Präparierkurse in der Charité. Weiterbildung mit historischem Flair. In: *Oralchirurgie Journal*, 2018, (1), 32.

10 The "Wickersheimer solution" consisted of 3,000 g of boiling water, in which 100 g of alum, 25 g of salt, 12 g of saltpeter, 60 g of potassium carbonate, and 10 g of arsoric acid were dissolved. To 10 quarts of this colorless, neutral, and odorless liquid, 4 quarts of glycerol and 1 quart of methyl alcohol were added. [(see Keil, Ernst (1879): Die Gartenlaube. Das Wickersheimers'sche Conservierungsverfahren. Hg. v. Verlag von Ernst Keil. Leipzig. Available online at jttps://de.wikisource.org/w/index.php?title=Seite:Die_Gartenlaube_(1879)_844.jpg&oldid= (Version vom 21.05.2018), (last visited September 11, 2019). In the Kingdom of Prussia, 1 quart = 1.145 liters (author's note)].

11 The Zoologisches Museum (Zoological Museum) of the Friedrich-Wilhelms-Universität zu Berlin was renamed the Museum für Naturkunde (Museum of Natural History) Berlin after 1945. On January 1, 2009, it was transferred to a foundation and has since been called Museum für Naturkunde – Leibniz Institute for Evolution and Biodiversity Research.

12 August Bernhard Brauer was from 1906 until his death in 1917 director of the Zoological Museum of the Friedrich Wilhelm University in Berlin.

13 Instructions for collecting, preserving, and packaging animals for the Zoological Museum in Berlin (translation by the author).

14 Arsenic soap or arsenic potassium to poison skins (translation by the author).

15 The mixture consisted of 50 cm³ glycerol and 70 cm³ distilled water, which was mixed with two to five drops of sublimate solution (Cf. Brauer 1907, 76).

References

Archive material

SMB-PK EM (1884): I/MV 0611, Pars I B, Bd. 4, E. Nr. 2165/84. Acta betreffend die Erwerbung ethnologischer Gegenstände aus Australien. Loseblattsammlung. Angebot von Ethnografica. Scharf & Kayser. 1 Liste ohne Datum sowie 3 Briefe vom 16.08.1884, 4.9.1884 und vom 20.11.1884, Blatt 1–2, 4 Seiten.

Other sources

Achsel, Bettina (2012): Das Manuale von Giovanni Secco Suardo von 1866/1894. Zugl.: Dresden, Hochschule für Bildende Kunst, Dissertation, 2011 u.d.T.: Achsel, Bettina: Kommentierte Übersetzung von Giovanni Secco Suardo, "Il manuale ragionato per la parte meccanica dell'arte del ristauratore dei dipinti" (1866) und "Il Restauratore dei dipinti" (1894). Göttingen: V & R Unipress.

Behrhol, Hans; Henckendorrf, Evelyn; Schnalke, Thomas; Wilcke, Günther (2018): Präparierkurse in der Charité. Weiterbildung mit historischem Flair. In: Oralchirurgie Journal, 2018, (1), 32.

Böttcher, first name unknown (author's note) (1927): Entmottungsanlagen nach dem Zyklonverfahren. *Zeitschrift für Desinfektion und Gesundheitswesen*, 19(4), 143–144.

Boyle, first name unknown (author's note) (1665): A way of preserving birds taken out of the Egge, and other small Faetus's. *Philosophical Transactions*, 1665–1666, (1), 199–201.

Brauer, August, Bernhard (1907): Anleitung zum Sammeln, Konservieren und Verpacken von Tieren für das Zoologische Museum in Berlin. Dritte vermehrte Auflage. Berlin: A. Hopfer in Burg b. M. I–VI.

Eibner, Alexander (1928): Entwicklung und Werkstoffe der Tafelmalerei. München: B. Heller.

Giere, Peter; Bartsch, Peter; Quaisser, Christiane (2018): BERLIN: From Humboldt to HVac – The Zoological Collections of the Museum für Naturkunde Leibniz Institute for Evolution and Biodiversity Science in Berlin. In: Beck, Lothar A. (Hg.). Zoological Collections of Germany. Cham: Springer International Publishing (Natural History Collections), 89–122.

Homolka, Martina (2015a): Eulan – ein Biozid gegen Keratin-Schädlinge und seine Relevanz in musealen Sammlungen. Berlin: Stiftung Deutsches Historisches Museum (1. Produktgeschichte).

Homolka, Martina (2015b): Eulan – ein Biozid gegen Keratin-Schädlinge und seine Relevanz in musealen Sammlungen. Berlin: Stiftung Deutsches Historisches Museum (2. Lexikalischer Produktschlüssel).

Keil (1879): Keil, Ernst. Die Gartenlaube. Das Wickersheimers'sche Conservierungsverfahren. Hrsg. v. Verlag von Ernst Keil. Leipzig. Available online at jttps://de.wikisource. org/w/index.php?title=Seite:Die_Gartenlaube_(1879)_844.jpg&oldid= (Version vom 21.05.2018).

Kuckhan, T.S. (1770): Four Letters from T.S. Kuckhan, to the President and Members of the Royal Society, on the Preservation of Dead Birds. *Philosophical Transactions*, (60), 303–304.

Lehmann, Detlev (1964): Die EULAN-Behandlung von Textilien und zoologischen Präparaten. Staatliche Museen Berlin. Islamische Abteilung. In: *Ergänzungsbände des Berliner Jahrbuchs für Vor- und Frühgeschichte*, (Band I).

Marte, Fernando; Péquignot, Amandine; von Endt, David W. (2006): Arsenic in Taxidermy Collections: History, Detection, and Management. *Collection Forum*, 21(1–2), 146–158.

Murray, Andrew (1877): The Museum Mite. *The American Naturalist*, Band 8, (11), 479–482.

Réaumur de, René-Antoine Ferchault (1748): Divers Means for Preserving from Corruption Dead Birds, Intended to Be Sent to Remote Countries, so That They May Arrive There in a Good Condition. Some of the Same Means May Be Employed for Preserving Quadrupeds, Reptiles, Fishes, and Insects. *Philosophical Transactions*, (45), 304–320.

Richter, Christian Gottlieb (1829): Anweisung Vögel auszustopfen, nebst Angabe aller dazu erforderlicher Hülfsmittel. Mit einem Vorwort von Brehm, Pastor in Renthendorf. Mit 2 Kupfertafeln. Jena: August Schmid.

Rookmaaker, L.C.; Morris, P.A.; Glenn, I.E.; Mundy, P.J. (2006): The Ornithological Cabinet of Jean-Baptiste Bécoeur and the Secret of the Arsenical Soap. *Archives of Natural History*, 33(1), 146–158.

Schiessl, Ulrich (1984): Historischer Überblick über die Werkstoffe der schädlingsbekämpfenden und festigkeitserhöhenden Holzkonservierung. *Maltechnik/Restauro*, (2), 9–40.

Tatershall, Creassy Edward Cecil (1924): To Preserve Woolen Textiles from Moth. *Museum Journal*, (23), 199–200.

Thon, Theodor (1827): Handbuch für Naturaliensammler oder gründliche Anweisung die Naturkörper aller drei Reiche zu sammeln, im Naturalienkabinet aufzustellen und aufzubewahren, namentlich Thiere aller Arten, Säugethiere, Vögel, Reptilien, Fische,

Conchylien, Crustaceen, Insekten, Zoophyten und Eingeweidewürmer auszustopfen, zuzubereiten, zu versenden, so wie Pflanzen zu trocknen, Herbarien, Fruchtkabinette, Holzbibliotheken und Mineraliensammlungen anzulegen, einzurichten und in vollkommner Schönheit zu erhalten. Frei nach dem Französischen bearbeitet und vervollständigt. Ilmenau: Verlag von Bernhard Friedrich Voigt.

Titschak, Erich (1937): Der Schutz von Museumsgegenständen gegen Mottenfraß. Mit 6 Abbildungen auf Tafel I. *Museumskunde, Zeitschrift für Verwaltung und Technik öffentlicher und privater Sammlungen*, Band 26(9), 19–24.

Troschel, Ernst (1916): Handbuch der Holzkonservierung. Berlin: Springer.

Unger, Achim (2012): "Eulanisierte" Textilien – eine Gefahr für Mensch und Material? *Beiträge zur Erhaltung von Kunst- und Kulturgut*, (2), 25–39.

Vanhöffen, Ernst (1918): Zur Erinnerung an August Brauer. *Mitteilungen aus dem Zoologischen Museum in Berlin*, 9. Band (1. Heft), 1–12.

Weber, Jörg; Unger, Achim (2018): Experimente zur Entfernung alter Holzschutz- und Holzfestigungsmittel mit Methyl-tert-butylether (MTBE) aus ungefassten und gefassten Holzproben. *VDR Beiträge zur Erhaltung von Kunst- und Kulturgut*, (2), 60–73.

6 Protective and human toxic effects of historical pesticides and their suitability tests

Chemical agents and means for protecting people and properties during expeditions

Researchers were dependent on available means and have accepted their use. Karl von den Steinen,[1] who embarked on his second Schingú expedition in 1887, carried a substantial quantity of arsenic pills in his first-aid kit (Cf. von den Steinen 1894, 1). In addition, ammonia, carbolic acid (phenol), alcohol, thein (teein), nicotine, quinine, and beijú[2] were added. A prophylactic intake of 10–14 arsenic pills at 0.002 g per dose and some quinine was not uncommon for protection against, for example, malaria. Skin diseases have occasionally been treated with Fowler's arsenic solution (Cf. Tello 2006, 35; von den Steinen 1894, 134–135). Further expeditions, which led from Germany and Belgium to Africa between 1874 and 1884, were described in detail by Fabian. The list of already mentioned drugs is extended as follows: Opium in the form of laudanum or morphine, zinc oxide (zinc (II) oxide), lead sulfate (lead (II) sulfate), tannin, citric acid, mustard plasters, and many more. Fabian even mentioned the death of a researcher from an overdose of quinine. However, the self-medication used in these circumstances and its often unintentionally drastic consequences led some researchers to plead for complete abstinence (Fabian 2000, 284; 65–66). Rare evidence of the equipment of an expedition to Abyssinia in 1905 included the payment of a drug suitcase, the contents of which cost 29 marks (SMB-ZA 1905). Theodor Koch-Grünberg[3] used naphthalene during his expedition to the Rio Negro to remove sand fleas from his own toes (Koch-Grünberg 1903–1905, 146–147). The intake of active substances and agents during non-European trips has sometimes been viewed critically. This raises compelling questions from today's perspective. The subject of the following investigations is the period from which and in which areas of society severely toxic or chronic effects on humans were investigated.

Toxic effects of pesticides on humans

Evidence to sensitivity to health hazards in museum collections due to pesticides can be traced back to individual cases as early as the 18th century for specific active ingredients and agents. As de la Varenne (1730–1792) noted,[4] sublimate (today referred to as mercury (II) chloride; see glossary) is a highly toxic substance, and

DOI: 10.4324/9781003407607-9

improper handling can result in harm, even from mere contact (Farber 1977, 556). In 1887, according to Hough,[5] handling toxic substances, such as sublimate, can cause skin diseases as well as can break off fingernails. Furthermore, he attributed arsenic-containing dusts to diseases, such as catarrh, which is the inflammation of the mucous membranes (see Part III, chapter 13.; Hough 1889, 552). A link between the use of pesticides and human toxicity effects can also be demonstrated for individual active substances and agents, as shown by the example of lead, which was used previously as a lead arsenate in plant protection products. The toxicity of lead was recognised early on by the Kaiserliches Gesundheitsamt/Reichsgesundheitsamt (KGA/RGA) (Imperial Health Office/Reich Health Office) in Berlin. The objective was to improve general hygiene standards for society (Cf. Hüntelmann 2008, 55–75). Drafts of several bills were created to safeguard the welfare and protection of both young and adult workers. In 1898, an expert opinion pointed out the dangers of lead, which can cause human organ damage (Ibid. 211). Additionally, the policy article *"Über die idealen und praktischen Aufgaben der ethnographischen Museen"*[6] delves into the health impairments caused by toxic substances. Richter devoted himself to the hygienic workplace conditions in museums during the conservation of museum objects (Richter 1907, 14–25, 99–120), specifically in the island states of Indonesia and Oceania. As a museum professional, he worked for seven years at the Anthropologisch-Ethnographisches Museum Dresden (Anthropological-Ethnographic Museum Dresden) and then from 1904 for a short time at the KMfV in Berlin (Richter 1908, 92–93). His early death was most likely caused by severe blood poisoning, which he had contracted while handling a poisoned arrow (Ibid. 158). This and his knowledge of the various methods for combating insect pests at both museums may have led him to be one of the few museum experts at the beginning of the 20th century to focus on the topic of health hazards. He clearly emphasised that ethnological museums were still in the experimental stage and urged his colleagues to share their knowledge and make it accessible to others (Ibid. 159). He considered the prevalent poor hygiene facilities at the workplaces of museum employees to be deplorable, citing symptoms of illness, such as eye inflammation and severe colds, suffered by colleagues during winter months (Ibid. 158). However, he did not provide a clear cause for these symptoms.

The professional activity of Richter and the official requirements of the KMfV/MfV in Berlin may have resulted in increased sensitivity toward health protection when handling chemical substances. On November 12, 1913, the museum received an official order from the then Minister of Culture, Hermann von Chappuis (Winter und Grabowski 2014, 44), instructing the use of substitutes for lead paints, such as lithopone and zinc white, to protect the occupational group of painters due to recurrent lead diseases. The context of these tests involved finding a substitute for lead white for exterior coatings, as lead paints, especially lead white, were only permitted if they could not be replaced for technical or artistic reasons

"und die Farben dabei nicht in Pulverform, sondern mit Leinöl verrieben beschafft werden".[7]

(SMB-PK EM 1913)

At the Königliche Museen (KM) in Berlin, the insecticide agent "Globol" was evaluated in a very different manner. In 1913, Eduard Krause[8] introduced it at the KMfV under its original name "Dichlorobenzene Agfa" (see Part III, chapter 10). However, the further procurement of the product was prevented in 1919 by the department heads of the museum because of its strong smell (see Part III, chapter 10). It was unclear from the files when this attitude changed, but exactly ten years later, the dangers posed by "Globol" to humans were addressed. Serious incidents of illness in a Dutch museum and the early death of Willy Foy[9] had alerted Walter Lehmann.[10] In a letter to Kurt Stubenrauch,[11] he requested that Director General Wilhelm Waetzoldt[12] be appointed to draw attention to the so-called "Globol disease", as both scientific professionals and the assistant conservator, Erich Zorn, must handle this pesticide extensively in the cupboards of the storage facilities in Berlin-Dahlem. When Carl Brittner[13] accidentally inhaled the substance, he experienced general malaise and headaches, prompting him to consult relevant toxicologists and health authorities (Archiv des Museums am Rothenbaum – Kulturen und Künste der Welt 1929a). A complex, yet unclear, response was provided by the Chemisches Laboratorium (chemical laboratory) on behalf of Brittner from his predecessor Rathgen. The response referred to the Museum für Völkerkunde (Museum of Ethnology) in Hamburg, where the agent "Globol" has been used for some time. From his point of view, Foy seemed to be very ill long before the application of the agent in the Rautenstrauch-Joest-Museum (R-J-M) in Cologne, which was why his early death was not associated with his handling of toxic substances. He recommended the agent "Areginal", which was used in Eberswalde.[14] However, he saw danger in every chemical agent. His statement ended rather resignedly, as he considered fumigation with "Areginal" in the in-house plant in Berlin-Dahlem as the only possibility. Heat treatment of the collection objects at about 70–80 °C was not feasible because the Staatliche Museen zu Berlin (National Museums in Berlin) (SMB) did not have a suitable facility (Archiv des Museums am Rothenbaum – Kulturen und Künste der Welt 1929b). From the Museum für Völkerkunde in Hamburg, Waetzold received a statement on "Globol" from the then director Georg Thilenius.[15]

He confirmed the symptoms of illness, such as discomfort and headaches, in his employees when they work on open cabinets equipped with "Globol". Even just being in the storage areas leads to these symptoms. The personal disposition of the individual is crucial, but he is looking for a substitute for "Globol" and was very interested in medical examinations from Berlin, if one intends to have them carried out there (Archiv des Museums am Rothenbaum – Kulturen und Künste der Welt 1929c). The company Fritz Schulz jun. Aktiengesellschaft, based in Leipzig, as well as its representative in Hamburg, Hugo Senger, tried to convince Thilenius in writing about the non-toxicity of "Globol". They cited 17 years of experience in dealing with this agent and the fact that hundreds of thousands of households used "Globol" to protect against moths. However, they were unable to provide scientific evidence (Archiv des Museums am Rothenbaum–Kulturen und Künste der Welt 1929d). Two months later, a confidential letter from Zacher circulated within the I. G. Farbenindustrie Aktiengesellschaft, in which he also spoke out against the pesticide "Globol" (LASA 1930). In the 1930s, another agent caused difficulties in the storage units of the Africa Department of the Museum für Völkerkunde in Hamburg. There, a wooden shelf with "Xylamon" had been painted to protect against wood-destroying insects. A female draftsman who worked in the storage facility complained the next day of discomfort and headaches. Despite leaving the site, she experienced the same symptoms when she returned to her workplace. Subsequently, the premises were examined by a building official of the City of Hamburg and an official from the Ministry of Work and Technology. As a result, it was stated that rooms for treatment with "Xylamon" must not be ventilated for months so that the wood preservative can work effectively. Consequently, such premises are unsuitable as workplaces (Archiv des Museums am Rothenbaum – Kulturen und Künste der Welt 1934).

The most commonly used fumigant in museum collections during the 19th century was carbon disulfide. This liquid was placed in bowls and positioned on the lower shelves of cabinets. In Berlin's Museum für Naturkunde (Museum of Natural History), employees were required to distribute 120 kg of carbon disulfide in about 50 large cabinets twice a year. These actions often resulted in acute toxic reactions, such as nausea, headache, and vomiting (Arndt 1932a, 54–55). Walther Arndt[16] reported 16 taxidermists, scientific staff, and unskilled workers who sustained health issues due to the impact of carbon disulfide. Only the use of respirators minimised the symptoms (Cf. Arndt 1932a, 55–56). Carbon disulfide was also applied in gaseous form in specially built fumigation plants.[17] There is currently no evidence to suggest that museum employees have experienced any adverse effects from handling these facilities. Richter's reflections and questions from members of the Deutsche Museumsbund (German Museums Association)[18/19] prompted Arndt to conduct a medical study on occupational diseases and damage in natural, zoological, botanical, and anthropological museum collections both in Germany and abroad in the year 1932. Thanks to a fortunate circumstance that he was employed as a doctor with academic qualifications in the Zoologisches Museum (Zoological Museum) of the Friedrich-Wilhelm-Universität Berlin in 1921 (Komander 2004). His study was the first of its kind conducted for employees in museum institutions. For this purpose, he interviewed scientists, taxidermists, unskilled workers, as well as civil servants and employees from German, Austrian, and other European museums. He divided his findings into three parts and first addressed those poisonings caused by arsenic, sublimate, formalin, fumigants, various preservatives, and chemicals (Cf. Arndt 1932a, 47–61). Furthermore, he described damage caused by poisons and hair of animal origin (Ibid. 61–66). In the second part of his study, he described damage that can occur through the transmission of pathogens (Arndt 1932b, 103–105). In the final and third part of his investigation, he discussed damage caused by inhalation of dusts (Ibid. 105–106). The results clearly show that arsenic poisoning occurs most frequently in the museums that he examined (Cf. Arndt 1932a, 49). Acute toxic effects include skin inflammation associated with suppuration, vomiting, diarrhea, and jaundice (Ibid. 52–53). Chronic poisoning includes headaches, fatigue, browning of the skin, conjunctivitis, eyelid rim inflammation, stomach complaints, liver diseases, gallbladder inflammation, nervous disorders, and general nervousness (Ibid. 53–54). For zoological, botanical, and anatomical purposes, "formalin" was introduced as a preservative in relevant museums in 1893. Subsequently, damage to the skin, especially the hands in connection with protracted eczema and irritation of the nasal mucosa, conjunctiva, and respiratory tract occurred as clinical pictures (Ibid. 57–60). In summary, 41.4 % taxidermists and senior taxidermists, 27.1 % scientific officials, 25.6 % other civil servants and employees, and 5.9 % voluntary unskilled workers were affected by occupational diseases and injuries caused by pesticides (Cf. Arndt 1932b, 109–110).

Arndt's investigations and Richter's observations on the handling of hazardous substances and pesticides in museum collections have been groundbreaking. Therefore, both are credited with significant pioneering achievements. From today's perspective, this offers numerous opportunities for further research in museums in the field of occupational health and safety during the period under investigation. This is underlined by the fact that Arndt's studies incorporated information from numerous natural history museums on both national and international levels. Therefore, a threat to active substances and agents for pest control in the first quarter of the 20th century was already known and widely discussed. Based on this fact, it is unclear why chemical substances for pest control have been used without scrutiny in museum collections for many decades. The question also inevitably arises as to the extent to which both scientific and cultural institutions have conducted thorough tests on pesticides to determine their harmful effects on health.

Suitability testing of historical pesticides

Initially, pesticide tests regarding their insecticidal effect were conducted in state-funded institutions at the start of the 20th century. Experiments were conducted at the Kaiserliche Biologische Anstalt für Land- und Forstwirtschaft (KBA) (Imperial Biological Institute for Forestry and Agriculture) in Berlin. Industry showed great interest in these investigations, as they were crucial for the approval of products in the free market. In 1916, the agent "Dichlorbenzin Agfa" was investigated at the institution and was found to be effective for controlling harmful insects in textile fabrics and furs (Schwartz 1916, 15–16). Further evaluations and assessments of plant protection products were documented in the Naumburg branch of the Biologische Reichsanstalt in 1920. Blunck presented his research results at a general meeting of the Gesellschaft für Entomologie (Society for Entomology), which involved using various chemical agents to combat coal fleas (*Phyllotreta* sp.), rapeseed fleas (*Psylliodes* sp.), rapeseed beetles (*Brassicogethes* sp.), and cabbage weevils (*Ceutorhynchus* sp.) (Blunck 1922, 40–41). Oilseed rape and beet, which served as oleaginous fruits and fodder crops, were tested in the study. These crops were used to evaluate the insecticidal properties of industrially manufactured products from de Haën and Hoechst, as well as solutions found in literature and proprietary formulations of liquid and powder preparations. Furthermore, Urania Green (Schweinfurter Grün/Schweinfurt green), lead arsenate, powdered arsenic preparations, nicotine preparations, naphthalene, and insect powder were investigated. The suspended capacity, wetting ability, resistance to atmospheric gases, effect on fodder plants, percentage of dead coal fleas (*Phyllotreta* sp.), and "feeding" on plants were evaluated (Ibid. 42–48). In principle, Blunck stated that none of the liquid arsenic preparations could guarantee the minimum requirement of a lethal effect within eight days with complete protection of the fodder plants. Powdered "Sturmsche Mittel"[20] performed the best. Similar to Richter and Walther, Blunck warned against the use of arsenic-containing agents and preparations due to their extreme toxicity, human health risks, severe malformations, and plant burns (see also LASA 1929). Therefore, for the time being, he prioritised the mechanical interception of beetles with a specially designed apparatus (Cf. Blunck 1922, 49–55.

Investigations on a scientific basis with the sole aim of testing the insecticidal effect of pesticides in museums could be proven in two cases in cooperation between the KMfV and the k.k. l-c-V. After Friedrich Rathgen and Johann Bolle[21] met personally in Vienna during a conference in 1904 (see Part III, chapter 13.), the latter undertook a study trip to Berlin to visit the KM and their Chemisches Laboratorium in 1910. Felix von Luschan, at that time director of the Afrikanisch-Ozeanische Abteilung (African-Oceanic Department) at the KMfV, supported Bolle's efforts to evaluate insecticides for their effectiveness in killing of insect pests in collection cabinets. During his stay in Berlin, Bolle received three soldered zinc boxes containing insects in worm-like wood from Gorizia, which were used as test objects in three collection cabinets of the KMfV (SMB-PK EM 1910/1911). The efficacy of 100 g of turpentine was tested in cabinet 17, 100 g camphor in cabinet 58, and carbon disulfide (without quantity, author's note) in cabinet 67 during a six-month experiment. Each cabinet had a volume of 6 m³ (SMB-PK EM 1911a).

Figure 6.1 Johann (Giovanni) Bolle. Director of the kaiserlich königlichen land-wirtschaftlich-chemischen Versuchsstation in Gorizia, Italy, from 1891 to 1912.

Figure 6.2 Felix von Luschan. Director of the Afrikanisch-Ozeanische Abteilung at the Königliches Museum für Völkerkunde from 1904 to 1911.

Bolle tended to put only carbon disulfide in the cabinets, as turpentine was only effective if the insects were not too deep in the wood. Camphor, on the other hand, did not cause mortality in insects (SMB-PK EM 1911b). August Eichhorn, von Luschan's successor, also supported Bolle's further experiments on the collection cabinets of the museum (SMB-PK EM 1911c), so that at the beginning of 1911 a second examination consisting of three experiments was carried out on the third floor of the museum. The cabinets had a capacity of approximately 9 m³ and were successively equipped with 450 cm³ and 225 cm³ carbon disulfide and 450 cm³ carbon tetrachloride. The fumes acted on the collection items for 16 days at a time (SMB-PK EM 1911d). After evaluating this experiment, Bolle reported that in the experiments with carbon disulfide, a 100 % mortality of the insects had occurred. The amount of carbon tetrachloride used did not kill insects. Nevertheless, Bolle advised against a higher dosage, as he worried that an unpleasant smell might spread throughout the hall, which would be particularly unfavorable to visitors (SMB-PK EM 1911e). To minimise odor nuisance, carbon disulfide was blended with small amounts of nitrobenzene and rosemary oil (SMB-PK EM 1912). Bolle obviously felt committed exclusively to the paradigms of the natural sciences in his evaluation of the experiments by not going into the odor nuisance caused by carbon disulfide. From today's perspective, carbon disulfide belongs to the so-called CMR substances (**C**arcinogenic, **M**utagenic, and toxic to **R**eproduction), whose harmful effect on human health has been clearly proven (Taeger 1941).

Notes

1 About Karl von den Steinen (see Part III, chapters 8 and 12).
2 Beijú is a small cake made from cassava or tapioca flour that ferments together with water.
3 About Theodor Koch-Grünberg (see chapter 8).
4 Pierre-Jean-Claude Mauduyt de la Varenne was a physicist, naturalist, and coauthor of the *Encyclopédie méthodique, ou par ordre de matières*.
5 Walter Hough was chief curator of the anthropology department at the Smithsonian Museum of Natural History, Washington D.C. from 1910 to 1935.
6 On the ideal and practical tasks of ethnographic museums (translation by the author).
7 And the paints are not procured in powder form but rubbed with linseed oil (translation by the author).
8 Eduard Krause, b. in 1847, d. 30.10.1917, was from 01.04.1884 until his death the first conservator at the KMfV in Berlin (see Part III, chapter 8).
9 Willy Foy, b. November 27, 1873, d. March 1, 1929, was, from 1901 to 1925, founding director of the R-J-M in Cologne.
10 Walter Lehmann, b. 1878, d. 1939, was, from 1927 to 1934, the director of the African, Oceanic and American collections at the Museum of Ethnology at the National Museums in Berlin. (See: Grabowski et al. (Hrsg.), 2010, 155).
11 Kurt Stubenrauch was at that time an assessor in the general administration of the National Museums in Berlin.
12 Wilhelm Waetzold, b. 1880, d. 1945, was Director General of the National Museums in Berlin from 1928 to 1934. (See: Grabowski et al. (Hrsg.), 2010, 151).
13 Carl Brittner, b. 1883, d. 1958, was, from 1928 to 1948, head of the Chemisches Laboratorium at the National Museums in Berlin. (See Grabowski et al. (Hrsg.), 2010, 165).
14 Further information that would indicate a specific institution in Eberswalde cannot be found in Friedrich Rathgen's letter (author's note).
15 From 1904 to 1935, Georg Thilenius was the first director of the Museum für Völkerkunde in Hamburg, founded in 1879 (see Part III, chapter 13).
16 Walther Arndt, b. January 8, 1891, d. June 26, 1944, was a German zoologist and physician. He was denounced for his critical statements about national socialism, sentenced by the People's Court on May 11, 1944, and executed in the Brandenburg-Görden penitentiary. A memorial plaque commemorates him at the Museum für Naturkunde, Invalidenstraße 43, 10115 Berlin.
17 The use of carbon disulfide is discussed in detail in Part III, chapter 8.
18 German Museums Association (translation by the author).
19 The Deutscher Museumsbund was founded on November 23, on the suggestion of the museum directors Karl Koetschau, Gustav Pauli, and Georg Swarzenski at the Städelsches Kunstinstitut in Frankfurt am Main in 1917.
20 See glossary: "Sturmsches Mittel".
21 About Johann Bolle (See Part III, chapter 8).

References

Archive material

Archiv des Museums am Rothenbaum – Kulturen und Künste der Welt (1929a): Findbuch. 101–1, Nr. 281. Lehmann, Walter. "Globolkrankheit". Abschrift des Briefes vom 16.02.1929 an die Generalverwaltung der Staatlichen Museen zu Berlin, 2 Seiten, ohne Paginierung.

Archiv des Museums am Rothenbaum – Kulturen und Künste der Welt (1929b): Findbuch. 101–1, Nr. 281. Rathgen, Friedrich. Stellungnahme zu "Globol". Abschrift des Briefes vom 18.2.1929 an die Generalverwaltung der Staatlichen Museen zu Berlin, 2 Seiten, ohne Paginierung.

Archiv des Museums am Rothenbaum – Kulturen und Künste der Welt (1929c): Findbuch. 101–1, Nr. 281. Thilenius, Georg. Stellungnahme zu "Globol". Brief vom 27.02.1929 an den Generaldirektor der Staatlichen Museen zu Berlin, 2 Seiten, ohne Paginierung.

Archiv des Museums am Rothenbaum – Kulturen und Künste der Welt (1929d): Findbuch. 101–1, Nr. 281. Weiterleitung einer Stellungnahme zu "Globol" von der Firma Fritz Schulz jun. Aktiengesellschaft. Senger, Hugo. Brief vom 08.05.1929 an die Direktion des Museums für Völkerkunde Hamburg, 1 Seite, ohne Paginierung.; Ebd. Fritz Schulz jun. Aktiengesellschaft, Leipzig. Brief vom 03.05.1929 an die Direktion des Museums für Völkerkunde Hamburg, 2 Seiten, ohne Paginierung.

Archiv des Museums am Rothenbaum – Kulturen und Künste der Welt (1934): Findbuch. 101–1, Nr. 281. Anonymus. Gesundheitliche Beeinträchtigung durch "Xylamon". 3 Berichte vom 19.05.1934, vom 24.05.1934 und vom 09.06.1934, 2 Seiten, ohne Paginierung.

LASA (1929): I 532, Nr. 399. I.G. Farbenindustrie Aktiengesellschaft, Höchst a. M. Anonymus. Untersuchungen über Pflanzenverbrennungen durch Schweinfurter Grün. Bericht vom 02.05.1929, Blatt 1–9, 9 Seiten.

LASA (1930): I 532 Nr. 600. I.G. Farbenindustrie Aktiengesellschaft, Abteilung Z III, Frankfurt am Main 28.07.1930. Stellungnahme vom 28.07.1930 zu "Globol", "Schädlingsnaphthalin", "Areginal" und "Areginal U", 1 Seite, ohne Paginierung.

SMB-PK EM (1910/1911): I/MV 0075, Pars II c, Vol. 1. E. Nr. 1360/10. Königliches Museum für Völkerkunde zu Berlin und kaiserlich königliche landwirtschaftlich-chemische Versuchsstation 1910–1911 in Görz, 1910–1911. Luschan von, Felix. 2 Aktennotizen vom 04.07.1910 und vom 07.07.1910 auf Brief vom 02.07.1910 des k.k. Inspektors (Name unbekannt, Anmerk. d. Verf.) aus Görz, 1 Seite, ohne Paginierung.

SMB-PK EM (1911a): I/MV 0075, Pars II c, Vol. 1. E. Nr. 1360/10. Luschan von, Felix. Brief an Johann Bolle vom 24.01.1911, Blatt 3–4, 2 Seiten.

SMB-PK EM (1911b): I/MV 0075, Pars II c, Vol. 1. E. Nr. 1360/10. Luschan von, Felix. Aktennotiz vom 24.01.1911 auf Brief von Johann Bolle vom 19.01.1911, Blatt 2, 1 Seite.

SMB-PK EM (1911c): I/MV 0075, Pars II c, Vol. 1. E. Nr. 1360/10. Eichhorn, August. Brief an Johann Bolle vom 01.02.1911, Blatt 5, 1 Seite.

SMB-PK EM (1911d): I/MV 0075, Pars II c, Vol. 1. E. Nr. 298/11. Bolle (09.02.1911): Bolle, Johann. Brief vom 09.02.1911 an die Generaldirektion der Königlichen Museen zu Berlin, 1 Seite, ohne Paginierung.; Bolle, Johann. Brief vom 02.03.1911 an den Direktor (Name is not mentioned, author's note), Blatt 1, 2 Seiten.

SMB-PK EM (1911e): I/MV 0075, Pars II c, Vol. 1. E. Nr. 298/11 Bolle (14.04.1911): Bolle, Johann. Brief vom 14.04.1911, 1 Seite, ohne Paginierung.

SMB-PK EM (1912): I/MV 0075, Pars II c, Vol. 1. E. Nr. 467/12. Versuche zur Schädlingsbekämpfung im Königlichen Museum für Völkerkunde zu Berlin mit der kaiserlich königlichen landwirtschaftlich-chemischen Versuchsstation in Görz. Bolle (04.03.1912): Bolle, Johann. Brief vom 04.03.1912, 2 Seiten, ohne Paginierung.

SMB-PK EM (1913): I/MfV 0034, Pars Ia, Bd. 9, E. Nr. 1886/13. Acta betreffend Dienstbestimmungen und Instruktionen. Loseblattsammlung. Chappuis von, Hermann. Anordnung über den Einsatz von bleihaltigen Farben für den Innen- und Außenbereich vom 12.11.1913, 1 Seite, ohne Paginierung.

SMB-ZA (1905): GV 696, Nr. 1–19. Loseblattsammlung. Königliche Museen zu Berlin (1905): Belege für das Etatjahr 1905. Zur Exordinarien Rechnung. Zusammenstellungen aus dem Jahr 1905, Blatt 16, 1 Seite.

Other sources

Arndt, Walther (1932a): Die Berufskrankheiten an naturwissenschaftlichen Museen. I. Vergiftungen. *Museumskunde, Neue Folge* IV, (2), 47–66.

Arndt, Walther (1932b): Die Berufskrankheiten an naturwissenschaftlichen Museen. II. Schädigungen durch Übertragung von Krankheitserregern. *Museumskunde, Neue Folge* IV, (3), 103–117.

Blunck, Hans (1922): Über die Wirkung arsenhaltiger Gifte auf Ölfruchtschädlinge nach Beobachtungen an der Naumburger Zweigstelle der Biologischen Reichsanstalt. Dritte Mitgliederversammlung zu Eisenach vom 28–30 September 1921. *Verhandlungen der Deutschen Gesellschaft für Entomologie*, 40–55.

Fabian, Johannes (2000): Out of Our Minds. Reason and Madness in the Exploration of Central Africa. The Ad. E. Jensen Lectures at the Frobenius Institute, University of Frankfurt. Berkeley: University of California Press. Available online at http://site.ebrary.com/lib/academiccompletetitles/home.action.

Farber, Paul Lawrence (1977): The Development of Taxidermy and the History of Ornithology. *ISIS*, 68(244), 550–566.

Grabowski et al. (Hrsg.) (2010): Grabowski, Jörn; Winter, Petra; Ebelt, Beate; Pilgermann, Carolin. Kunst recherchieren. 50 Jahre Zentralarchiv der Staatlichen Museen zu Berlin. Staatliche Museen zu Berlin. Berlin: Deutscher Kunstverlag, 155.

Hough, Walter (1889): The Preservation of Museum Specimens from Insects and the Effects of Dampness. For the Year Ending June 30, 1887. Report of the National Museum, Washington, DC.

Hüntelmann, Axel Cäsar (2008): Hygiene im Namen des Staates. Das Reichsgesundheitsamt 1876–1933. Diss.-Bremen. Univ., 2005. Göttingen: Wallstein. Available online at http://deposit.d-nb.de/cgi-bin/dokserv?id=3099685&prov=M&dok_var=1&dok_ext=htm.

Koch-Grünberg, Theodor (1903–1905): Tagebuch Rio Negro-Expedition 1903–1905. Völkerkundliche Sammlung der Philipps-Universität Marburg, VK Mr KG-B-I.2. Heft 1. Loseblattsammlung. Nachlass Theodor Koch-Grünberg, Unveröffentlichte Quelle, 4-seitig.

Komander, Gerhild H.M. (2004): Die Geschichte Berlins. Arndt, Walther. Hg. v. Verein für die Geschichte Berlins e.V., gegr. 1865. Available online at www.diegeschichteberlins.de/geschichteberlins/persoenlichkeiten/persoenlichkeiteag/434-arndt.html.

Richter, Oswald (1907): Über die idealen und praktischen Aufgaben ethnographischer Museen (Fortsetzung). *Museumskunde, Zeitschrift für Verwaltung und Technik öffentlicher und privater Sammlungen*, (Band III), 99–120.

Richter, Oswald (1908): Über die idealen und praktischen Aufgaben ethnographischer Museen. *Museumskunde, Zeitschrift für Verwaltung und Technik öffentlicher und privater Sammlungen*, (Band IV), 92–235.

Schwartz, N., first name unknown (author's note.) (1916): Versuche zur Bekämpfung tierischer Schädlinge mit Giften. In: Mitteilungen Kaiserliche Biologische Anstalt, Berlin: Verlagsbuchhandalung Paul Parey und Julius Springer, 1916, (16).

Taeger, Harald (1941): Die Klinik der entschädigungspflichtigen Berufskrankheiten. Berlin, Heidelberg: Springer.

Tello, Helene (2006): Investigations on Super Fluid Extraction (SFE) with Carbon Dioxide on Ethnological Materials and Objects Contaminated with Pesticides. Diplomarbeit. Fachhochschule für Technik und Wirtschaft, Berlin. Fachbereich 5, Gestaltung, Studiengang Restaurierung/Grabungstechnik.

von den Steinen, Karl (1894): Unter den Naturvölkern Zentral-Brasilien. Reiseschilderung und Ergebnisse der zweiten Schingú-Expedition, 1887–1888. Berlin: Geographische Verlagsbuchhandlung von Dietrich Reimer [Mit 30 Tafeln (1 Heliogravüre, 11 Lichtdruckbilder, 5 Autotypien und 7 lithogr. Tafeln) sowie 160 Text-Abb.].

Winter, Petra; Grabowski, Jörn (Hrsg.) (2014): Zum Kriegsdienst einberufen. Die Königlichen Museen zu Berlin und der Erste Weltkrieg. Staatliche Museen zu Berlin. Köln: Böhlau (Schriften zur Geschichte der Berliner Museen, 3).

7 Typological recording of pesticides

Table 7.1 Typological recording of natural active substances

Inorganic substances

Chronology	Active substance	Use	Source
c. 484–424 BC	Aluminum potassium sulfate dodecahydrate (alum)	Fire retardant	Unger et al. 2001
1742	Potassium carbonate (potash)	Admixture in arsenic soap	Rookmaaker et al. 2006
1770 1875	Potassium nitrate (saltpeter)	Combating of insect pests in natural history collections; preservation of medicinal preparations	Kuckhan 1770; Keil 1879
1877	Boric acid	Wood preservative	Unger et al. 2001
ca. 1920	Carbon dioxide	Stock protection	Römpp Lexikon Chemie 1992 und 1996

Organic substances

Chronology	Active substance	Use	Source
1600 BC* 200 A.D. ca. 1900	Coloquints	Medical purposes; control of rodents; control of insect pests in museum collections	Ebers 1875; Wink et al. 2008; Königliches Museum für Völkerkunde zu Berlin 1901
15th century 17th century	Aloe	Antimicrobial ingredients; healing effect; wood impregnation	Lehmann und Lehmann 1985; Schiessl 1984
16th century end of 19th century	Laudanum (opium)	Drug; intoxicant; agents in first-aid kits of explorers	Ball 2014; Fabian 2000

(*Continued*)

DOI: 10.4324/9781003407607-10

Table 7.1 (Continued)

Organic substances			
Chronology	Active substance	Use	Source
Mid-17th century End of 19th century	Quinine	Medical application; malaria control; food industry; remedy in first-aid kits of explorers	Müller-Jahncke et al. 2005; Römpp Lexikon Chemie 1992 und 1996; Fabian 2000
1682 1770	Musk	Medical healing power; aphrodisiac; dry preparation of birds	Schröck et al. 1682; Kuckhan 1770
17th century	Lavender oil	Repellent against insects; fragrance; sedative	Schiessl 1984; Rochussen 1920
1770	Tansy	Dry preparation of birds; repellent against insect pests	Kuckhan 1770; Rochussen 1920
1770 1909	Wormwood leaves	Remedy; preparation of dead birds; protection against wood-destroying insects	Römpp Lexikon Chemie 1992 und 1996; Kuckhan 1770; Aberle und Koller 1968
ca. 1800	Pyrethrum (Dalmatian and Persian insect powder)	Insecticide	Unger et al. 2001
1819 1884	Teein (caffeine)	Discovery by Friedlieb Ferdinand Runge; remedy in first-aid kits of explorers	Gossauer 2006; Steinen, von den 1884
1830	Camphor	For medical and cosmetic preparations; disinfection; prevention and combating of insect pests in entomological and ethnological collections	Römpp Lexikon Chemie 1992 und 1996; Odegaard et al. 2005
1842	Thymol	Fungicide; combating mold; antiseptic	Lüllmann et al. 2010; Burns 1941
1846 ca. 1920	Coumarin	Narcotic effect; perfume; medical application; repellent against rodents; stock protection	Lowe 2017; Römpp Lexikon Chemie 1992 und 1996; Zacher 1924

Organic substances

Chronology	Active substance	Use	Source
1875	Rosemary oil	Preservation of taxidermy and plants	Neumayer und Ascherson 1875; Rochussen 1920
ca. 1900	Patchouli oil	Remedy; repellent against insects	Wiesner 1927
End of 19th century ca. 1900	Pine oil	Detergent; solvents for resins; technical use; control of insect pests in museum collections	Birnbaum und Merck 1884; Wiesner 1927; Königliches Museum für Völkerkunde zu Berlin 1901
End of 19th century	Quendel oil	Repellent against insect pests in natural history collections	Rochussen 1920; Pfister 2008
End of 19th century	Strychnine	Combating insect pests in anthropological collections; rat poison	Mason 1889
End of 19th century	Tannin	Technical use in leather production; medical use; remedies in first-aid kits of explorers	Römpp Lexikon Chemie 1992 und 1996; Fabian 2000
1924	Dammar resin	Conservation of damaged wooden objects	Unger et al. 2001
1924	Tung oil (Chinese wood oil)	Conservation of damaged wooden objects	Unger et al. 2001
Early 20th century	Spike oil	Repellent against textile moths	Rochussen 1920; Aal 1925

Note: * In the case of multiple mentions of an active substance in the sources, the oldest time indication in the respective column is highlighted in bold in the first place.

Table 7.2 Typological recording of synthetic active substances

Inorganic substances

Chronology	Active substance	Use	Source
1452	Arsenic (III) oxide (arsenic)	Wood-destroying insects	Unger et al. 2001
1452	Mercury (II) chloride (sublimate)	Insecticide; wood preservative	Unger et al. 2001

(Continued)

Table 7.2 (Continued)

Organic substances			
Chronology	*Active substance*	*Use*	*Source*
16th century*	Sulfuric acid	Insecticide	Römpp Lexikon Chemie 1992 und 1996
1720	Copper (II) sulfate	Insecticide; wood preservative	Unger et al. 2001
1742	Arsoric acid (Arsenic (III) acid)	Component in plant protection products; admixture in arsenic soap	Römpp Lexikon Chemie 1992 und 1996; Rookmaaker et al. 2006
1770	Sulfur bloom (flowers of sulfur)	Dry preparation of birds	Kuckhan 1770
1774	Ammonia	For technical syntheses; stock protection	Gmelin 1827; Römpp Lexikon Chemie 1992 und 1996; Zacher 1924
1815	Zinc (II) chloride (zinc chloride)	Wood preservative	Unger et al. 2001
1818 1919	Hydrogen peroxide	Bleaching and disinfectant; fungicide; conservation of polychrome sculptures	Römpp Lexikon Chemie 1992 und 1996; Bolle 1919
Mid-19th century	Lead hydrogen arsenate	Pesticide; feeding poison	Römpp Lexikon Chemie 1992 und 1996
1861 ca. 1940	Sodium fluoride	Wood preservative; control of insect pests; admixture in bookbinding paste	Unger et al. 2001; Burns 1941
1875 ca. 1900	Potassium cyanide	Industrial use; preservation of taxidermy and plants; room fumigation in museums	Römpp Lexikon Chemie 1992 und 1996; Neumayer und Ascherson 1875; Toothaker 1908
1880 1915	Hydrogen cyanide (hydrogen cyanide)	Plant protection; stock protection; fumigation against wood-destroying insects	Unger et al. 2001
1885	Arsoric acid (arsenic (V) acid)	Pesticide; insecticide in museum collections	Römpp Lexikon Chemie 1992 und 1996; Hough 1889

Organic substances

Chronology	Active substance	Use	Source
End of 19th century	Lead (II) sulfate	Painting color; remedies in first-aid kits of explorers	Römpp Lexikon Chemie 1992 und 1996; Fabian 2000
End of 19th century	Zinc (II) oxide (zinc oxide)	Antiseptic; pigment; remedies in first-aid kits of explorers	Römpp Lexikon Chemie 1992 und 1996; Fabian 2000
ca. 1900	Barium hexafluorosilicate	Conservation of herbaria	Purewal 2001
ca. 1900 1951	Sodium arsenite	Dry preparation of animal skins; herbicide; contact poison	Wray 1908; Odegaard et al. 2005
1914 1919	Hydrogen sulfide (hydrogen sulfide)	Browning of organic and inorganic materials	Römpp Lexikon Chemie 1992 und 1996; SMB-PK EM 1914h; Bolle 1919
ca. 1920	Sodium silicate	Industrial use; control of wood-destroying insects	Römpp Lexikon Chemie 1992 und 1996; Department of Scientific and Industrial Research 1926
ca. 1920	Peroxodisulfates	Strong oxidizing agents; conservation of paper	Römpp Lexikon Chemie 1992 und 1996; Scott 1922
ca. 1920	Sulfurous acid	Stock protection	Zacher 1924
ca. 1920	Chlorine water	Stock protection; disinfectant	Zacher 1924; Römpp Lexikon Chemie 1992 und 1996
1934	Ammonium arsenate (V)	Admixture to adhesives during bookbinding	Plenderleith 1934

Organic substances

Chronology	Active substance	Use	Source
1774	Ammonia	For technical syntheses; stock protection	Gmelin 1827; Römpp Lexikon Chemie 1992 und 1996; Zacher 1924

(*Continued*)

Table 7.2 (Continued)

Organic substances			
Chronology	Active substance	Use	Source
1796 1875	Ethanol (synthetic)	Solvents and extractants; preservation of taxidermy and plants	Römpp Lexikon Chemie 1992 und 1996; Neumayer und Ascherson 1875
Early 19th century ca. 1900	Benzene	Additive for motor fuels, solvents; control of insect pests in museum collections	Römpp Lexikon Chemie 1992 und 1996; Königliches Museum für Völkerkunde zu Berlin 1901
1830	Nicotine	Against insect pests in ethnological collections	Odegaard et al. 2005
1848 1916 1945	Chloropicrin (active ingredient: trichloronitromethane)	Textile moths, beetles, cockroaches; chemical warfare agent in the First World War; gassing of objects made of organic materials	Odegaard et al. 2005; Römpp Lexikon Chemie 1992 und 1996; Lehmann 2005
1st half 19th century 1884 1889	Chloroform	Anesthetic; against bacon beetles; against insect pests in objects with organic materials	Römpp Lexikon Chemie 1992 und 1996; Smith 1884; Hough 1889
1875	Methanol (methyl alcohol)	Component for the Wickersheimersche Lösung	Keil 1879
1877	Diethyl ether (ether)	Anesthetic (no longer in use); against insect pests in natural history collections	Römpp Lexikon Chemie 1992 und 1996; Murray 1877
1887	Naphthalene	Combating textile moths	Odegaard et al. 2005
1887	Naphthole	Manufacture of paints; antiseptic effect; combating textile moths	Römpp Lexikon Chemie 1992 und 1996; Odegaard et al. 2005
1888	Cresol (cresol); trade name carbolineum	Wood preservative	Unger et al. 2001

Organic substances

Chronology	Active substance	Use	Source
1890	Phenol (carbolic acid)	Disinfection; wood preservation	Muter 1890; Wray 1908
1893	Methanal (formaldehyde)	Fungicide; fumigants in intensive animal husbandry and restricted in museum collections	Römpp Lexikon Chemie 1992 und 1996
ca. 1900	Carbon disulfide	Fumigation of wood-destroying insects	Unger et al. 2001
ca. 1900	Tetrachloromethane (carbon tetrachloride)	Fumigation of wood-destroying insects	Unger et al. 2001
1909 1912	Nitrobenzene (mirbane oil)	Disinfection; against insect pests in natural history collections	Lueger 1904; Deschka 1987
1910	Pentylacetate	Food industry; solvents for Zaponlack	Römpp Lexikon Chemie 1992 und 1996; Chemische Fabrik Griesheim-Elektron 1910
1916	Dichlorodiethyl sulfide (mustard gas, LOST group)	Chemical warfare agent in the First World War	Schnedlitz 2008
ca. 1920	1-Butanol	Solvents, component of detergents; stock protection	Römpp Lexikon Chemie 1992 und 1996; Zacher 1924
ca. 1920	1,2-Butylene oxide (1,2-Epoxybutane)	Stock protection	Römpp Lexikon Chemie 1992 und 1996
ca. 1920	*n*-Butyl formate	Stock protection	Römpp Lexikon Chemie 1992 und 1996
ca. 1920	Glacial acetic acid	Stock protection	Zacher 1924
ca. 1920	Isovaleric acid	Stock protection; remedy	Römpp Lexikon Chemie 1992 und 1996
ca. 1920	Carbon monoxide	Syngas; industrial use; stock protection	Römpp Lexikon Chemie 1992 und 1996; Zacher 1924

(Continued)

Table 7.2 (Continued)

Organic substances			
Chronology	*Active substance*	*Use*	*Source*
1928	Ethylene oxide	Fumigation of wood-destroying insects; kills bacteria, viruses, fungi	Unger et al. 2001
1930 1935/1936	Polyvinyl acetate	Wood consolidation; conservation of paintings	Plenderleith 1934; Schiessl 1984
1930	Tetrachlorobenzene	Solid substance to produce insecticides and herbicides	I.G. Farbenindustrie 1930
1932	Bromomethane (methyl bromide)	Discovery of insecticidal activity	Unger et al. 2001
1939	Dichlorodiphenyldi-chloroethane (DDD) metabolite of dichlorodiphenyl-trichloroethane (DDT)	Insecticide; wood preservative	Simon 1999; Unger et al. 2001
1939	Dichlorodiphenyldi-chloroethylene (DDE) metabolite of DDT	Insecticide; wood preservative	Simon 1999; Unger et al. 2001
1939	DDT	Insecticide; wood preservative	Simon 1999; Unger et al. 2001
1941	1,2-Dichloroethane	Fumigation of objects made of organic materials	Burns 1941
1951	Dichlorvos	Insecticide	Römpp Lexikon Chemie 1992 und 1996
End of 1950s	Illo-Spezial-T (active ingredient: tetrachloroethylene or obsolete perchloroethylene)	Fumigation of ethnological objects; solvents in the textile industry	Scharn[1] GESTIS-Stoffdatenbank 2017
1960	Chlorpyrifos	Insecticide	Römpp Lexikon Chemie 1992 und 1996

Note: * In the case of multiple mentions of an active substance in the sources, the oldest time indication in the respective column is highlighted in bold in the first place.

Table 7.3 Typological recording of preparations containing active substances in industry

Organic substances

Chronology	Active substance	Use	Source
1805	Schweinfurt green (active ingredient: copper (II) arsenite acetate)	Paint; insecticide; wood preservatives	Liebig 1822, 447; Schiessl 1984
1915	Antorgan (woodworm antorgan) (active ingredient: zinc fluoride)	Wood conservation in house, cellar, and garden; control of wood-destroying insects	Unger, verbal communication, 2017
1924	Chlorinated lime	Disinfectant against bacteria and viruses; stock protection	Zacher 1924
1925	Sturmsches Mittel (Esturmit; active substance: calcium (II) arsenate)	Pesticide	Stellwaag 1927
1926	Diametan (active ingredient: sulfur)	Fumigation of premises against domestic harmful insects	Kolbe und Haug 1979; Römpp Lexikon Chemie 1992 und 1996
1928	"Calcid" (active ingredient: calcium cyanate)	Fumigation of citrus trees; combating rats	Kalthoff und Werner 1998
1929	"Zyklon B" (active ingredient: hydrogen cyanide)	Fumigation of wood-destroying insects	Unger et al. 2001

Organic substances

Chronology	Active substance	Use	Source
c. 2000 BC	Tar	Against wood-destroying insects; organic materials	Unger et al. 2001
Early 18th to early 20th century	Kerosene	Wood preservative	Zinke 1802
1781	Turpentine oil	Wood preservative	Unger et al. 2001
1850	Petroleum ether	Dry cleaning	Karlsch und Stokes 2003
1877	Petrol	Combating of mites in natural history collections; combating of insect pests on objects containing organic materials	Murray 1877; Rathgen 1908

(Continued)

Table 7.3 (Continued)

Organic substances			
Chronology	Active substance	Use	Source
1877	Naphtha	Combating of mites in natural history collections	Murray 1877
1888	Carbolineum (active ingredients: cyclic hydrocarbons)	Wood preservative	Unger et al. 2001
End of 19th century	Beech tar oil	Combating of insect pests in natural history collections	Pfister 2008
End of 19th century	Perthan (active ingredient: chlorinated hydrocarbon, related to DDD)	Stock protection; textile moths and carpet beetle	Sirois et al. 2010; GESTIS-Stoffdatenbank 2017
End of 19th century	Crude benzene	Against plant pests in stock protection	Schwartz 1925
ca. 1900	Linseed oil varnish	Impregnation of wooden objects	Unger et al. 2001
1906	Autan (active ingredients: paraformaldehyde and barium peroxide)	Disinfectants against bacteria, microorganisms, and rodents	Eichengrün 1906
1914	Globol (active ingredient: 1,4-dichlorobenzene)	Insecticide; textile moths	AIFM Wolfen 1912–1918; Odegaard et al. 2005
1920	Eulan (various active substances)	Insecticide; control of insect pests on textiles and carpets	Unger 2012
ca. 1920	Tillantin R (active ingredient: nitrophenolmercury)	Stock protection	Taeger 1941
1923	Flit (active ingredient: DDT)	Insecticide; control of insect pests on textiles, skins, and feathers	Wilhelmi und Kunike 1927
1923	Xylamon (active ingredients: chloronaphthalenes)	Wood preservative	Unger et al. 2001
1923	Xylamon-LX Hell (Wirkstoffe: chloronaphthalenes)	Wood preservative	Unger et al. 2001
1924	Paraffin	Conservation of damaged wooden objects	Unger et al. 2001

Organic substances

Chronology	Active substance	Use	Source
1926	Eryl (active ingredient: unknown)	Stock protection; control of the grain beetle	Kleine 1926
1930	T-gas (active ingredient: ethylene oxide)	Housing fumigation; control of bacteria, microorganisms, and rodents	Kalthoff und Werner 1998

Table 7.4 Typological recording of prescriptions of pharmacists and doctors

Inorganic and organic substances

Chronology	Active substance	Use	Source
1742	Arsenic soap (active ingredients: arsoric acid, arsenic (III) oxide, camphor, soap, potassium carbonate, and lime powder)	Dry preparation of animal skins and ethnographic objects	Rookmaaker et al. 2006
1786	Fowler's solution (active ingredient: potassium arsenite)	Preservation of animal skins	Neumüller 1973
ca. 1900	Antisekt (active ingredient: unknown)	Insecticide for objects made of organic materials	Germann 1933
ca. 1900	Sodium arsenicosum (sodium arsenate)	Insecticide for objects made of organic materials	SMB-PK EM 1904

Table 7.5 Typological recording of prescriptions of individuals

Inorganic and organic substances

Chronologie	Active substance	Use	Source
ca. 1800	Mustard plaster	Home remedies and remedies for various ailments	Pierer 1857–1865
1840	"Zacherlin" (active ingredient: pyrethrum)	Insecticide	Sotriffer 1996; Unger et al. 2001; Offenthaler 2013

(Continued)

Table 7.5 (Continued)

Inorganic and organic substances			
Chronologie	Active substance	Use	Source
ca. 1850	Phosphorus paste	Control of rodents	Hertwig 1847; Meyers Konversationslexikon, 5. Aufl. 1896
ca. 1850	Chinese moth tincture (active ingredients: peel of red pepper, alcohol, and camphor)	Insecticide	Lange 1923
1875	Wickersheimersche solution (active ingredients: water, alum, salt, potassium carbonate, arsoric acid, glycerin, and methyl alcohol)	Wet preparation of medicinal preparations, animal skins, and plants	Keil 1879
ca. 1900	Moth ether (active ingredients: camphor, naphthalene, clove, and lavender oil)	Insecticide	Arends et al. 1927
ca. 1900	Arsenous jelly	Preparation of pathological samples of animal organs	Delépine 1914

Note

1 Verbal communication on August 1, 2018 by Klaus Scharn, conservator and chief conservator at the Ethnological Museum from 1966 to 1999.

References

Archive material

AIFM Wolfen, ohne Signatur (1912–1918): Loseblattsammlung. *AGFA Jahresgeschäftsberichte. Actien-Gesellschaft für Anilin-Fabrikation. Blatt*, 105–106; 109–114, 12 Seiten.

HAStK (1910/1913): Best. 614, Nr. 438, Chemische Fabrik Griesheim-Elektron; Chemische Industrie-Gesellschaft, Ankauf von Imprägniermitteln, Akte. Loseblattsammlung. Brief von der Chemischen Fabrik Griesheim-Elektron vom 25.06.1910, Blatt 1, 1-seitig; 2 Briefe von der Chemischen Industrie-Gesellschaft Berlin vom 15.11.1913 und 08.12.1913; Blatt 2–3, je 1-seitig.

LASA (1930): I 532, Nr. 600, I.G. Farbenindustrie Aktiengesellschaft, Abteilung Z III, Frankfurt am Main (28.07.1930): Stellungnahme zu Globol – Schädlingsnaphtalin – Areginal – Areginal U. Brief, 1- seitig, ohne Paginierung.

Königliches Museum für Völkerkunde zu Berlin (01.01.1901 bis 30.04.1903): Acta betreffend den Umzug und die Aufstellung der Sammlungen des Museums. Aktenzeichen: I/MV 0057, Bd. 5, Pars I c. Umzugsakte.

SMB-PK EM (1904): I/MV 730, Vol. 30, Pars I. B., E. Nr. 578/04. Acta betreffend die Restauration von Alterthümern. Loseblattsammlung. Anweisung für den Einsatz von Schädlingsbekämpfungsmitteln. Luschan von, Felix. Aktennotiz vom 14.06.1904. Blatt 224, 4 Zeilen.

SMB-PK EM (1914h): I/MV 932, E. Nr. 324/14. Königliche Museen zu Berlin 1914. Personalakte Eduard Krause von 1914. Eichhorn, August. Schreiben an Wilhelm Bode vom 24.02.1914, 1 Seite, rechte Spalte, ohne Paginierung.

Other sources

Aall, Hans (1925): Arbeide og ordning i kulturhistoriske Museer. Kort Veiledning. Oslo: Utgitt med Statsbidrag.

Aberle, B.; Koller, M. (1968): Konservierung von Holzskulpturen. Probleme und Methoden. Wien: Institut fiir Osterreichische Kunstforschung des Bundesdenkmalamtes, 43 Seiten.

Arends, Georg; Frerichs, Georg; Zörnig, Heinrich (Hrsg.) (1927): Naphthalinum. In: Hagers Handbuch der Pharmazeutischen Praxis. Berlin, Heidelberg: Springer.

Ball, Philip (2014): The Devil's Doctor. Paracelsus and the World of Renaissance Magic and Science. London: Cornerstone Digital.

Birnbaum, Carl; Merck, Klemens (Hrsg.) (1884): Klemens Merck's Warenlexikon für Handel, Industrie und Gewerbe. Beschreibung der im Handel vorkommenden Natur- und Kunsterzeugnisse unter besonderer Berücksichtigung der chemisch-technischen und anderer Fabrikate, der Droguen- und Farbewaren, der Kolonialwaren, der Landesprodukte, der Material- und Mineralwaren. 3., gänzlich umgearb. Aufl., 2., rev. Abdr. Leipzig: Gloeckner.

Bolle, Johann (1919): Die Ermittlung der Wirksamkeit von insektentötenden Mitteln egen die Nagekäfer des verarbeiteten Werkholzes. *Zeitschrift für angewandte Entomologie*, (Band 5), 105–117.

Burns, Ned J. (1941): Field Manual for Museums. Washington, DC: United States Government Printing Office.

Delépine, Sheridan (1914): On the Arsenious Acid-Glycerin-Gelatin ("Arsenious Jelly") Method of Preserving and Mounting Pathological Specimens with Their Natural Colours, and on the Use of New Forms of Receptacles for Keeping Museum Specimens. From the Public Health Department, University of Manchester. *The Museums Journal*, (13), 322–329.

Department of Scientific and Industrial Research (ed.) (1926): The Cleaning and Restoration of Museum Exhibits. Third Report Upon Investigations Conducted at the British Museum. London: Published under the Authority of his Majesty's Stationery Office.

Deschka, Gerfried (1987): Die Desinfektion egent Insektensammlungen nach neueren Gesichtspunkten. *Steyrer Entomologenrunde*, 21, 57–61.

Ebers, Georg (Hrsg.) (1875): Papyros. Das Hermetische Buch über die Arzneimittel der alten Ägypter in hieratischer Schrift. Available online at http://digi.ub.uni-heidelberg.de/diglit/ebers1875bd2.

Eichengrün, Arthur (1906): Über das neue Autan-Desinfektionsverfahren. *Pharmazeutische Zeitschrift*, LI (77), 852.

Fabian, Johannes (2000): Out of Our Minds. Reason and Madness in the Exploration of Central Africa; The Ad. E. Jensen Lectures at the Frobenius Institute, University of Frankfurt.

Berkeley: University of California Press. Available online at http://site.ebrary.com/lib/academiccompletetitles/home.action.

Germann, Paul (1933): Bekämpfung der Museumsschädlinge. *Museumskunde*, Neue Folge, V; Sonderdruck (I), 9–11.

GESTIS-Stoffdatenbank (2017): Available online at www.dguv.de/ifa/gestis/gestis-stoffdatenbank/index.jsp.

Gmelin, Leopold (1827): Handbuch der theoretischen Chemie. Frankfurt: Franz Varrentrapp.

Gossauer, Albert (2006): Struktur und Reaktivität der Biomoleküle. Eine Einführung in die organische Chemie. Zürich: Verl. Helvetica Chimica Acta.

Hertwig, Heinrich Carl (1847): Praktische Arzneimittellehre für Thierärzte. 3. vermehrte Auflage Berlin: Veith & Comp., Anmerkung auf S. 520.

Hough, Walter (1889): The Preservation of Museum Specimens from Insects and the Effects of Dampness. For the Year Ending June 30, 1887. Report of the National Museum, Washington, DC.

Kalthoff, Jürgen; Werner, Martin (1998): Die Händler des Zyklon B. Tesch & Stabenow; eine Firmengeschichte zwischen Hamburg und Auschwitz. Hamburg: VSA-Verlag.

Karlsch, Rainer; Stokes, Raymond G. (2003): Faktor Öl. Die Mineralölwirtschaft in Deutschland 1859–1974. München: Beck. Online verfügbar unter http://www.gbv.de/dms/faz-rez/FD1200304101799547.pdf.

Keil, Ernst (1879): Die Gartenlaube. Das Wickersheimers'sche Conservierungsverfahren. Hrsg. v. Verlag von Ernst Keil. Leipzig. Available online verfügbar at jttps://de.wikisource.org/w/index.php?title=Seite:Die_Gartenlaube_(1879)_844.jpg&oldid=-(Version vom 21.05.2018).

Kleine, R. (1926): (First name unknown, author's note). Bekämpfungsversuche von Calandra granaria mit Eryl. In: Mitteilungen der Gesellschaft für Vorratsschutz e.V., Friedrich Zacher (Hrsg.). Berlin-Steglitz: Heft 6, 2. Jahrgang, November, 69–80.

Kolbe, Wilhelm; Haug, Gustav (1979): Rückblick auf 65 Jahre Bayer-Pflanzenschutz-Abteilung (1914–1979) und 50 Jahrgänge Pflanzenschutz-Nachrichten (1922–1979). Verlag: Leverkusen, (Bayer AG).

Kuckhan, T.S. (1770): Four Lletters from T.S. Kuckhan, to the President and Members of the Royal Society, on the Preservation of Dead Birds. *Philosophical Transactions*, (60), 303–304.

Lange (1923): Lange, Otto. Mottenmittel. In: Chemisch-Technische Vorschriften, III. Band: Harze, Öle, Fette. Berlin Heidelberg GmbH: Springer-Verlag, 754.

Lehmann, Dieter; Lehmann, Andrea (1985): Zwei wundärztliche Rezeptbücher des 15. [fünfzehnten] Jahrhunderts vom ObLerrhein. Zugl.: Würzburg, Univ., Diss. 1983. Pattensen/Han.: Wellm (Würzburger medizinhistorische Forschungen, 34).

Lehmann, Jirina (2005): Geschichte der Konservierung und Restaurierung in Russland und in der Sowjetunion. Im Buch von Professor M.W. Farmakowskij. *VDR Beiträge zur Erhaltung von Kunst- und Kulturgut*, (2), 47–62.

Liebig, Justus von (1822): Darstellung der unter dem Namen Wienergrün im Handel vorkommenden Malerfarbe. *Repertorium für die Pharmacie*, 13, 446–457.

Lowe, Derek B. (2017): Das Chemiebuch. Vom Schießpulver bis zum Graphen, 250 Meilensteine in der Geschichte der Chemie. Kerkdriel: Librero IBP.

Lueger, Otto (Hrsg.) (1904): Lexikon der gesamten Technik und ihrer Hilfswissenschaften. Im Verein mit Fachgenossen herausgegeben von Otto Lueger. Zweite, vollständig neu bearbeitete Auflage. 8 Bände. Stuttgart, Leipzig: Deutsche Verlagsanstalt.

Lüllmann, Heinz; Mohr, Klaus; Heinz, Lutz (2010): Pharmakologie und Toxikologie: Arzneimittelwirkungen verstehen – Medikamente gezielt einsetzen: Ein Lehrbuch für

Studierende der Medizin, der Pharmazie und der Biowissenschaften, eine Informationsquelle für Ärzte, Apotheker und Gesundheitspolitiker. 17. vollst. überarb. Stuttgart: Auflage, Verlag Thieme.

Mason, Otis Tufton (1889): Report Upon the Work in the Department of Ethnology in the U.S. National Museum. For the Year Ending June 30, 1886. Part II. [S.l.]: Government Printing Office.

Meyers Konversationslexikon (1896): 5. Auflage, Leipzig und Wien.

Müller-Jahncke, Wolf-Dieter; Friedrich, Christoph; Meyer, Ulrich (2005): Arzneimittelgeschichte. 2., überarb. und erw. Aufl. Stuttgart: Wiss. Verlag-Ges.

Murray, Andrew (1877): The Museum Mite. *The American Naturalist*, Band 8, (11).

Muter, John (1890): The Analysis of Carbolic and Sulphurous Disinfecting Powders. *The Analyst*, 63.

Neumayer, Georg von; Ascherson, Paul (1875): Anleitung zu wissenschaftlichen Beobachtungen auf Reisen. Mit besonderer Rücksicht auf die Bedürfnisse der kaiserlichen Marine. Berlin: Verlag von Robert Oppenheim. Available online at http://data.onb.ac.at/ABO/%2BZ102227706.

Neumüller, Otto-Albrecht (Hg.). (1973): Römpps Chemie-Lexikon. 7. Aufl. Stuttgart: Franckh'sche Verlagshandlung, 1185–1186.

Odegaard, Nancy; Sadongei, Alyce (2005): Old Poisons, New Problems. A Museum Resource for Managing Contaminated Cultural Materials. Walnut Creek: AltaMira Press.

Offenthaler, Eva (2013): "Zacherlin wirkt staunenswert!" – Johann Zacherl und sein Pulver. Hrsg. v. Verlag der Österreichischen Akademie der Wissenschaften. Österreichisches Biographisches Lexikon, Biografie des Monats, 6. Available online at www.oeaw.ac.at/inz/forschungen/oesterreichisches-biographisches-lexikon/, last visited March 14, 2017.

Pfister, Aude-Laurence (2008): L'Influence des Biocides sur la Conservation des Naturalis. Diplomarbeit. Haute École Arts, Appliqués-La Chaux-de-Fonds, Filière Conservation-Restauration.

Pierer's Universal-Lexikon der Vergangenheit und Gegenwart (1857–1865): 4. Auflage, 1. Band, Altenburg: Verlagsbuchhandlung Heinrich August Pierer.

Plenderleith, Harold James (1934): The Preservation of Antiquities. Oxford: University Press (I).

Purewal, Victoria Jane (2001): Analysis of the Pesticide Residues Present on Herbarium Sheets within the National Museums and Galleries of Wales. Proceedings of 2001: A Pest Odyssey. In: Kingsley, Helen; Pinniger, David; Xavier-Rowe, Amber; Winsor, Peter (eds.). Integrated Pest Management for Collections. London: James & James, 144.

Rathgen, Friedrich (1908): Mitteilungen aus dem Laboratorium der Königlichen Museen zu Berlin. IV. Die Verwendung von Tetrachlorkohlenstoff in der Konservierungspraxis. In: *Museumskunde, Zeitschrift für Verwaltung und Technik öffentlicher und privater Sammlungen*, Karl Koetschau (Hrsg.), Berlin: Georg Reimer, Band IV, 90–91.

Rochussen, Frank (1920): Ätherische Öle und Riechstoffe. 2. umgearb. Auflage. Berlin, Leipzig: Vereinigung wissenschaftlicher Verleger (Sammlung Göschen).

Römpp Lexikon Chemie (1992 und 1996): Jürgen Falbe und Manfred Regitz (Hrsg.). 9. und 10. Aufl. Stuttgart, New York: Thieme.

Rookmaaker, L.C.; Morris, P.A.; Glenn, I.E.; Mundy, P.J. (2006): The Ornithological Cabinet of Jean-Baptiste Bécoeur and the Secret of the Arsenical Soap. *Archives of Natural History*, 33(1), 146–158.

Schiessl, Ulrich (1984): Historischer Überblick über die Werkstoffe der schädlingsbekämpfenden und festigkeitserhöhenden Holzkonservierung. *Maltechnik/Restauro*, (2), 9–40.

Schnedlitz, Markus (2008): Chemische Kampfstoffe. Geschichte, Eigenschaften, Wirkung. [Studienarbeit. Zugl.: Wiener Neustadt, FH, Seminararbeit, 2008. 1. Aufl. München: Grin-Verlag.

Schroeck, Lucas; Franz, Anselm; Hafner, Melchior. Historia Moschi (1682): Ad normam Academiae Naturae Curiosorum. Augustae Vindelicorum: Theophilius Göbelius.

Schwartz, Martin (1925): Die reichsgesetzlichen Pflanzenschutzbestimmungen für die Einfuhr lebender Pflanzen und frischer Pflanzenteile nach Deutschland. *Der Deutsche Erwerbsgartenbau*, (45), Berlin: Verlag der Gärtnerischen Verlagsgesellschaft m.b.h. 653–654.

Scott, Alexander (1922): The Restoration and Preservation of Objects at the British Museum. *Journal of the Royal Society of Arts*, 24 (LXX; No. 3618), 327–339.

Simon, Christian (1999): Kulturgeschichte einer chemischen Verbindung. Basel: Christoph-Merian-Verlag.

Sirois, Jane; Poulin, Jennifer; Stone, Tom (2010): Detecting Pesticide Residues on Museum Objects in Canadian Collections. A Summary of Surveys Spanning a Twenty-year Period. *Collection Forum*, 24(1–2), 28–45.

Smith, John (1884): Some Observations on Museum Pests. *Proceedings of the Entomological Society*, 1884–1889, 1.

Sotriffer, Kristian (1996): Die Blüte der Chrysantheme. Die Zacherl – Stationen einer anderen Wiener Bürgerfamilie. Wien u.a.: Böhlau Verlag.

Stellwaag, Fritz (1927): Der Gebrauch der Arsenmittel in deutschem Pflanzenschutz. Ein Rückblick und ein Ausblick unter Verwertung der ausländischen Erfahrungen. *Zeitschrift für angewandte Entomologie, Bd.*, 12(1), 35–36.

Taeger, Harald (1941): Die Klinik der entschädigungspflichtigen Berufskrankheiten. Heidelberg: Springer Berlin.

Toothaker, Charles Robinson (1908): Fumigation. *Proceedings of the American Association of Museums*. Baltimore: The Waverly Press, vol. II, 1908, 119–123.

Unger, Achim (2012): Eulanisierte" Textilien – eine Gefahr für Mensch und Material? *Beiträge zur Erhaltung von Kunst- und Kulturgut*, (2), 25–39.

Unger, Achim; Schniewind, Arno P.; Unger, Wibke (2001): Conservation of Wood Artifacts. A Handbook. Berlin: Springer (Natural Science in Archaeology). Available online at www.loc.gov/catdir/enhancements/fy0815/2001020310-d.html.

Wiesner, Julius (1927): Die Rohstoffe des Pflanzenreiches. 4. Auflage. Leipzig: Wilhelm Engelmann (I, II).

Wilhelmi, Julius; Kunike, Hugo (1927): Versuche und Untersuchungen über die Wirksamkeit des Petroleum-Raffinates „Flit" bei der Fliegen- und Stechmückenbekämpfung. *Zeitschrift für Desinfektion und Gesundheitswesen*, 19(3), 98–99.

Wink, Michael; van Wyk, Ben-Erik; Wink, Coralie (2008): Handbuch der giftigen und psychoaktiven Pflanzen; mit 13 Tabellen. Stuttgart: Wiss. Verlag-Ges.

Wray, L. (first name unknown, author's note) (1908): The Preservation of Mammal Skins. *Museums Journal*, (8), 207–208.

Zacher, Friedrich (1924): Methoden der Vorratsschädlingsbekämpfung. Vierte Mitgliederversammlung zu Frankfurt a.M. vom 10. bis 13. Juli 1924. *Verhandlungen der Deutschen Gesellschaft für angewandte Entomologie*, 45–50.

Zinke, Georg Gottfried (1802): Kunst allerhand natürliche Körper zu sammeln, auf eine leichte Art für das Kabinett aufzubereiten und vor der Zerstörung feindlicher Insecten zu sichern. Jena: Göpferdt, J. C. G.

Part III

Collecting and preserving cultural assets in Berlin and beyond from the end of the 19th century to the beginning of the 20th century

8 Spatial conditions and personnel requirements for the preservation of collections at the Königliches Museum für Völkerkunde in Berlin from the end of the 19th century to the beginning of the 20th century

The storage and spatial conditions of the Königliches Museum für Völkerkunde/Staatliches Museum für Völkerkunde at the Koeniggraetzer Straße and in Berlin-Dahlem

The collections of the Ethnologisches Museum (EM) (Ethnological Museum) have their origins in the Kuriositätenkabinette (Electoral Cabinets of Curiosities) of the 17th century, where they ended up as individual exotic objects in the Kunstkammer (Royal Chambers of Art) of the Great Elector Friedrich Wilhelm in his former Stadtschloss (City Palace) (Cf. Westphal-Hellbusch 1973, 6–8; Haas 2002, 16; Bolz 2001, 13–14, 2007, 174–183). Subsequently, ethnographic collections were exhibited publicly for the first time in 1859 at the newly opened Neues Museum in Berlin (Ledebur 1869, 193–204). As early as December 27, 1873, permission was granted to build a separate building for ethnological collections, as the spatial capacities in the Neues Museum were completely exhausted. Other storages, cellars, and rooms of the Königliche Museen (KM) (Royal Museums) had already been provisionally filled with ethnological objects. As early as 1885, before the completion of the building at the Königgrätzer Straße, there was an urge to release rooms for ethnological objects (SMB-PK EM 1885a). Adolf Bastian immediately had ethnological collection items removed from rented premises at the Dorotheenstraße 5 (SMB-PK EM 1885b). The ceremonial opening of the KMfV in Berlin took place one year later on December 18, 1886 (Cf. Haas 2002, 17). However, setting up ethnological exhibitions took several years (Anonymous 1888, 1). The newly constructed building in the former Königgrätzer Straße in Berlin-Mitte originally housed prehistoric collections and the Heinrich Schliemann collection on the ground floor of the building. Ethnological collections from Africa, Oceania, and America were on the first floor. The second floor housed Asian collections (Anonymous 1889, 15). At the opening of the museum, the contemporary witness Friedrich von Hellwald expressed himself effusively in an article:

"Seitdem Berlin zur Metropole des Deutschen Reiches emporgestiegen, ist es nicht bloß Weltstadt, sondern auch ein Centrum der Wissenschaftspflege geworden".[1]

(Hellwald 1887, 1)

DOI: 10.4324/9781003407607-12

Figure 8.1 Königliches Museum für Völkerkunde, Königgrätzer Strasse 120, Berlin-Mitte, 1886.

Thus, Berlin has done

". . . einen Schritt, wie ihn unseres Wissens noch keine europäische Haupt-stadt zu verzeichnen hat".[2]

(Ibid., 101)

Hellwald's euphoric exclamations stemmed primarily from his personal convictions. The construction of its own building with collection objects from non-European cultures and the associated opening of the then KMfV in Berlin is empathetically regarded by him as a major milestone in the cultural landscape.

Barely 30 years later, the First World War had considerable consequences for KM in Berlin. On August 2, 1914, all museums were closed immediately. At that time, one feared less the imminent danger of an air raid but was forced to take this step *ad hoc* because numerous employees were drafted for military service. Only one and a half weeks later were the museums reopened on August 12, 1914, with significant restrictions. In the KMfV, the remaining guards were distributed over the halls, and entire exhibition areas were closed (Cf. Winter und Grabowski 2014, 9–10). With the end of the monarchy in Germany in 1918, the state's self-image also changed, which resulted in name changes for the KMfV in Berlin and numerous other institutions. The KMfV was transformed into the Staatliches Museum für Völkerkunde (MfV) (National Museum for Ethnology) in Berlin (Ibid. 49–50), and in 2000, it was renamed to EM in Berlin (Cf. Bolz 2001, 11–12). The approaching loss of German colonies (see Part I, chapter 1) had a significant impact on collecting activities from 1876 onward, when Bastian took office. Under his leadership, the collections expanded in an unusual way within a short period of time (Cf. Bolz 2007, 185–188). The preservation of an extensive collection, such as that of the EM in Berlin, brought about numerous problems. The control of material pests and preventive protection against them became of great importance. However, neither specialised facilities for researching agents and methods of material and object protection nor their own specialists were available in this field. As a result, museum staff encountered significant challenges in taking care of the collected objects while handling them in inadequate storage facilities. Thus, in 1904, Felix von Luschan[3] was no longer able to accommodate objects from Africa offered by the Hamburg medical officer Fülleborn[4] to the KMfV. Even the cellars there were so overcrowded that the government offered to build a shed in Berlin-Dahlem to house additional ethnographic objects by the end of 1904 (SMB-PK EM 1904). In the autumn of 1905, the shed was completed, so that in 1906 collection items from the Königgrätzer Straße could be transferred there. The shed was constructed at a low cost and resulted in considerable damage to the objects from the beginning, prompting von Luschan to return the objects to the KMfV. It was only after the renovation of the shed that it was temporarily used as a storage for collection items (Westphal-Hellbusch 1973, 29). Nevertheless, the cramped storage conditions were not resolved. According to a report by the media in 1909, the building was overcrowded and resembled a junk store more than a scientific collection (Viktor 1909).

In 1914, a new building was constructed in Berlin-Dahlem for the Asiatisches Museum (Asian Museum), the Asiatische Abteilung (Asian department) of the KMfV, and the Islamische and Ostasiatische Abteilung (Islamic and East-Asian department). After von Bode postponed the construction for a short time,[5] work on the building came to a complete stop in November 1916 due to a lack of manpower. The end of the First World War and the new political conditions finally led to the decision that the Dahlem building was only provisionally completed as a storage facility, thus marking the first separation of permanent collections and the objects to be stored (Cf. Westphal-Hellbusch 1973, 20–34). The storage building at the new location had no heating facilities when the move there began and was designed and furnished as a pure study collection (Ibid. 17). The relocation of collection objects, which were no longer intended for permanent collections in the Königgrätzer Straße, began in 1923 and extended over a longer period of time (Ibid. 32). In May 1925, a clerk from the Preußisches Ministerium für Wissenschaft, Kunst und Erziehung (Prussian Ministry of Science, Art, and National Education) visited the newly built storage building in Berlin-Dahlem. He then submitted a report to his superior about boxes and packaging materials that he had found in the immediate vicinity of the storage rooms. The Generalverwaltung of the Staatliche Museen (General administration of the National Museums) was then instructed to remove the packaging materials that had been placed in the basement of the building to prevent the spread of vermin through contaminated packaging materials to collection items (SMB-PK EM 1925a). August Eichhorn[6] urgently appealed to the Generalverwaltung to set up an iron cabinet in Berlin-Dahlem, which had been disassembled on site and was intended for moth-infested textiles. At the same time, Eichhorn immediately rejected any further responsibility for the collection items entrusted to him if not remedied immediately. His appeal had an effect and led to the cabinet being set up by the locksmith Kurzhals within ten days (SMB-PK EM 1925b). Eichhorn also pointed out that efficient housing in the new storage was not possible because the door passages were too narrow for handling long boxes, which would require more workers and therefore additional wages (Ibid. E. Nr. 550/25).

Personnel requirements for the preservation of the collections of the Königliches Museum für Völkerkunde/Staatliches Museum für Völkerkunde in Berlin from the end of the 19th century to the beginning of the 20th century

The former employees who were responsible for receiving and caring for these large collection of objects can now be regarded as pioneers in the field of ethnographic conservation. They had a diverse range of vocational training to perform their tasks. From April 1, 1884, the KM in Berlin employed the architect and chemist Eduard Krause[7] as the first conservator at the KMfV (Born et al. 2004/2005, 488; Peltz 2017, 55). In 1873, Krause had already met Rudolf Virchow and Albert Voß, who inspired him to apply for a job in 1879 at the Ethnological and Nordic Department of the Neues Museum (New Museum) of the KM in Berlin (Seler 1917, 212–213). The contributions of Krause to the conservation and restoration efforts of EM were investigated, focusing on his role in the establishment of the institution's current preservation practices. His responsibilities, now assumed by curators and collection managers, were dedicated to the field until the end of his career. In March 1885, Bastian asked him not only to examine the collection of the lawyer Gühler,[8] but also to select relevant pieces for the museum (SMB-PK EM 1885d). As a conservator, he was responsible for the entire collection of the museum and dealt equally with the conservation and restoration of objects made of inorganic and organic materials. In 1887, he was sent on a business trip to Mainz and Worms to study the methods of conservation and restoration of other museums (SMB-PK EM 1887). For this purpose, funds were made available from the relocation of the newly built KMfV (SMB-PK EM 1885c).

Krause's publications document his extensive preoccupation with archaeological and ethnological objects, ranging from the preservation of iron-based artifacts to comparative studies of flints and drums from various regions around the world. In addition, he specialised in the preservation of prehistoric artifacts, including the treatment of red pigmentation on human skeletal remains, as well as the creation of prehistoric clay vessels. His conservation measures were lauded in several publications, including his work on the skull of *Homo moustériensis Hauseri*. His monograph on fishing equipment, which he wrote in 1896 for the German fisheries exhibition during the Berlin trade exhibition, remains a standard work today. The

Figure 8.2 Wilhelm Eduard Julius Krause. First conservator at the Königliches Museum für Völkerkunde in Berlin from 1884 to 1917.

combination of his conservational and curatorial interests is evident in two publications on "Wunderliche Heilige",[9] in which he described the extraordinary practices of Indian fakirs (Krause 1882, 533–537, 1883, 361, 1897, 37–40, 1899, 576–579, 1900, 193–196, 1901, 361–367, 1903, 317–323, 1904; Schuchhardt 1912a, 443–446, 1912b, Spalte 4–10; Hoffmann 1997, 7–16; Cziesla 2000, 173–186; Wegner und Hoffmann 2002, 218–221).

Employees at the Königliches Museum für Völkerkunde/Staatliches Museum für Völkerkunde in Berlin with conservation and restoration tasks

After Krause's death, individuals from diverse professional backgrounds joined the KMfV/MfV in Berlin to work on conservation and restoration efforts. These included, among others, house inspectors, technical assistants, and museum guards. The fact that these professional groups were responsible for the conservation of collection objects is inconceivable from today's perspective. However, it shows how poorly defined the profession of conservators was during that period. With all the advantages and disadvantages, pioneering work was also carried out in this area. The aforementioned occupational groups were integrated into the promotion system of "Wartestandsbeamte"[10] of the Prussian state, where museum guards could pursue better-paying positions with conservation activities by preparing for a waiting period of 20 to 25 years during their years of service. Therefore, it was not the professional qualification that was decisive for a higher grouping but the personal patience and perseverance of the individual. Occasionally, there were disagreements between the Generalverwaltung of the KM (General administration of the Royal Museums) and the Preußisches Ministerium für Wissenschaft, Kunst und Erziehung. The ministry was quite interested in giving its low-ranking officials from the institution the opportunity to work in museums, which the Generalverwaltung viewed as a disadvantage for those who had started their careers from the beginning at the KM (SMB-ZA 1934–1936).

For example, the collection manager Strebe,[11] who had previously served as Krause's unskilled assistant, was tasked with some activities (see chapter 8). More examples are provided as follows to illustrate which individuals were employed for the conservation and restoration of collection objects at the Staatliche Museen (SMB) (National Museums) and especially at the MfV in Berlin.

Regarding Mr. Schellin,[12] there is no information about his vocasional training, but it is known that he worked as a cashier at SMB. He was proposed by the General Direktion (General Management) as a technical assistant, and from 1919 he worked as a civil servant at the Museum für Deutsche Volkskunde. There, he was among others responsible for the

"Beseitigung von Holz- und Textilschädlingen, Staub- und Metallpilzen, die eine genaue Kenntnis der einschlägigen Chemikalien erfordert".[13/14]

(Ibid., 94–95)

Rudolf Kuhn, who was born on January 8, 1892, served as a museum guard from June 1, 1925. He later became a technical assistant in the Chemisches Laboratorium (Chemical Laboratory) of SMB, starting from June 1, 1929. His job description provides an insight into occupational safety at the end of the 1920s:

"Restaurierung und Konservierung von Museumsgegenständen, namentlich Ausgrabungsfunden. Die bei Ausübung dieser Tätigkeit stets vorhandenen Gefahrenmomente erfordern erhöhte Aufmerksamkeit. Es wird von dem Beamten verlangt, daß er genaue Kenntnis der Chemikalien, mit denen er arbeitet, besitzt und sie in ihrer Anwendung und Wirkung vollkommen beherrscht. Umgang mit konzentrierten Säuren und Alkalien, mit feuer- und gesundheitsgefährlichen Lösungsmitteln oder Giften, weiter die Bedienung eines Schamottebrennofens bei einer Temperatur von 600 °C verlangen ein selbständiges, verantwortungsbewußtes Handeln und erschweren die Arbeit, die in Schutzkleidung und unter besonderen Vorsichtsmaßnahmen ausgeübt werden muß. Die Durchführung der Konservierungen setzen Kenntnisse in chemischen Arbeitsverfahren voraus. Zu nennen sind besonders die umfassenden Konservierungs- und Restaurierungsarbeiten an der assyrischen und babylonischen Tontafelsammlung. Allein im Jahr 1932 sind von K u h n mehrere Tausend Tontafeln und Bruchstücke behandelt worden. Auch Reduzierungen und Freilegung von Ornamenten an antiken Funden wurden von ihm ausgeführt. Die Verschiedenheit des Materials und der Gegenstände machen ein öfteres Einarbeiten in neue Gebiete erforderlich. Beispiele: Metalle, besonders Silber, Bronze, Kupfer, Eisen, ferner Steine, Holz, Elfenbein, Knochen, Mumien, Textilien, usw. Daß die gewöhnlichen technischen Arbeiten, wie Photographieren, Beschriften und Zeichnen von dem Beamten ausgeführt werden, ist selbstverständlich. Seine besondere Eignung zu technischen Arbeiten geht aus seiner früheren Tätigkeit im Maschinenbetrieb hervor. Die im Chemischen Laboratorium auszuführenden Arbeiten dienen dem Schutze des Museumsgutes und somit der Erhaltung großer Kulturwerte".[15]

(Ibid., 70–71)

Hermann Siebert, who was born on May 29, 1884, began his career at the museum as a guard on April 1, 1919. From 1926 until 1932, he worked mainly in the departments of the African, Oceanic, and American collections in the MfV with the following tasks:

"Hilfe bei den Katalogzetteln der Süd- und Nordamerikanischen Sammlung. Führung des Plattenarchivs. Betreuung des Inhalts der Sammlungsschränke und Führung der Kartothek. Neuaufstellung von Schränken und tapeziermäßigen Verrichtungen dabei (als gelernter Tapezierer und Dekorateur). Neuaufstellung der peruanischen Stoffsammlung".[16]

(Ibid., 71–72)

In 1936, August Lösekrug, a technical assistant, was appointed as the assistant conservator by the Ministerial Chancellery Secretary of the Preußisches Ministerium für Wissenschaft, Kunst und Erziehung. However, the introduction of the professional title "assistant conservator" could not be clarified in time, but Lösekrug had already been working in the departments of the African, Oceanic, and American collections since May 1922:

"*Restaurieren* von Büchern, Druckschriften, Karten, Zeichnungen, Plänen, Photographien und Bildern etc."[17]

(Ibid., 85)

The selection of conservation and technical assistants presented here was based not only on their already acquired skills in caring for collections but also on other factors. From 1933, more criteria were added for promotions, including membership in the NSDAP or qualification as a frontline fighter under the April 7, 1933 professional civil service law, as well as individual merits from the First World War. These criteria were applied to every appointed official:

" . . . die Gewähr dafür, daß er jederzeit rückhaltlos für den nationalen Staat eintreten wird".[18]

(Ibid., 70–96)

The collection structure in the EM, the on-site presentation and storage in Berlin, and the conservational care of the holdings by former employees are characterised less by a systematic approach and more often by regrettable circumstance of purely pragmatic decisions. After Krause's death in 1917, the position of conservator was not filled by an academically educated person for a long period of time, according to sources. With this background, the next focus will be on analyzing our methods for pest control and the role of the Chemisches Laboratorium in this regard.

Notes

1 Since Berlin became the metropolis of the German Reich, it has become not only a cosmopolitan city but also a center of science cultivation (translation by the author).
2 A step that, to our knowledge, no European capital has yet taken (translation by the author).
3 Felix von Luschan was director of the African–Oceanic Department at the KMfV in Berlin from 1904 to 1911.
4 First name unknown (author's note).
5 See about Wilhelm von Bode (Part I, chapter 1).
6 August Eichhorn was head of the Oceanic Department at the KMfV from 1916 and at the MfV from 1918 to 1929. Vgl. Grabowski et al. 2010, 155.
7 Eduard Wilhelm Krause was born in 1847 and died on October 30, 1917. Friendly oral communication from the Zentralarchiv oft the Staatliche Museen zu Berlin, February 2017.
8 First name unknown (author's note).
9 Whimsical saints (translation by the author).
10 Officers on hold (translation by the author).
11 First name unknown (author's note).
12 First name unknown (author's note).
13 Elimination of wood and textile pests, dust, and metal fungi, which requires precise knowledge of the relevant chemicals (translation by the author).
14 The term "metal mushrooms" is not a technical term in conservation. Presumably, this refers to chlorine corrosion (cf. Rathgen 1924, 1–7.).
15 Restoration and conservation of museum objects, namely, excavation finds: The moments of danger that are always present when carrying out this activity require increased attention. The official is required to have precise knowledge of the chemicals with which he works and to have a complete command of their application and effect. The operation of a fireclay kiln at 600 °C, which involves handling concentrated acids, alkalis, solvents, or hazardous substances that pose risks to both fire and health, requires independent, responsible actions and must be carried out under special precautions and protective clothing. The implementation of preservatives necessitates an understanding of chemical processes. Particularly noteworthy are the extensive conservation and restoration work on the Assyrian and Babylonian clay tablet collections. In 1932 alone, several thousand clay tablets and fragments were treated by Ku h n. He also carried out reductions and exposure of ornaments on ancient finds. The diversity of material and subject matter necessitates more frequent exposure to new areas. Examples are metals

(especially silver, bronze, copper, and iron), stones, wood, ivory, bones, mummies, textiles, etc. It goes without saying that the usual technical work, such as photographing, labeling, and drawing, is carried out by the official. His aptitude for technical work is evident from his previous work in machine operation. The work performed in the Chemical Laboratory aims to safeguard the museum's assets and thereby preserve significant cultural values (translation by the author).

16 Help with the catalogue notes of the South and North American collections. Management of the photographic plate archive. Supervision of the contents of the collection cabinets and management of the map library. Reinstallation of cabinets and wallpapering operations (as a trained upholsterer and decorator). Realignment of the Peruvian fabric collection (translation by the author).

17 *Conservation* of books printed materials, maps, drawings, plans, photographs, and pictures etc. (translation by the author).

18 The guarantee that he will at all times stand up wholeheartedly for the national state (translation by the author).

References

Archive material

SMB-PK EM (1885a): I/MV 0053, I c, Vol. 1, E. Nr. 105/85, Acta betreffend den Umzug und die Aufstellung der Sammlungen des Museums. Loseblattsammlung. Anonymus. Bericht vom 21.5.1885 über die Unterbringung von Sammlungsgut des Königlichen Museums für Völkerkunde vor Bezug in das neue Gebäude in der Königgrätzer Straße, 2 Seiten, je 1 Spalte, ohne Paginierung.

SMB-PK EM (1885b): I/MV 0053, I c, Vol. 1, Ebd. E. Nr. 112/85, Bastian, Adolf. Brief vom 24.05.1885, 1 Seite, ohne Paginierung.

SMB-PK EM (1885c): I/MV 0053, I c, Vol. 1. E. Nr. 180b/1885. Acta Umzug und Aufstellung der Sammlungen. Loseblattsammlung. Bastian, Adolf. Antrag auf Renumeration für den Konservator Eduard Krause vom 14.08.1885, 1 Seite, ohne Paginierung.; Genehmigung durch den Generaldirektor am 15.08.1885, 1 Seite, ohne Paginierung.

SMB-PK EM (1885d): I/MV 0611, Pars I B, Bd. 4. zu E. Nr. 66/85, Begutachtung der Sammlung Gühler. Krause, Eduard. Bericht vom 10.03.1885, Blatt 1–4, 6 Seiten.

SMB-PK EM (1887): I/MV 0075, Pars II c, Vol. 1. E. Nr. 584/1887. Dienstreise zum Studium der Konservierung nach Mainz und Worms. Krause, Eduard. Bericht vom 06.09.1887, Blatt 1–13, 15 Seiten.

SMB-PK EM (1904): I/MV 730, Pars I B 30, zu E. Nr. 203/04. Acta betreffend die Erwerbung ethnologischer Gegenstände aus Afrika. Loseblattsammlung. Luschan von, Felix. Brief an den Stabsarzt Herrn Fülleborn vom 13.02.1904, Blatt 79, 1 Seite.

SMB-PK EM (1925a): I/MV 0015, Band 1, Pars I 3, E. Nr. 550/25. Akte betreffend Umzug und Aufstellung der Sammlungen im Staatlichen Museum für Völkerkunde zu Berlin. Loseblattsammlung. Besuch im Magazingebäude in Berlin-Dahlem. Nentwig, (Vorname unbekannt, Anmerk. d. Verf.). Bericht vom Preußischen Ministerium für Wissenschaft, Kunst und Volksbildung vom 20.05.1925, 1 Seite, ohne Paginierung.

SMB-PK EM (1925b): I/MV 0015, Bd. 1, Pars I 3. E. Nr. 600/25. Aufstellung eines Schrankes in der ozeanischen Abteilung. Eichhorn, August. Bericht vom 02.06.1925, 1 Seite, ohne Paginierung.

SMB-ZA (1934–1936): I/GV 0043, Kaderleitung der Generalverwaltung bei den Staatlichen Museen zu Berlin vom 04.08.1934–30.07.1936. Das Problem der Wartestandsbeamten im Beförderungssystem der Staatlichen Museen. Personallisten und Korrespondenzen an das Preußische Ministerium für Wissenschaft, Kunst und Volksbildung, Blatt 67–146, 44 Seiten.

Other sources

Anonymous (1888): Übersicht über die Amerikanischen Sammlungen des Königlichen Museums für Völkerkunde. Zusammengestellt für die 7. Tagung des internationalen Amerikanisten-Kongresses. Berlin: H.S. Hermann.

Anonymous (1889): Übersichtlicher Abriß der Sammlungen im Königlichen Museum für Völkerkunde. Den Mitgliedern des Deutschen Geographentages in ihrer 8. Sitzung überreicht. Berlin.

Bolz, Peter (2001): Ethnologisches Museum: Neuer Name mit traditionellen Wurzeln. Die Umbenennung des Berliner Museums für Völkerkunde. *Baessler-Archiv. Beiträge zur Völkerkunde* ausgegeben am 25. February 2003. Sonderdruck aus Bd. 49, 11–16.

Bolz, Peter (2007): From Ethnographic Curiosities to the Royal Museum of Ethnology. Early Ethnological Collections in Berlin. In: Fischer, Manuela; Bolz, Peter; Kamel, Susan (eds.). Adolf Bastian and His Universal Archive of Humanity. The Origins of German Anthropology, Hildesheim: Georg Olms, 173–190.

Born, Hermann; Hausdörfer, Ute; Thieme, Franziska (2004/2005): Die Restaurierungswerkstätten. In: Das Berliner Museum für Vor- und Frühgeschichte; Festschrift zum 175-jährigen Bestehen, (36/37).

Cziesla, Erwin (2000): Spätpaläolithische Widerhakenspitzen aus Brandenburg. Eine Forschungsgeschichte. *Archäologisches Korrespondenzblatt*, 173–186.

Grabowski et al. (2010): Grabowski, Jörn; Winter, Petra; Ebelt, Beate; Pilgermann, Carolin (Hg.). Kunst recherchieren. 50 Jahre Zentralarchiv der Staatlichen Museen zu Berlin. Staatliche Museen zu Berlin. Berlin: Deutscher Kunstverlag.

Haas, Richard (2002): Brasilien an der Spree. Zweihundert Jahre ethnographische Sammlungen in Berlin. In: Deutsche am Amazonas Forscher oder Abenteurer? Expeditionen in Brasilien 1800–1914, Anita Hermannstädter (Hrsg.), Begleitbuch zur Ausstellung im Ethnologischen Museum, Berlin-Dahlem in Zusammenarbeit mit dem Brasilianischen Kulturinstitut in Deutschland. IX; Neue Folge 71. Berlin: Ethnologisches Museum, 2., unveränd. Aufl. Münster: LIT. Veröffentlichungen des Ethnologischen Museums Berlin Fachreferat Amerikanische Ethnologie, 17.

Hellwald, Friedrich von (1887): Das Berliner Museum für Völkerkunde. *Vom Fels zum Meer*, (2), 101.

Hoffmann, Almut (1997): Zur Geschichte des Fundes von Le Moustier. In: *Acta Praehistorica et Archaeologica*, Bd. 29, 7–16.

Krause, Eduard (1882): Hr. Ed. Krause berichtet über ein neues Verfahren zur Conservierung der Eisen-Alterthümer. Sitzung am 11. November 1882. *Verhandlungen der Berliner anthropologischen Gesellschaft*, 533–537.

Krause, Eduard (1883): Mittheilungen über trapezförmige Feuersteinscherben. *Zeitschrift für Ethnologie und der Verhandlungen der Berliner Gesellschaft für Anthropologie, Ethnologie und Urgeschichte*, 15, 361.

Krause, Eduard (1897): Wunderliche Heilige. *Illustrierte Familienzeitschrift "Für alle Welt"*, 1897, Heft 1, 8.; Heft 2, 37–40.

Krause, Eduard (1899): Die Verwendung von Celluloid-Lack zur Conservirung von Althertümern, sowie von Holz, Stoffresten und Papier, namentlich alten Zeichnungen, Drucken, Acten in Archiven usw. Verhandlungen der Berliner Gesellschaft für Anthropologie, Ethnologie und Urgeschichte. *Zeitschrift für Ethnologie*, Jahrgang, 31, 576–579.

Krause, Eduard (1901): Zur Frage der Rotfärbung vorgeschichtlicher Skelettknochen. *Globus*, Sonder-Abdruck aus Bd. 83(23), 361–367.

Krause, Eduard (1903): Über die Herstellung vorgeschichtlicher Tongefässe. *Zeitschrift für Ethnologie*, 35, Heft II und III, 317–323.

Krause, Eduard (1904): Vorgeschichtliche Fischereigeräte und neuere Vergleichsstücke. Eine vergleichende Studie als Beitrag zur Geschichte des Fischereiwesens. Berlin: Borntraeger.

Krause, Eudard (1900): Die ältesten Pauken. *Globus*, Bd. 78, 193–196.

Ledebur, Leopold Freiherr von (1869): Aus der Ethnologischen Sammlung des Königlichen Museums zu Berlin. *Zeitschrift für Ethnologie und ihre Hülfswissenschaften als Lehre vom Menschen in seinen Beziehungen zur Natur und zur Geschichte*, Erster Jahrgang (III), 193–204.

Peltz, Uwe (2017): Das Chemische Laboratorium bis zur Gründung als "Zwillingsinstitute" im geteilten Berlin. *Berliner Beiträge zur Archäometrie, Kunsttechnologie und Konservierungswissenschaft*, (25), 55–94.

Rathgen (1924): Rathgen, Friedrich. Die Konservierung von Altertumsfunden. Mit Berücksichtigung ethnographischer und kunstgewerblicher Sammlungsgegenstände. 2. Aufl. Berlin, Leipzig: Walter de Gruyter & Co. Handbücher der Staatlichen Museen zu Berlin, II. und III. Teil.

Schuchhardt, Carl (1912a): Die neue Zusammensetzung des Schädels vom Homo Mousteriensis Hauseri. *Praehistorische Zeitschrift*, Bd. IV, 443–446.

Schuchhardt, Carl (1912b): Die neue Zusammensetzung des Schädels vom Homo Mousteriensis Hauseri. *Amtliche Berichte aus den Königlichen Kunstsammlungen*, 34(1), Spalte 4–10.

Seler, Eduard (1917): Sitzung vom 17. November 1917. *Zeitschrift für Ethnologie; Organ der Berliner Gesellschaft für Anthropologie, Ethnologie und Urgeschichte*, (49), 212–213.

Viktor, Adolf (1909): Tausend Topfscherben. Berliner Tagblatt vom 21.09.1909. Quoted in Menghin (2005): Das Berliner Museum für Vor- und Frühgeschichte. Festschrift zum 175-jährigen Bestehen. Staatliche Museen zu Berlin-Preußischer Kulturbesitz, Menghin, Wilfried (Hrsg.). Acta Praehistorica et Archaeologica 36/37, 2005, 130.

Wegner, Dietrich; Hoffmann, Almut (2002): Der Schädel vom Combe Capelle im Museum für Vor- und Frühgeschichte wiederaufgefunden. *Archäologisches Nachrichtenblatt*, 7(3), 218–221.

Westphal-Hellbusch, Sigrid (1973): Zur Geschichte des Museums. Hundert Jahre Museum für Völkerkunde. *Baessler-Archiv*, XXI, 1–99.

Winter, Petra; Grabowski, Jörn (Hrsg.) (2014): Zum Kriegsdienst einberufen. Die Königlichen Museen zu Berlin und der Erste Weltkrieg. Staatliche Museen zu Berlin-Preußischer Kulturbesitz, Ethnologisches Museum. Köln: Böhlau Verlag (Schriften zur Geschichte der Berliner Museen, 3).

9 Explorers, collectors, and adventurers at the Königliches Museum für Völkerkunde in Berlin from the end of the 19th century to the beginning of the 20th century

Research and expedition trips

During James Cook's early expeditions between 1768 and 1780, when Georg Forster accompanied him on his second voyage, items collected from North America and Alaska were brought to Berlin for the Kurfürstlichen Kunstkammern (Electoral Chambers of Art) resp. in the Königlichen Kunstsammlungen (Royal art collections) (Cf. Ledebur 1869, 203–204). It is widely acknowledged that Alexander von Humboldt's journey to South America between 1799 and 1804 sparked a great deal of scientific interest in exploring the Americas and other continents (Beck 1985, 300–304; Hermannstädter 2002, 26). For example, the Royal Prussian merchant ship "Prinzess Louise" was also used to transport items from Alaska and North America during several voyages between 1829 and 1837 (Ledebur 1869, 197–198, 202; Meyen 1834, 1–493).

During the period of 1800 to 1831, the Berlin ethnological collections grew as a result of natural history research trips. The naturalist Count Johann Centurius von Hoffmannsegg was a key figure in this expansion. Thanks to his financial independence, he was able to pursue his interests in natural history and science immediately after studying these subjects (Cf. Hermannstädter 2002, 27). On his behalf, his valet Friedrich Wilhelm Sieber, the botanist Friedrich Sellow, and then Prussian diplomat Ignaz Maria von Olfers collected items in Brazil (Ibid. 26–27). In 1815, Friedrich Sellow, the zoologist Georg Wilhelm Freyreiss, and Prince Maximilian zu Wied met at the home of Baron Georg Heinrich von Langsdorff, who was in Russian service, in Rio de Janeiro. In the same year, they set off together on a research trip to the hitherto little-known east coast of Brazil, which resulted in additional specimens that were added to the Berlin museum's collection (Ibid. 31–34).

With the help of physician and psychiatrist Adolf Bastian, a new era of collecting ethnographica began. Together with Rudolf Virchow, he catered to the growing interest among educated citizens in foreign cultures and founded the Berliner Gesellschaft für Anthropologie, Ethnologie und Urgeschichte (Berlin Society for Anthropology, Ethnology, and Prehistory) in 1869 (Ibid. 44–45). Bastian and Virchow are credited with establishing ethnology as an independent subject within the sciences. The other founding members were Leopold Freiherr von Ledebur[1] and

DOI: 10.4324/9781003407607-13

Albert Voß.[2] Together, they took an interdisciplinary, scientific approach, in which anthropology was seen as the study of human beings who were related to nature and history (Junker 2004/2005, 425). From 1850 to 1903, Bastian undertook nine global expeditions, from which he brought back numerous objects. Based on these journeys, he disseminated his teachings on non-European peoples and taught as a private lecturer and, from 1868, as an associate professor of Ethnology in Berlin, thus establishing Ethnology as a humanities discipline in the university sector (Fischer et al. 2007, 297). During the second half of the 19th century, Bastian's generous acquisition budget, as well as private donations, were generously endowed by Crown Prince Friedrich Wilhelm (Cf. Hermannstädter 2002, 47–48).[3] As a traveler, scholar, and museum expert, he had a global network of contacts in his time and was constantly on the lookout for people he could inspire to support his cause. This led to a chance encounter with Karl von den Steinen, which turned his professional career around, which he described as follows:

> "Ich wäre einen völlig anderen Lebensweg gegangen, wenn ich nicht eines Tages – es sind nun 25 Jahre – im Fremdenbuch des Hotels von Honolulu den Namen "Dr. Bastian-Berlin"gelesen hätte. – Er eroberte den Menschen sofort, indem er ihm hohe, seltene Aufgaben stellte, und zwar mit einem Vertrauen, einer Zuversicht, daß man sich förmlich selbst wachsen fühlte. Das Geheimnis seiner erstaunlichen, suggestiven Kraft war kein anderes, als daß er immer nur an die besten Instinkte appellierte und selbst keine anderen besaß".
>
> (von den Steinen 1905, 168)[4]

After completing his medical degree, Bastian switched to ethnology and began collecting ethnographica for the KMfV in Berlin. Between 1882 and 1888, he took part in three extensive and elaborately organised expeditions and was thus significantly involved in the exploration of the Rio Xingú in central Brazil (Ibid. 67–86; Kraus 2004a, 30–32).

Theodor Koch-Grünberg, a teacher from Grünberg in Hesse, was already interested in the scientific literature of South American Indians during his time in school. Presumably, through his geography professor, Wilhelm Sievers, contact was made with Herrmann Meyer from Leipzig,[5] who took him on a research trip to the Rio Xingú in 1899 as a photographer and scientific companion. He followed von den Steinen in the newly developing ethnology for the study of indigenous peoples and undertook a total of three major expeditions to the Amazon. On his return, Koch-Grünberg resigned from his teaching position in 1901 and, with the support of Bastian and von den Steinen, was employed first as a volunteer and later as a research assistant at the KMfV in Berlin until 1909. From his research trip in 1903 to the Amazon region and the Rio Negro, he brought back 1300 objects and about 1000 photographs for the American collections of the KMfV. This trip brought him not only scientific fame but also the reputation of the "Indian friend". From 1911 to 1913, he traveled Orinoco. On his last research trip, Koch-Grünberg died on October 8, 1924, in Vista Alegre on the Rio Branco due to malaria. He bequeathed

Figure 9.1 "Four of my carriers". Koch-Grünberg, Theodor, 1917.

to the Berlin MfV an extensive collection and photographs from the lowlands of Amazonia, as well as numerous publications (Cf. Hermannstädter 2002, 87–106; Cf. Kraus 2004a, 35–36, 2004b). Just as significant to this day is the expedition financed by Wilhelm Kissenberth[6] on behalf of KMfV. From 1907 onward, he was employed as a trainee in the America department of the museum and brought back 1164 objects and about 300 photographs from Rio Araguaya in Brazil between 1908 and 1910 (Cf. Hermannstädter 2002, 106–131; Cf. Kraus 2004b, 40–42).

Researchers and adventurers

Large-scale expeditions and research trips are often closely associated with the names of those who have broken new ground and subsequently gained public recognition and scientific fame. Although traveling to non-European countries at the end of the 19th and beginning of the 20th century was extremely strenuous and extended over very long periods of time, such important names as James Cook and Alexander von Humboldt also attracted less well-known people who went to little or even unknown areas outside Europe as explorers or for the pleasure of adventure. Among them were showmen, taxidermists, sailors, doctors, merchants, and politicians. Through their ethnological, scientific, or purely mercantile interests, they helped build up the collections of the EM Berlin from the early 19th century onward. By the end of the 19th century, the acquisition budget of Adolf Bastian for

KMfV resulted in a vast collection of items. This thesis is substantiated and illustrated for the American collections by example of individual persons.

A big boost began for the collections of the Königlichen Kunstkammern at the Stadtschloss in Berlin in 1829 under the scientific direction of Leopold Freiherr von Ledebur. During this time, the North American collections included objects belonging to the American captain and showman Samuel Hadlock and Ferdinand Deppe, who first worked as gardeners at the royal court and later became taxidermists (Bolz und Sanner 1999, 26). Among the politicians and diplomats was Friedrich Ludwig von Roenne, Prussian Prime Minister and later Reich Envoy to the USA, who donated some ethnographic objects to the Kunstkammer. The Prussian consul Hebenstreit sent collector's items to Berlin, and the physician Georg Engelmann, who worked in St. Louis, also sold objects from North America to the Kunstkammer. Furthermore, the collection was supplemented by Frederick Röver, a merchant from St. Louis (Ibid. 27–28). Long before his time as the second director general of the KM in Berlin, Ignaz Maria von Olfers brought back some ethnographica from North and South America from his travels to America as a diplomat and legation secretary and handed them over to the Kunstkammer (Cf. Ledebur 1869, 194–200, 202–203).[7]

Adolf Bastian spent many years traveling, but he still was able to devote himself intensively to his directorial duties between 1880 and 1889 (Cf. Hermannstädter 2002, 44–46). Whether he was an ethnologist, researcher, adventurer, or even just an observing traveler is the subject of recent studies in the history of science (Fiedermutz-Laun 2007, 55–74). In the present context, only his unusually high acquisition budget and the money he received from private sponsors are significant. As a result, he was able to win over merchants, diplomats, travelers, and other Germans living abroad as suppliers for the development of ethnological collections in Berlin. In 1880, a man named Mr. Wilhelm Pietzker sent Afro-Brazilian cult objects to the museum of the Rio Grande do Sul. In the same year, the German Consul General Hermann Haupt in Rio de Janeiro brought typical fishing gear to the international fisheries exhibition in Berlin. All pieces exhibited there were handed over to the KMfV after the end of the exhibition. A targeted acquisition policy began as early as the 1870s, which is why traveling laymen were provided with instructions on how to acquire ethnographica. The Berliner Gesellschaft für Anthropologie, Ethnologie und Urgeschichte even distributed questionnaires with research instructions to German officers traveling overseas. The ambassador to Argentina, Baron Theodor von Holleben, systematically arranged suitable collectors for Bastian on site. Among them was Richard Rohde, son of a sergeant in the Lithuanian Dragoon Regiment. From 1882 to 1884, he traveled to the border area between Paraguay and Brazil on behalf of the KMfV in Berlin, where he collected hundreds of objects for the museum (Cf. Hermannstädter 2002, 47–53).

The core group of scholars who built up the subject of ethnology in Germany and founded scientific research in Amazonia includes the following people: Karl von den Steinen and Paul Ehrenreich, both born in 1885 and trained as physicians, were inspired for ethnology by Bastian and Virchow. Ehrenreich also turned to physical anthropology. As a young historian and ethnologist, Konrad Theodor

Preuss entered the service of the KMfV, where he remained until retirement. Starting as a research assistant, he held the post of director of the American department until the end of his service in 1934. During this time, he made two major research trips to Mexico and Colombia. Koch-Grünberg, who came from a protestant pastor's family, switched from his teaching service to the subject of ethnology, as described. His contemporary Max Schmidt was also interested in ethnology after completing his doctoral thesis in law and began his career as a volunteer at the KMfV. After several research trips, he quit his service at the Berlin Museum and finally settled in South America in 1929 (Cf. Kraus 2004a, 36–37). Fritz Krause (1881–1960) is the last to be counted among this group of people. His academic career, which was geared toward the teaching profession, had already changed at the University of Leipzig, where he studied ethnology, geography, and geology. In 1908, he took part in an expedition to Brazil and in 1927 became the director of the GRASSI Völkerkundemuseum (Museum of Ethnology) in Leipzig. At the same time, he taught ethnology at the University of Leipzig (Ibid. 37–40).

Other notable individuals who have contributed to Amazonia research include Herrmann Meyer (1871–1932), Wilhelm Kissenberth (1878–1944), and Felix Speiser (1880–1949). Although all three had different academic backgrounds, they were distinct from the first group in that they either did not work for the KMfV/MfV in Berlin or did not teach. Because of the financial independence of Herrmann Meyer and Felix Speiser, both were able to carry out their research trips on their own initiative and with their own funds (Ibid. 40–46). The development of young scientists at the KMfV began with Bastian. In his spirit, young academics from various disciplines devoted themselves enthusiastically to non-European ethnology after their studies. They were pioneers in their time and, as a rule, had to bring a university degree in another subject area with them as the only requirement. They were doctors, lawyers, and historians. Other selection criteria for employment at the museum, which are now good scientific practice, were still unknown in the founding years of anthropology. The situation was different for people who were active as museum collectors. They were shaped very differently by their origins and their social status.

Tradesman for ethnographica

During the period studied, it was common practice for travelers to act as dealers of ethnographica for several museums and institutions. For instance, Richard Rohde (see subchapter: Researchers and adventurers) not only supplied the Berlin MfV but also the Hamburg animal dealer Carl Hagenbeck (Cf. Hermannstädter 2002, 47–55). On a much larger scale, museums were offered ethnographica from distant continents by import and export companies. The company Scharf & Kayser from Hamburg traded from overseas Coprah (Kopra).[8] In August 1884, they sent the General Direktion of the KM in Berlin a list of various ethnographica of the Rapanui from Easter Island. The items on offer, such as fishing nets, wooden idols, and sewing needles made of bone or headgear, represented an additional source of income for the Hanseatic Trading Company SMB-PK EM (1884). According to

Andrew Zimmermann, the science of ethnology in Germany is primarily based on the collection of objects, unlike in other countries. According to this, ethnologists referred to the artifacts they collected as

" . . . natürliche wissenschaftliche Präparate, die über das Grundwesen der menschlichen Natur Aufschluss geben"

(Cf. Zimmermann 2013, 247–258).

Far fewer ethnologists traveled to non-European countries to collect objects for their museums than objects that were obtained from dealers, officials, and soldiers that were sent through post. Many professions found collecting and donating ethnographica to be a popular practice, as it offered the possibility of being awarded the Preußischen Roth Adler-Ordens (Prussian Order of the Roth Adler) or the Königlichen Krone-Orden (Royal Order of the Crown) after the order had been fulfilled (Ibid. 253). The Roth Adler-Orden, founded in 1705 by Hereditary Prince Georg Wilhelm of Brandenburg-Bayreuth, passed to the Kingdom of Prussia in 1791 and has since been the second highest order in the Prussian state (Hoeftmann 1868, 37–39). By collecting objects from non-European countries, the order made a significant contribution to society as a whole, which was also allowed to be shown publicly by a medal.

Notes

1 Leopold Freiherr von Ledebur, born in 1799 and died in 1877, was director of the Ethnographischen Sammlung and director of the Museum Vaterländischer Altertümer of the KM Berlin from 1829 to 1873.
2 Albert Voß, born in 1837 and died in 1906, was a physician and a student of Rudolf Virchow. Through his mediation, Voß became in 1874 first a scientific assistant of the Nordische Sammlungen (Nordic collections) and then from 1886 to 1906 head of the Vorgeschlichtlichen Abteilung (pre-graduation department) of the KMfV. The term "auxiliary worker" corresponds to the current term "research assistant" (author's note).
3 Crown Prince Friedrich Wilhelm became German Emperor through the death of his father Wilhelm I on March 9 until his death on June 15, 1888, and thus received the nickname "99-day emperor".
4 I would have taken a completely different path in life if one day – it is now 25 years – I had not read the name "Dr. Bastian-Berlin" in the stranger's book (guestbook; author's note) of the Honolulu Hotel. He immediately conquered man by giving him high, rare tasks, and he did so with a trust, a confidence that one could literally feel oneself growing. The secret of his astonishing, suggestive power was none other than that he always appealed only to the best instincts and had no others but himself (translation by the author).
5 Herrmann August Heinrich Meyer, born on January 11, 1871 and died on March 17, 1932, was a German publisher, geographer, and explorer.
6 Wilhelm Kissenberth, born in 1878 and died in 1944. From 1907–? he was a volunteer at the KMfV Berlin and between 1922 and 1924 employed in the entwickllungsgeschichtliche Abteilung (Department of Evolutionary History) at the MfV Berlin.
7 From 1839 to 1864, Ignaz Maria von Olfers was the second director general of the KMfV Berlin (Cf. Grabowski et al. 2010, 167.).
8 Coprah (copra) is the core meat of coconuts from which coconut oil is extracted (author's note).

References

Archive material

SMB-PK EM (1884): I/MV 0611, Pars I B, Band 4, E. Nr. 2165/84. Acta betreffend die Erwerbung ethnologischer Gegenstände aus Australien. Loseblattsammlung. Angebot von Ethnografica. Scharf & Kayser. 1 Liste ohne Datum sowie 3 Briefe vom 16.08.1884, 4.9.1884 und vom 20.11.1884, Blatt 1–2, 4 Seiten.

Other sources

Beck, Hanno (1985): Alexander von Humboldts Amerikanische Reise. Stuttgart: Thienemann Ed. Erdmann (Alte abenteuerliche Reiseberichte).

Bolz, Peter; Sanner, Hans-Ulrich (1999): Indianer Nordamerikas. Die Sammlungen des Ethnologischen Museums Berlin. SMPK, G+H Verlag Berlin 1999.

Fiedermutz-Laun, Annemarie (2007): The Scientific Legacy of Adolf Bastian (1826–1905). Compilation, Evaluation and Significance of Knowledge About the Life and Work of the Scholar. In: Fischer, Manuela; Bolz, Peter; Kamel, Susan (eds.), Adolf Bastian and His Universal Archive of Humanity. The Origins of German Anthropology, Hildesheim: Georg Olms, 55–74.

Fischer, Manuela; Bolz, Peter; Kamel, Susan (eds.) (2007): Adolf Bastian and His Universal Archive of Humanity. The Origins of German Anthropology. Ethnological Museum Berlin, Hildesheim: Georg Olms.

Hermannstädter, Anita (ed.) (2002): Deutsche am Amazonas – Forscher oder Abenteurer? Expeditionen in Brasilien 1800 bis 1914. Begleitbuch zur Ausstellung im Ethnologischen Museum, Berlin-Dahlem in Zusammenarbeit mit dem Brasilianischen Kulturinstitut in Deutschland. Ethnologisches Museum Berlin, Ausstellung. 2., unveränd. Aufl. Münster: LIT. Veröffentlichungen des Ethnologischen Museums Berlin Fachreferat Amerikanische Ethnologie, Neue Folge, 2002, (71).

Hoeftmann, Friedrich Wilhelm (1868): Der Preußische Ordens-Herold. Zusammenstellung sämmtlicher Urkunden, Statuten und Verordnungen über die Preußischen Orden und Ehrenzeichen. Berlin: Königliche Buchhandlung von Mittler & Sohn.

Junker, Horst (2004/2005): Zur Dokumentation archäologischer Sammlungen und Archivierung von Quellenmaterial am Museum für Vor- und Frühgeschichte. In: Staatliche Museen zu Berlin-Preußischer Kulturbesitz, Menghin, Wilfried (Hrsg.). Das Berliner Museum für Vor- und Frühgeschichte. Festschrift zum 175-jährigen Bestehen. Acta Praehistorica et Archaeologica 36/37, 415–471.

Kraus, Michael (2004a): Bildungsbürger im Urwald. Die deutsche ethnologische Amazonienforschung (1884–1929). Marburg/Lahn: Curupira (Reihe Curupira, Bd. 19).

Kraus, Michael (Hrsg.) (2004b): Koch-Grünberg, Theodor. Die Xingu-Expedition (1898–1900). Ein Forschungstagebuch. Köln: Böhlau.

Ledebur, Leopold, Freiherr von (1869): Aus der Ethnologischen Sammlung des Königlichen Museums zu Berlin. *Zeitschrift für Ethnologie und ihre Hülfswissenschaften als Lehre vom Menschen in seinen Beziehungen zur Natur und zur Geschichte*, Erster Jahrgang (III), 193–204.

Meyen, Franz Julius Ferdinand (1834): Reise um die Erde. Ausgeführt auf dem königlich preussischen Seehandlungs-Schiffe Prinzess Louise, commandirt von Capitain W. Wendt, in den Jahren 1830, 1831 und 1832. Theil 1: Historischer Bericht. Berlin in der Sanderschen Buchhandlung. C.W. Eichhoff.

von den Steinen, Karl (1905): Gedächtnisfeier für Adolf Bastian. Am 11. März 1905. *Zeitschrift der Gesellschaft für Erdkunde zu Berlin*, Sonderabdruck, (3), 168.

Zimmermann, Andrew (2013): Bewegliche Objekte und globales Wissen. Die Kolonialsammlungen des Königlichen Museums für Völkerkunde in Berlin. In: Habermas, Rebekka; Przyrembel, Alexandra (Hrsg.). Von Käfern, Märkten und Menschen. Kolonialismus und Wissen in der Moderne. 1. Aufl. Göttingen: Vandenhoeck & Ruprecht.

10 Active ingredients and agents for the protection of persons and goods on expeditions

A valuable foundation for travelers in non-European countries was the book *Anleitung zu wissenschaftlichen Beobachtungen auf Reisen*,[1] created and published in 1875 by Georg Neumayer[2] (Neumayer und Ascherson 1875). The manual served as a solid foundation for scientists during their fieldwork, enabling them to orient themselves in unknown areas and fields. After the reader becomes familiar with the continents explored at that time, the state of art in astronomy, geography, oceanography, topography, meteorology, and geology is presented in individual chapters (Ibid. 1–332). At the end of the monograph, several authors devote themselves in great detail to the collection of plants, reptiles, fish, marine invertebrates, arthropods, birds, and mammals (Ibid. 384–515). The treatment of dried plants and dead animals is described in detail, including the necessary equipment and means of preservation.[3] On a theoretical level, this was done to ensure that the collected items were transported intact to their new locations (Ibid. 333–515). It became clear that in Neumayer's time, there was already a recognition that the preservation of plants and animals for museum collections necessitated specialised knowledge. Another important handbook was written by Albert Voß in 1888 *Merkbuch, Alterthümer aufzugraben und aufzubewahren*[4] (Voß 1888). This manual, initiated and published by the Preußischen Ministerium der geistlichen, Unterrichts- und Medizinalangelegenheiten (Prussian Ministry of Spiritual, Educational and Medical Affairs), was sent to museums, private collectors, libraries, pastors, teachers, mayors, and administrative authorities along with a questionnaire (Born et al. 2004/2005, 487). As a result, it became widely used and was on everyone's lips. Originally written for excavations, the handbook summarised the state of the art in conservation technology at the time in an exemplary manner. It was a practical companion during archaeological excavations, as well as a guide for the conservation of finds and collected objects in non-European countries. Its pocket-handy format corresponded to a "do it yourself" handbook for the local practitioner and received much recognition from anthropologists, prehistorians, and ethnologists (Lissauer 1906, 761–762).

The extent to which researchers, collectors, and adventurers resorted to funds from their own first-aid kits to protect collected objects can only be discussed to a limited extent here. Therefore, great importance is attached to the urgent statement made by Wilhelm Kissenberth (see Part III, chapter 8, researchers and adventurers)

DOI: 10.4324/9781003407607-14

regarding the preservation of future museum objects made of organic materials. In a letter to Eduard Seler, he wrote:

"Es ist eine grässliche Zeit, in der alles schimmelt, die Kisten von Kakerlaken wimmeln, Naphthalin, Pfeffer, Zacherlin ignorierend. Ich hoff nur sehnlichst, dass, was ich mit Fleiss und Sorgfalt gesammelt habe, auch einigermaßen wohlbehalten nach Berlin gelangt. Wie sich der Transport meiner Sammlung in Zukunft gestalten wird, das ist mir vorläufig noch völlig unklar".

(Cf. Kraus 2004a, 171–172).[5]

Notes

1 Guide to scientific observations while traveling (translation by the author).
2 Georg Balthasar Neumayer, born on June 21, 1826, in Kirchheimbolanden and died on May 24, 1909, in Neustadt an der Haardt, was a geophysicist and polar explorer from Bavaria-Palatinate.
3 For preservation, the following have been proposed: ether (ether), sodium arsoric acid (sodium arsenate), arsenic soap, alum, carbolic acid (phenol), potassium cyanide (potassium cyanide), potassium double chromic acid (potassium dichromate), glycerol (propane-1,2,3-triol), naphthalene, nicotine, osmium acid, petroleum, sulfur ether (diethyl ether), sublimate (mercury(II) chloride), and alcohol (ethanol). The following repellents were proposed: camphor, cajeput oil, carbolic acid (phenol), crushed pepper, and rosemary oil.
4 Memo book to dig up and store antiquities (translation by the author).
5 It is a ghastly time when everything is moldy; the boxes are teeming with cockroaches, ignoring naphthalene, pepper, and Zacherlin. I just hope that what I have collected with diligence and care will reach Berlin reasonably safely. How the transport of my collection will be in the future is still completely unclear to me for the time being (translation by the author).

References

Other sources

Born, Hermann; Hausdörfer, Ute; Thieme, Franziska (2004/2005): Die Restaurierungswerkstätten. In: Das Berliner Museum für Vor- und Frühgeschichte; Festschrift zum 175-jährigen Bestehen, (36/37).
Kraus, Michael (2004a): Bildungsbürger im Urwald. Die deutsche ethnologische Amazonienforschung (1884–1929). Marburg/Lahn: Curupira (Reihe Curupira, Bd. 19).
Lissauer, Abraham (1906): Sitzung vom 21. Juli 1906. *Zeitschrift für Ethnologie und der Verhandlungen der Berliner Gesellschaft für Anthropologie, Ethnologie und Urgeschichte*, (38), (IV und V).
Neumayer von, Georg; Ascherson, Paul (1875): Anleitung zu wissenschaftlichen Beobachtungen auf Reisen. Mit besonderer Rücksicht auf die Bedürfnisse der kaiserlichen Marine. Berlin: Verlag von Robert Oppenheim. Available online at http://data.onb.ac.at/ABO/%2BZ102227706.
Voß, Albert (1888): Merkbuch–Alterthümer auszugraben und aufzubewahren. Eine Anleitung für das Verfahren bei Ausgrabungen, sowie zum Konservieren vor- und frühgeschichtlicher Alterthümer. Unter Mitarbeit von Gustav von Gossler (Hrsg.). Berlin: Ernst Siegfried Mittler und Sohn.

11 Developments and experiments on pest control at the Königliches/Staatliches Museum für Völkerkunde in Berlin

Preserving colonial collections at the Königliches/Staatliches Museum für Völkerkunde and the Königliche/Staatliche Museen in Berlin

Eduard Krause, as the first conservator of the KMfV, was integrated into the hierarchical structure of the museum and played a role in the practical implementation of pest control experiments. He was involved in the execution of experiments through the curators of the museum as well as directly through the General Direktion of the KM in Berlin. However, he always required a director's permission if he wanted to conduct his own pest control experiments (SMB-PK EM 1912/1913). This also applied to recommendations for pest control by Friedrich Rathgen of the Chemisches Laboratorium (chemical laboratory).[1] This laboratory occupied an important position in the field of pest control and was the world's first exclusive scientific conservation facility for museum objects (Bracchi 2013, 105–120). Rathgen (1862–1942) became the laboratory's first chemical director at the age of 26. Among his staff members was Carl Brittner, who became his successor after his retirement in 1928. The renaming of Rathgen-Forschungslabor did not take place until 1975, following the end of the Second World War, when it reopened in Berlin-Charlottenburg (Maertins 2005, 11).

Rathgen's initial primary task was to develop procedures and methods for the conservation of archaeological finds and artifacts made of inorganic materials. On behalf of the Generalverwaltung of KM, he visited numerous museums in Germany and abroad to become knowledgeable about the methods used for the conservation of museum objects (Rathgen 1896, 125–127). As a result, he published his standard work in 1898 "*Die Konservirung von Alterthumsfunden*".[2] The chemical substances described herein for the conservation of museum objects were also ordered by him, so it is assumed that they were also used (Cf. Bracchi 2013, 261). During his tenure, he was one of the few specialists in museum industry and actively participated in national and international meetings and conferences relevant to his field. The presentation of his research to an expert audience was one of his official tasks as the director of the laboratory (Bracchi 2014, 6). Family connections within the European nobility even enabled Rathgen in one case to accept an invitation from the Duchess of Mecklenburg to her castle Bogensperk in Slovenia,

DOI: 10.4324/9781003407607-15

Figure 11.1 Friedrich Rathgen. Head of the Chemisches Laboratorium at the Königliche/
Staatliche Museen zu Berlin from 1888 to 1927.

accompany her professionally during her excavations, and look after her finds in
terms of conservation (Nemecek 2013).

The Chemisches Laboratorium attracted both admiring and envious attention
from abroad. During the 13th meeting of the Royal Society of Arts in London
in 1922, Walter Reid reported on an invitation from the SMB and his visit to the
Chemisches Laboratorium, which he described as an institution with impressive
technical and scientific equipment. He was so impressed that he expressed public
shame for the backwardness of his native Great Britain compared to this institution (Scott 1922, 337). Therefore, it is not surprising that German was the leading
language of science at the beginning of the 20th century (Reinbothe 2011, 49–52).
Rathgen's manual can also be perceived internationally soon after its publication.
In addition, interest in his standard work was so high that it was translated into
English as early as 1905 (Rathgen 1898). Rathgen's area of responsibility soon
expanded to include pest control, primarily for KMfV/MfV. The primary objective
was to safeguard and preserve the collections, which consist mainly of organic
materials, in cooperation with the conservator Krause (Otto 1979, 48–49). Consequently, Rathgen published a second and third part of his manual, as well as an
addition to the three parts in 1924 and 1926, which contained instructions on how
to preserve objects made of organic substances (Rathgen 1924, 1926).

The pesticides offered by industry and trade were obviously the reason Rathgen
examined industrially applied substances in his Chemisches Laboratorium regarding their use in conservation to control insect pests. Therefore, he experimented
with carbon tetrachloride, among others. Despite its higher price compared to

petrol, he saw the great advantage that this substance is neither flammable in liquid nor in a vaporous state (Rathgen 1908, 90–91). To verify whether the vapors of carbon tetrachloride have a damaging effect on sensitive surfaces of works of art, he conducted a study with 64 mineral colors. He found that there were changes in the bronze colors of gold,[3] and cobalt yellow changed its color to brownish. In several samples of linseed oil varnish that had been applied with paints, changes were observed due to exposure to carbon tetrachloride. Rathgen attributed this to a chemical reaction between the vapors and the also browned, fresh linseed oil. As a result, he advised against the use of carbon tetrachloride in paints containing resins or varnish as binders. For dyed wool and cotton, he conducted the same experiment without detecting color changes. He concluded by pointing out that common carbon tetrachloride always contains small amounts of carbon disulfide, which was why he advised using only technically pure carbon tetrachloride. However, this gas could only be applied to a limited extent for pest control on art and cultural property (Cf. Rathgen 1911, 219–220). He also recommended the use of airtight museum cabinets and showcases, which he had jointly developed with the Dresden company, August Kühnscherf & Söhne, for the display of objects with organic materials, such as pieces of fabric, furs, or feathers. As a preventive measure, a shell with carbon disulfide or carbon tetrachloride should also be set for such objects in airtight containers to protect against attacks by harmful insects (Rathgen 1908, 97–102). The renunciation of fumigants, such as carbon disulfide and carbon tetrachloride, was documented by Rathgen's successor, Carl Brittner. On February 1, 1936, he certified the correctness of a delivery of 10 liters of "Areginal" for use in the "Schädlingsbekämpfungsanstalt in Berlin-Dahlem" (SMB-ZA 1935).

During practical measures to combat insect pests on collection items, unforeseen difficulties arose. In the case of the suspected improper handling of mercury(II) chloride (sublimate) on objects from the Oceania collection, there were intense discussions within KM in Berlin. This example also illustrates the internal hierarchical structures between the Chemisches Laboratorium, KMfV, and the General Direktion of KM. In 1914, Krause was tasked to conserve feather shields from the Richard Thurnwald collection. Shortly thereafter, Eichhorn[4] informed von Bode that feather shields had been returned discolored and deformed (SMB-PK EM 1914a). This matter was subjected to a close examination, with Rathgen's involvement.[5] The General Direktion finally issued a reprimand to Krause due to a poorly executed conservation, which was sent to him by Grünwedel on September 5, 1914 (SMB-PK EM 1914b). Interestingly, on the same day, an entry with the note "secret" was deleted (SMB-PK EM 1914g). The question of whether the pursuit of independence in the conservation of ethnological objects or simply negligence during the measures taken on the feather shields from Oceania resulted in a poor conservation outcome remains unanswered. However, it is evident that Rathgen did not agree with Krause on all technical issues. In a journal entry in 1903, he had already expressed disapproval of his colleague and his methods of preserving iron finds (Rathgen 1903, 704). In fact, it is remarkable how much attention was paid to the incident described here shortly before the outbreak of the First World War. This may also be regarded as an indication of the individual desire for normality or "business as usual".

However, KM/SMB also received external support in the development of pest control methods from a personality completely unknown in today's conservation science. The founding director of the Kaiserlich Königlichen Landwirtschaftlich–Chemischen Versuchsstation (k.k. l-c-V) (Imperial Royal Agricultural–Chemical Experimental Station) in Gorizia, Italy, Johann (Giovanni) Bolle, was a prominent figure in his time and of great importance for the textile industry and agriculture (VDLUFA 2013; Bolle 1882, 1892, 1898, 1899; Bolle und Mewis 1892).[6] He also made significant contributions to the preventive and control measures of insect pests in museums and became a valuable partner of KMfV/MfV. With his expertise, he helped establish agricultural research facilities to increase yields for an ever-growing population. Bolle was constantly expanding the scope of the experimental station's pest control for agriculture, which led to its restructuring in 1891, and he was appointed director of the k.k. l-c-V. During his tenure, he traveled to various places, including Germany, France, Switzerland, Asia, and North America, to research plant and tree pests (Anonymous 1913b, 39–40).

Bolle was not only interested in plant protection and improving agriculture. During his visits to many museums and natural science collections throughout Europe, he studied the methods used to control pests on site and ensured the dissemination of his knowledge. He pursued his goals with great commitment and was instrumental in establishing a network of agricultural and cultural institutions for pest control. This led to his decisive role during some experiments in KMfV, which made him a key figure in conservation efforts within the museum. Between 1911 and 1912, Bolle and von Luschan conducted experiments on combating insects at KMfV while understanding the effects of several pesticides in detail. Notable, they determined the total ineffectiveness of camphor, which was used previously as a preservative against harmful insects in the collection cabinets of EM in Berlin-Dahlem. This is a striking example and seen in fact as an indication that knowledge has been lost over the course of more than a hundred years.

Bolle's active involvement beyond his term of office is reflected in his detailed publication on *Die Ermittlung der Wirksamkeit von insektentötenden Mitteln egen die Nagekäfer des verarbeiteten Werkholzes*, where he summarised his experiences from years of experiments (Bolle 1919, 105–117). He emphasised that leaf metals or metal foils turn brown in conjunction with the vapors of carbon disulfide, as a chemical reaction produces hydrogen sulfide (known today as hydrogen sulfide; see glossary). Bolle thus independently confirmed Rathgen's expertise. He also used his expertise in evaluating browning on polychrome works of art, such as when lead white was used for their frames. In one case, he even worked as a conservator on two white-framed statues that had tanned faces after treatment. He covered the faces with strips of canvas and drizzled a solution of hydrogen peroxide over them, which, according to his description, made the face colors fresh and rosy again (Ibid., 107–108). In this publication, it advocated carbon disulfide as the most suitable of all active substances and agents. He considered the great advantage of the substance while using it under reduced air pressure. The liquid only must act on the collection items for two hours.

Dissemination of knowledge within museums

The existing sources were unable to provide evidence of a systematic approach to preserving museum collections made of organic materials. However, relevant inquiries from smaller museums provided important insights into the conservation methods of larger institutions at that time. Bolle became increasingly visible in his frequent travels and extensive correspondence. Tirelessly and with great commitment, he drew attention to the progress made in the field of pest control in various museum institutions throughout Europe. Consequently, knowledge of pest control within museums–both nationally and internationally–has been constantly expanding. Since its foundation in 1888, KM/SMB has had a unique selling point in terms of conservation with the Chemisches Laboratorium, which lies both in professional competence and in the technical equipment of this laboratory. Within the institution, this primarily benefited KMfV/MfV. In addition, museums and conservation laboratories were known far beyond the borders of Berlin. Smaller or distant museums sought advice from their Berlin colleagues. Their written questions were mainly concerned with methods for preserving mounted and unmounted wood. However, there are also questions regarding the protection of textiles against textile moths. To illustrate this, ten museums from the German, Western, and Eastern European regions show which paths individuals from external institutions took to enhance their own knowledge while also leveraging the "know-how" of others. These museums also benefited in the aftermath of specialist conferences, which were mainly attended by museum experts from large institutions. Similarly, a difference can be observed in the application of pesticides to individual objects or in the mass killing of insect pests. In addition, larger museums were more likely to raise funds for the construction of fumigation systems than smaller facilities.

However, the Städtisches Museum für heimatliche Altertümer in Herford (SMB-PK EM 1905a), Bormann) had to preserve archaeological wood, half of a tree coffin, in 1905. Through the "*Merkbuch*", one became familiar with the recommended treatment using kerosene and varnish, but wanted to make sure whether cracks in the now dried wood should be closed again before treatment. Krause recommended immersing the archaeological find in water for months to then coat it several times with the mixture of kerosene and varnish recommended in this manual, or alternatively, to apply a mixture of varnish and real Avenarius "Carbolineum".[7] He also mentioned that the treatment would result in a much darker appearance, which would fade over time (SMB-PK EM 1905b). Six years later, the Herford museum again asked for further advice on the conservation of a five and a half meter long dugout canoe.[8] In doing so, we learned that Krause had already distanced himself from the mixture recommended in the "*Merkbuch*" for some time in his choice of remedies. Petroleum and varnish, if not consumed quickly, formed a stiff gelatinous mass, which did not become liquid again even in various solvents. In contrast, he recommended a mixture of linseed oil varnish and Avenarius "Carbolineum" in a 1:1 ratio, which should be applied several times, if possible, when hot (SMB-PK EM 1911). Very important was the case where Rathgen visited the grave complex of the Duke Privy Councillor Count Samuel von Behr in the Doberan Minster in September 1912 (LAKD 1885–1953). The life-sized wooden horse and rider sculpture, which had undergone conservation in 1886, was once again infested with wood-destroying insects, as visible in the turmoil that ensued (LAKD 1912a).[9] After an extensive consultation with Bolle, Rathgen sent a detailed report to the local administration in November. He recommended applying and soaking the horse and the rider several times with kerosene, and treating the rider with Chinese wood oil to consolidate the wood (LAKD 1912b). Twenty years later, a new infestation occurred, which led to another treatment that used turpentine (LAKD 1933). Another request from the administrative district of Kassel reached the Kaiser-Friedrich Museum on December 19, 1913.[10] The local conservator Holtmeyer[11] inquired about conservation methods to combat wood-destroying insects (*Anobium* sp.) in the case of polychrome sculptures. Several employees came forward to offer their help. Theodor Demmler,[12] Otto von Falke,[13] the conservator Hauser,[14] as well as Rathgen all expressed themselves in a similar manner. General statements were not made, but individual cases had to be examined. The treatment of unprocessed wooden sculptures is generally easier than that of polychrome sculptures because there is no need to consider colors or gilding. There is no information available about companies that perform such work. It was recommended to immerse the wood in pure turpentine, and if necessary, strengthen it with a glue solution. Fumigation with carbon disulfide in a tin box or a sheet metal chamber was also recommended. However, Rathgen added that turpentine is not a safe agent for controlling wood pests and recommended a resin solution prepared in-house, as well as a specially formulated mixture from the Technical University in Dresden that was currently being tested in the laboratory. Skilled craftsmen can be entrusted with these conservation measures. Because no agent would protect against new infestation, Rathgen also recommended that the objects be monitored regularly (SMB-ZA 1913/1914).

Sometimes, it was necessary for pesticide manufacturers to engage in a little self-promotion, as evidenced by a letter from the conservator Paul Hübner of the Städtischen Sammlungen in Freiburg i. Br., to the Consolidirten Alkaliwerke, Hanover department, where the product "Xylamon-LX-Hell" was produced. He was not only enthusiastic about its effect in combating wood-destroying insects but also conducted his own experiments with it in his laboratory. Without being asked, he was willing to recommend "Xylamon-LX-Hell" to other museums and building authorities and to publish his results (SMB-ZA 1933).

Similarly, knowledge of museums has spread internationally. Sidney Harmer[15] also visited the Nordiska museet in Stockholm and admired the fumigation system "Lusknäppen" (see subchapter: Construction of plants for mass fumigation against insect pests). According to his contribution in 1922, this was the best he had ever seen in museum practice for pest control. Like many of his colleagues, he advocated the active ingredients carbon disulfide and hydrogen cyanide for the preservation of entomological collections (Harmer 1922, 333–334). At the beginning of the 20th century, problems with large spatial distances were solved through knowledge sharing. A correspondence between the Staatliche

Urallandesmuseum Yekaterinburg (Sverdlozk)[16] and MfV in Berlin revealed that experts at the Urals museum were interested in learning about the importance given by the Berlin colleagues to the product "Areginal" and the process of soaking with the product "Eulan", which has been newly developed against textile moths. Measures had to be taken to effectively protect Samoyad reindeer hides and Kyrgyz objects made of felt against moths (SMB-PK EM 1928). In Berlin, the toxic and flammable carbon disulfide was replaced by "Areginal". The Chemisches Laboratorium was in the process of experimenting with "Eulan"-soaked objects, which is why a conclusion on this agent could only be made the following year (Ibid., E. Nr. 356/28; Ibid., Serie 1928, Nr. 3026, Brief Nr. 356/28). Meanwhile, colleagues from Yekaterinburg reached out to the Scientific Museum Department in Moscow to request permission to maintain contact with Berlin museums. They also needed to obtain written permission to order "Eulan" from Germany (Ibid., May 17, 1928; Ibid., August 26, 1929). In its reply, the Chemisches Laboratorium mentioned only the results from laboratory tests with "Areginal" and "Eulan". Rathgen's successor, Brittner, recommended "Eulan M" for protecting against moths, but at the same time pointed out that before soaking, the dyes of oriental carpets, especially vegetable dyes, must be examined in general. Therefore, he urged caution regarding its application. However, he left no doubt that the efficacy of "Eulan" against textile moths had been scientifically proven, although he could not provide any information on the length of its effectiveness. Like his Russian colleagues, Brittner was not allowed to answer inquiries directly. He wrote originals, which were then sent to Yekaterinbrúrg in letter form by Otto Kümmel[17] (Ibid., Brief Nr. 356/28, Nr. 67). Manufacturers or suppliers of pesticides were also involved in the knowledge transfer process. They incorporated positive reviews from other museum experts into their promotional materials, highlighting their successful results.

Substantial progress has been made in the dissemination of knowledge as well as the advancement of research through relevant journals. The oldest journal is the *Journal of the Royal Society of Arts*, founded in 1852 (Stansbury et al. 1852) by the society of the same name that was founded in 1754. It published topics for the promotion of the arts, innovative products in industry, and trade. The journal *Museumskunde–Zeitschrift für Verwaltung und Technik öffentlichen und Privatsammlungen*, published in Germany by Karl Koetschau, is still the most important of its kind for museum institutions (Meyer 2014, 179). In the first issue in 1905, von Bode explained in detail the tasks of this journal and its inspiring effect on representatives of European museums. The international status of the journal was ensured solely by the publication of numerous articles in German and English (Ibid. 181). At the beginning of the 20th century, the exchange of ideas, opinions, and research among museum experts is widely comprehensible. Important topics, such as restoration, conservation, security measures against fire and theft, air conditioning, lighting, the choice of showcases for exhibitions and permanent collections, and different administrative structures in large and small museums, were discussed extensively in this journal (Ibid. 179–181). In 1901, another "specialist medium", the *Museums Journal*, was first published in Great Britain as a professional journal of the Museums Association, which was founded in 1889. This journal covered all topics related to museums, galleries, cultural monuments, and historical houses. Museum experts, such as Carl Bernhard Salin[18] from Sweden, challenged their colleagues at conferences to use the *Museums Journal* as a platform, where ideas regarding conservation issues should be exchanged intensively (Cf. Anonymous 1916, 269).

The importance of this platform is illustrated through examples. In the *Museums Journal*, Wray shared his experiences on skin and hide conservation, stating his own preparations and techniques. He described the treatment of killed mammals in the tropics and preparations made after their arrival in England (Wray 1908, 201–209, 201–206).[19] The production of mixtures was not possible in the tropics, so peeled fur was coated on both sides with salt and "carbolic disinfecting powder" and then rolled into a bundle. After applying a powdery solution of alum, the treated skins should be dipped into a tub containing alum, phenol, and salt, and then thoroughly cleaned with sufficient running water to prepare for the next step (Ibid. 207–208). Sheridan Delépine of the Public Health Department of the University of Manchester had originally reported in the *Journal of Pathology and Bacteriology* about the "arsenous jelly" he developed and used for the preparation of pathological samples of animal organs. It contained gelatin and arsoric acid, which had been previously dissolved in water. He recommended embedding, or permanent preservation in the jelly he has developed, for plants and animals. The publication of the article in the *Museums Journal* in 1914 led to the spread of this method of conservation among museum professionals. The precise information in the formulation made it possible for anyone to produce the product (Delépine 1914, 322–329). This exchange of knowledge between individual disciplines extended the application of remedies from the medical field to the field of conservation of art and cultural assets. The intensity with which the *Museums Journal* was perceived and served as inspiration for one's own experiments is also demonstrated by a 1916 publication from the Royal Albert Museum in Exeter. Frederick Richard Rowley, a museum enthusiast, tested Delépine's "arsenous jelly" for the wet preparation of seaweed. He praised several advantages of this method, including the better preservation of brown and red seaweed colors and the increased visibility of the objects prepared in this way. In addition, such preserved wet preparations can be easily displayed in desk showcases alongside other dry preparations. In particular, Rowley noted that the jelly-like mixture would not evaporate, unlike wet preparations stored in alcohol. Unfortunately, his attempts to preserve flowers in this way were not successful (Rowley 1916, 77–79).

To preserve the numerous objects in KMfV/MfV in Berlin, only two cases were reviewed for active ingredients and agents. The lack of "know-how" and limited resources within the museum sometimes led to the ongoing promotion of pest control products by industry, trade, and commerce. At the same time, the staff redirected their attention to similar developments in other museums, and this exchange took place either through conferences or by studying the writings of colleagues in the emerging specialist literature. This marked the commencement of a knowledge transfer within an expanding network of museum experts.

Table 11.1 Networking of museum experts at both national and international levels between 1874 and 1934.

Time	Name	Function	Department	City	Country	Correspondence
1874–1906	Adolph Bernhard Meyer	Director	Königliches Zoologisches und Anthropologisch-Ethnographisches Museum	Dresden	Germany	Johann Bolle
1884–1917	Eduard Krause	Conservator	Museum für Völkerkunde	Berlin	Germany	Johann Bolle Willy Foy Alfred Hackman Georg Thilenius
1888–1927	Friedrich Rathgen	Director	Chemisches Laboratorium	Berlin	Germany	Johann Bolle Willy Foy Alexander Scott Georg Thilenius Schulz/Klein (first names unknown)
1891–1912	Johann Bolle	Director	kaiserlich königliche landwirtschaftlich-chemische Versuchsstation	Gorizia	Italy	Willy Foy Felix von Luschan Adolph Bernhard Meyer Axel Nilsson Friedrich Rathgen Georg Thilenius
1894–1946	Hans Aall	Conservator/Director	Norsk Folkemuseum	Oslo	Norway	Friedrich Rathgen
1897–1924	Albert Frank Kendrick	Curator	Department of Textiles of the Victoria & Albert Museum	London	Great Britain	Johann Bolle
1901–1924	Willy Foy	Director	Rautenstrauch-Joest Museum	Cologne	Germany	Johann Bolle Friedrich Rathgen Georg Thilenius
1904–1911	Felix von Luschan	Director	African–Oceanic department Museum für Völkerkunde	Berlin	Germany	Johann Bolle

Years	Name	Position	Institution	City	Country	Correspondents
1904–1935	Georg Thilenius	Director	Museum für Völkerkunde Hamburg	Hamburg	Germany	Johann Bolle Willy Foy Axel Nilsson
1906–1914	Axel Nilsson	Curator	Nordiska museet	Stockholm	Sweden	Johann Bolle Sidney Harmer Georg Thilenius
1916–1929	August Eichhorn	Head of department	Oceanic department Museum für Völkerkunde	Berlin	Germany	Johann Bolle Georg Thilenius
1919–1927	Sidney Harmer	Director	Natural History department at the British Museum	London	Great Britain	Axel Nilsson
1919–1932	Alfred Hackman	Curator	Prehistoric Department at the National Museum	Helsinki	Finland	Eduard Krause Axel Nilsson
1919–1938 1924–1959	Alexander Scott Harold Plenderleith	Directors of natural science laboratory	British Museum	London	Great Britain	Friedrich Rathgen
1921–1934	Konrad Theodor Preuß	Director	American department Museum für Völkerkunde	Berlin	Germany	Willy Foy
1928–1948	Carl Brittner	Director	Chemisches Laboratorium	Berlin	Germany	Schulz/Klein (first names unknown)
1928–1933	Otto Kümmel	Director	Asian Collections Museum für Völkerkunde	Berlin	Germany	Schulz/Klein (First names unknown)
1928/1929	Schulz/Klein (first names unknown)	Unknown	Urallandesmuseum	Yekaterinburg		Carl Brittner Otto Kümmel

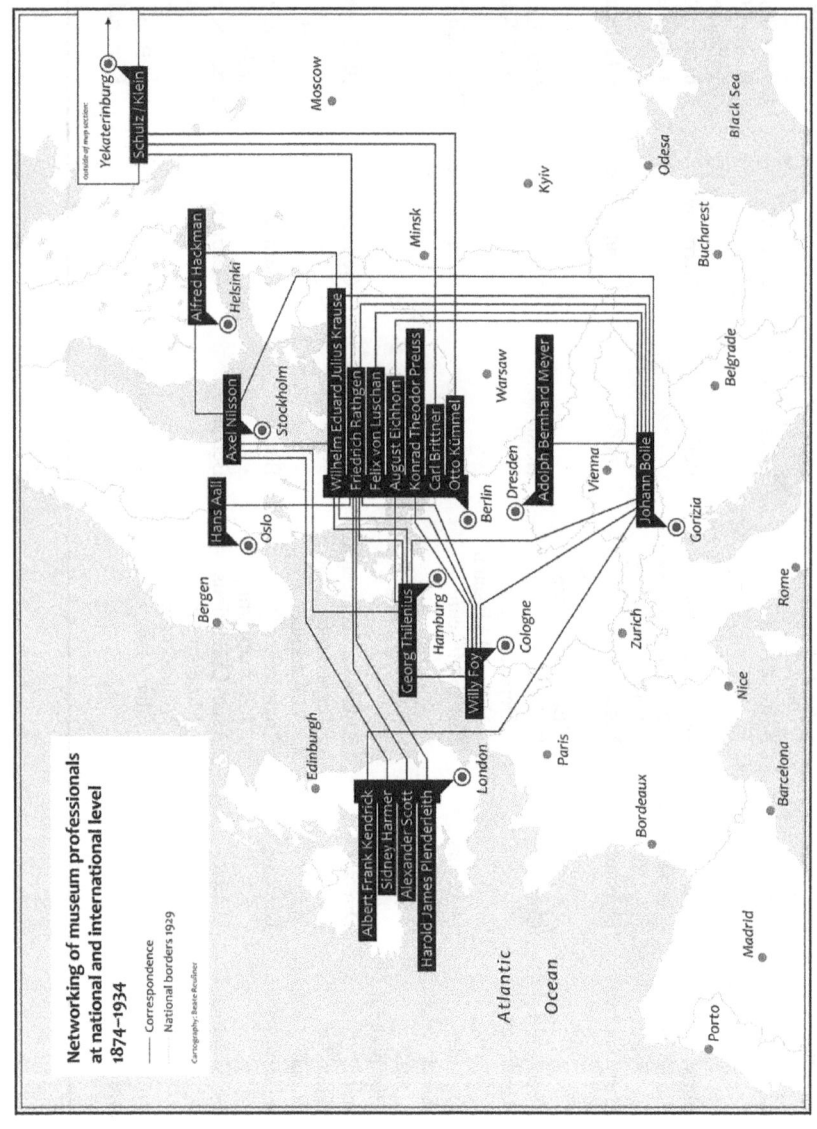

Figure 11.2 Networking of museum experts on both national and international levels from 1874 to 1934.

Construction of a plant for mass fumigation against insect pests at the Königliches Museum für Völkerkunde/Staatliches Museum für Völkerkunde in Berlin

Expensive fumigation systems were required to be able to counter the disintegration of extensive collections in KMfV/MfV. Bolle mentioned museums in Dresden, Stockholm, and Cologne, which already have such a facility (Ibid. 117). A short note in *Chemiker Zeitung* had announced the arrival of a new Swedish technology for combating insect pests using a fumigation system as early as 1907 (Nilsson 1907, 299). However, details about the plant and the expertise regarding it were not disseminated until August 1911, when Bolle shared the information from a study trip from Sweden (SMB-PK EM 1912a). The invention by Axel Nilsson from the Nordic Museum in Stockholm was thus made available to other museums in German-speaking countries (NMA 1902–1915c). It is unclear when and from where KM/SMB first learned of this new technology. Following the usual hierarchy, Eduard Seler[20] forwarded on November 4, 1911, a request from Konrad Theodor Preuss[21] on the procurement of a fumigation system to Felix von Luschan (SMB-PK EM 1911–1914). Four months later, Albert Grünwedel[22] mentioned that Preuss was familiar with the "Desinfektionsapparat" (disinfectant apparatus) of the Rautenstrauch-Joest-Museum (R-J-M) in Cologne and was willing to share his knowledge to support the construction of a similar facility at KMfV (Ibid., February 24, 1912). The planning, construction, and structural expansion of this novel technology at KMfV/MfV Berlin for mass disinfestation of collection objects during the First World War are the subjects of the following investigations.

Krause was immediately commissioned to obtain information about the details of the apparatus and procedure, and he contacted Bolle to obtain the details. In a letter dated January 28, 1912, he informed KMfV about the fumigation system already existing in the R-J-M and recommended that the museum send its construction plans for the fumigation system to Berlin. He also referred to a similar apparatus that was installed at the National Museum in Helsinki, which was based on models from the Nordiska museet in Stockholm. From his point of view, it was the most magnificent plant. He emphasised the importance of having a suitable pump to achieve a pressure of 1–2 atm within an hour,[23] and also emphasised that the boiler should be tightly sealed. In addition, Bolle offered his other services to review the Cologne plans (Ibid., January 28, 1912). Next, Krause contacted the director of the National Museum in Helsinki, Alfred Hackman, and requested on February 2, 1912, that he provide drawings, technical descriptions, and construction and installation costs for the Finnish plant. Hackman explained that the fumigation system at his museum was designed by Axel Nilsson based on the Swedish system from the Nordiska museet in Stockholm and set up by a local company. He expressed satisfaction with the results of fumigation at reduced air pressure but recommended that the plant be housed in a specially constructed building because of the risk of carbon disulfide explosion. From Swedish, he translated the exact details of the technical equipment for the Berlin Museum and provided a list of the total costs for a separate building, the machinery, customs, freight, and installation

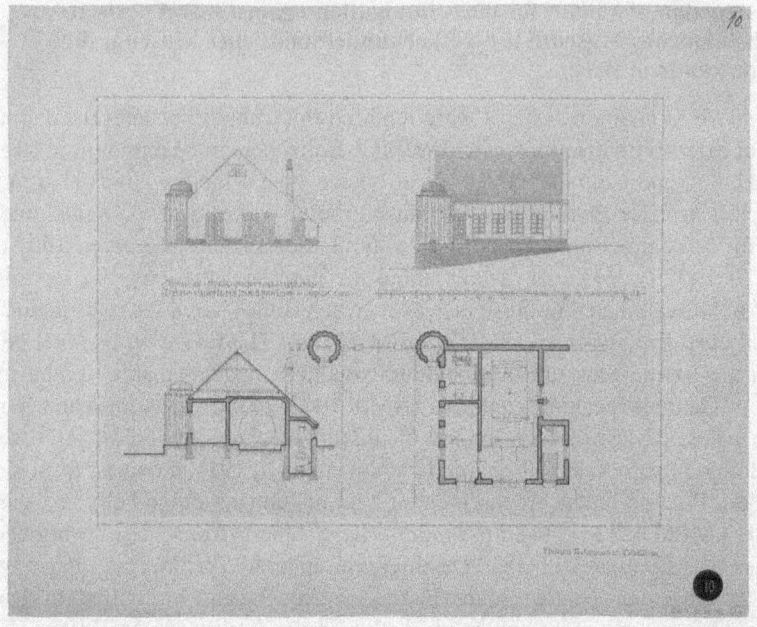

Figure 11.3 Fumigation system at the National Museum Helsinki, construction drawing.

of the plant, which amounted to a total of 32,716 Finnish marks (FHA 1912; SMB-PK EM 1912a, 1912b, 1912c).[24]

Krause then visited R-J-M in Cologne, where Willy Foy,[25] the director of the museum, said that he had no drawings but provided photographs of his plants instead. For economic reasons, he recommended the construction of large and small boilers. In Cologne, the system was not installed in a separate building but in a remote part of the museum. Grünwedel advocated a fumigation system for KMfV according to the model in Cologne (SMB-PK EM 1912d). As a future location, he requested room number 40 in the "Asian Museum" in Berlin-Dahlem (SMB-PK EM 1914a). On March 19, 1912, drawings and photographs of the fumigation system were received from Helsinki, as well as a brochure requested by Krause at the same time from the apparatus construction company Gustav Christ & Co. m.b.H. from Berlin-Weissensee (SMB-PK EM 1912e). However, plans for the extension of the museum in Berlin-Dahlem could not be executed due to the unavailability of necessary funds at that time (SMB-PK EM 1914b).

Nevertheless, the steadily growing number of collection items that arrived in Berlin required action. Following Eichhorn's appeal on January 7, 1914 (SMB-PK EM 1914c), Grünwedel asked Rathgen to inspect objects that had arrived in the Königgrätzer Straße from the Kaiserin-Augusta-Fluss-Expedition (SMB-PK EM 1914d). Two days later, he asked Krause to take care of the conservation of these objects (SMB-PK EM 1914e). Krause then had a box built for fumigation, which was set up in room 13 or the so-called Vergiftungsraum[26] (SMB-PK EM

1914f). This historical designation of a functional space, which was reserved for the treatment of collection objects preventively or acutely against pest infestation, indicates that the museum took precautions regarding the health care of its employees as early as 1892, in the sense of the official occupational safety and health regulations that have existed since that time. From this perspective, room 13 can be seen as a precursor to the establishment of work areas in the sense of the black-and-white principle practiced today (Bundesanstalt für Arbeitsschutz und Arbeitsmedizin. Februar 2010, 419–450). The work on the "Asian Museum", which began in 1914 according to the designs of the architect Bruno Paul, had to be stopped unfinished with the outbreak of the First World War.[27] After Krause died on October 30, 1917, efforts to locate relevant construction documents in his office were unsuccessful (SMB-PK EM 1912f). This lack of transparency and knowledge transfer among employees within SMB is further highlighted. Following Germany's defeat in World War I, the Weimar Republic was unable to proceed with the construction project as originally planned, and on July 23, 1921, it was decided to complete the building in Berlin-Dahlem using cost-effective methods. The building was used for another purpose and was made available to all departments of MfV as a storage area. In 1923, the relocation of collections began, which had to be largely financed by the sale of so-called duplicates from individual departments. The separation of show and study collections, which was completed for the first time, had a positive effect on operations in the Königgrätzer Straße (Cf. Westphal-Hellbusch 1973, 29–32). The construction of a fumigation plant in Berlin-Dahlem was granted permission on January 6, 1923 (SMB-PK EM 1923). The Chemisches Laboratorium was responsible for preparatory activities related to the construction of the fumigation plant. On January 18, 1923, the building administration of SMB ordered the delivery of "Apparatur"[28] for pest control from Friedrich Heckmann, which uses carbon disulfide to generate low air pressure. On May 19, 1923, the fumigation system was loaded by rail and sent from Breslau to Berlin.

After 12 years, the approval for the construction of a fumigation plant has finally been granted (SMB-ZA 1914, 1922–1923, 1928, 1915, 1919, 1922–1924, 1924–1928). Behind this fact hides a certain program, which was not only due to the circumstances of the First World War. The MfV in Berlin was—and still is—part of SMB with its General Direktion. The process of implementing decisions was often slow, which was why other ethnological museums in Germany were able to introduce fumigation systems to combat insect pests much earlier.

One week after the construction of the plant, the first test run was performed under the supervision of Rathgen. To use the fumigation system twice a week in the summer months, he calculated the required amount of carbon disulfide to be approximately 40 kg. He also requested a larger quantity of kerosene from the administration, seeking a barrel with a capacity of 125 or even 200 liters to eliminate insects from large wooden objects that were too big for the plant (Otto 1979, 248–250). During the test runs, Rathgen pointed out many technical defects. However, from February 6, 1924, onward, all new additions, especially those where pest infestation was visible, could finally be handed over to the

Figure 11.4 Fumigation system at the Museum für Völkerkunde der Staatlichen Museen zu
Berlin, closed state, 1923.

fumigation system and be treated there (SMB-PK EM 1937). Five months later,
Rathgen realised that the fumigation system was too small for many objects made
of wood and too large for others made of furs, feathers, etc. He therefore decided
to submit an application for the expansion of the fumigation plant in Berlin-
Dahlem. The "Asiatische Museum" and the storage facility for MfV had several
deficiencies, including the lack of heating. Because of the large climatic fluctua-
tions, there were many dissonances among the museum experts. The staff from
the East Asian collections even categorically decided against relocating their
objects to Berlin-Dahlem until a heating system had been installed in the room
of the fumigation system (Cf. Otto 1979, 262–267). The extension of this facility for the mass fumigation of collection objects in MfV extended painstakingly
over many years from 1925 onward. After a long process, the expanded fumigation plant in Berlin-Dahlem was delivered on August 21, 1926, but Rathgen was
again not satisfied. Finally, on June 14, 1927, a note was made indicating that
the fumigation system had been completed. In the 1930s, Rathgen's successor
Brittner had to take care of the technical safety of the fumigation system and the
smooth operation for the employees there (Cf. Otto 1979, 268–275). It is difficult
to understand the fact that the control of insect pests in this plant was carried out
preventively and combatively against harmful insects until 2003 in the Ethnological Museum in Berlin-Dahlem.

Figure 11.5 Fumigation system at the Museum für Völkerkunde der Staatlichen Museen zu Berlin, open state, 1923.

Construction of plants for mass fumigation against insect pests in German and European museums

The publications, as well as the participation in conferences of well-known directors and curators, prove the constant threat of harmful insects to ethnological or natural history collections. The desire to develop rational and effective working methods for large collections was significant. It was not the individual object that needed to be preserved but the entire collection. This has led to the development of equipment for mass fumigation at the international level. First, there is evidence in Germany that in 1902, at the Royal Zoological and Anthropological-Ethnographic Museum Dresden under the German scientist and anthropologist Adolph Bernhard Meyer,[29] the first plant for mass fumigation against insect pests was built and installed. However, a letter from Bolle to Foy stated that, according to employees of the Dresden Museum in 1904 to 1905, the facility would not work effectively (HAStK 1907–1914a).

In 1874, Meyer became the director of the Naturkundemuseum (Natural History Museum) in Dresden. Five years later, the museum was renamed the Königlich Zoologisches und Anthropologisch-Ethnographisches Museum (Royal Zoological and Anthropological-Ethnographic Museum). During his 30-year term of office, it was very important to him to combine the handicraft–technical and scientific–museum work. Many of his ideas were published in the "*Publicationen aus dem Königlich Zoologischen Museum zu Dresden"*, the *"Publicationen aus dem Königlich Ethnographischen Museum in Dresden"*, and the *"Abhandlungen und Berichten aus dem Zoologischen und Anthropologisch-Ethnographischen Museum zu Dresden"*. These were groundbreaking in the museum landscape and received international attention (Maertins 2005; Martin 2005). Extensive study trips took him through Europe and North America. As a result, he introduced many technical innovations in the Dresden Museum and, together with the Kühnscherf company in Dresden, developed dust-proof collection cabinets made of glass and steel (Ibid., without page number). However, it was not possible to clarify how Meyer came to the conclusion that he had a fumigation system against insect pests built and installed in his museum to preserve the museum's collections. During the turn of the century, hygienic purposes were served by stationary and mobile "disinfection devices" that were operated using steam, hot air, or even carbon disulfide (Ernst Keil's Nachfolger 1892, 661). However, he had the mechanical engineering company Herrmann & Ranft from Dresden install the first stationary fumigation system for pest control in his museum. The fumigation system consisted of a boiler, which was closed at its front with a gas-tight lid. The boiler contained a high-pressure water pipe, a vacuum gauge, and a ventilation tap. A bottle filled with carbon disulfide was placed next to it in a container of boiling water and was connected to the cauldron. Likewise, a water separator ensured that the used water was separated from the extracted air. Water was discharged into a separate pipe. The boiler had four iron rails on which wooden strips were mounted to hold collection objects. Once the boiler was filled and sealed, the blasting apparatus[30] evacuated the air from inside the boiler and controlled the air pressure using a vacuum gauge. Once the air pressure was sufficiently reduced, boiling water was poured into the water tank to quickly evaporate the carbon disulfide inside the boiler. Subsequently, the apparatus remained untouched for a few days. The vaporous carbon disulfide was then extracted from the boiler using a blasting apparatus and fed into the atmosphere via a chimney. After this step, it was possible to reintroduce air into the boiler via the ventilation tap and safely open the lid (Meyer 1903, 22).

The enthusiasm for new technologies, as well as Meyer's involvement in museum affairs, attracted numerous museum professionals from Germany and abroad to the Dresden Museum, seeking inspiration for their own home furnishings and interior design (Sächsisches Staatsarchiv–Hauptstaatsarchiv Dresden, 13842). Among them were three museum experts from Sweden: Gunnar Hazelius in 1902,[31] Hjalmar Stolpe in 1903,[32] and Carl Bernhard Salin in 1904. The study journeys by the Swedish museum experts were significant and groundbreaking for larger European museums, as will be shown as follows.

Although the first fumigation system for art and cultural assets is in Dresden, this technology was brought to maturity in the Nordiska museet in Stockholm and patented by its inventor. From 1904 to 1912, the art historian Axel Nilsson served as a curator of the museum and was responsible for collecting folklore objects from across Sweden. The concern for the preservation of objects, which consisted mainly of organic materials, led him to the idea of improving the method of controlling insect pests with carbon disulfide, which was widely used at the time. Through his

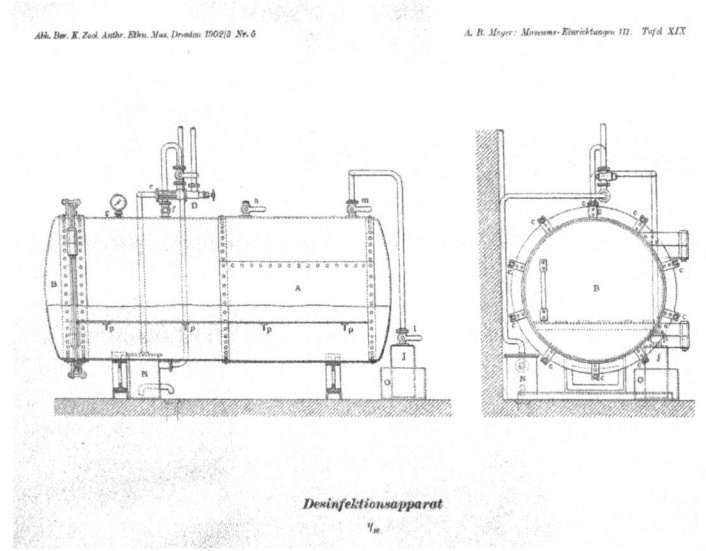

Figure 11.6 Fumigation system at the Königlich Zoologisches und Anthropologisch-Ethnographisches Museum Dresden, architectural drawing, 1903.

Figure 11.7 Museum objects in the disinfection room of the Königlich Zoologisches und Anthropologisch-Ethnographisches Museum Dresden, around 1920 (?).

colleagues' study journeys to Dresden, he also gained knowledge of the facility there. However, the peculiarity of his plant lies in its construction. His invention included a vacuum chamber with vacuum pumps connected in parallel. In contrast to the Dresden plant, this made it possible to create a larger vacuum. Once the required reduction in air pressure was achieved, carbon disulfide was introduced into the chamber. This allowed the fumes of carbon disulfide to penetrate deeper into the solid wooden structures (Brodin 2011, 57–73).

The museum's board of directors immediately took up this invention and worked hard to build the facility. The good cooperation between the Dresden and Stockholm museums is evidenced by a cost offer as well as a design drawing by the company Herrmann & Ranft from Dresden for the fumigation system they installed in the Königlich Zoologischen und Anthropologisch-Ethnographischen Museum Dresden (NMA 1903). The offer of the Swedish company Nya Aktiebolaget Atlas was realised in Stockholm. A system with a 4-meter-long and 2.8-meter-diameter cylinder, including accessories, was built and installed. Due to the high explosive tendency of carbon disulfide, a separate building next to the museum was designed for the fumigation plant. As a result, the museum management submitted an application to the Royal Building Authority in Stockholm to obtain permission to build a facility for combating insect pests behind the museum building. Its total cost was 6,000 Swedish kronor, and on May 4, 1903, the company Nya Aktiebolaget Atlas was awarded the contract for the construction of the entire plant with all the necessary accessories on the basis of the drawings and the calculated costs. At the beginning of 1904, the fumigation boiler was installed, and after lengthy tests, on April 6, 1904, they were satisfied with the functionality of the system. Carbon disulfide was supplied by Vasen Pharmacy in Stockholm at a price of 46 öre/kg (NMA 1904). Shortly thereafter, on November 25, 1905, Axel Nilsson received patent number 25800 of the class 26: c for his invention from the Royal Swedish Patent and Registration Office (Kunigl. Patent-och Registreringsamt. Patentschrift vom, November 7, 1908).

As early as April 1905, the Historical Museum in Stockholm asked whether the facility could be used for other collections. As a result, on April 12, 1907, the board of directors of the museum decided to rent out the facility for 25 crowns per filling, excluding loading and unloading. In addition, the Nordiska museet reserved the right to use the facility itself primarily for its own needs (NMA 1902–1915a). Later bottlenecks in the supply of carbon disulfide led to the museum obtaining permission to import carbon disulfide from Germany in 1915 (NMA 1915).

The collections of R-J-M in Cologne go back to the estate of the Cologne geographer and ethnologist Wilhelm Joest.[33] After his death, his sister Adele Rautenstrauch promoted the construction of his own museum, which was opened in 1906. The fact that the founding director Willy Foy was employed as an assistant at the Königlich Zoologischen und Anthropologisch-Ethnographischen Museum Dresden before his term of office may have contributed to the fact that R-J-M was one of the most innovative museums in Germany in its time regarding the control of insect pests. It has been proven that this museum is the second in Germany to have a facility for mass fumigation against insect pests since 1909. It is unclear whether the knowledge of this technology came to Cologne directly via the

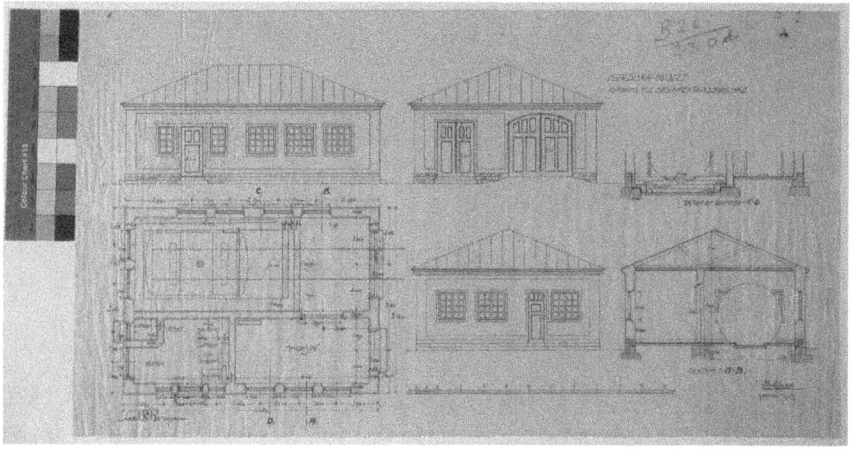

Figure 11.8 Building with fumigation system "Lusknäppen" at the Nordiska museet in Stockholm, construction drawing from October 3, 1903.

Nordiska museet or via Bolle. The plant engineer August Zerres from Cologne was entrusted with the construction of a fumigation plant and received an unspecified patent copy from the museum in 1909 (HAStK 1908–1912a). After the application for the acceptance of the plant for the Kölner Elektrizitätswerke von Foy was signed, it was put into operation two weeks later (Ibid., June 19, 1909). It consisted of small and large boilers in which the air pressure can be reduced.

Foy and Bolle began the first tests immediately after the plant was commissioned. They were ideal partners for testing the functionality of the Cologne plant as well as the effectiveness of different substances. Carbon disulfide and carbon tetrachloride were investigated, which, to date, have been used primarily in the protection of stocks (Part I, chapter 3). The museum was able to obtain these substances directly in Cologne from the wholesaler Duwalt, Korndoerfer & Co (HAStK 1909). For the first experiments, Bolle prepared test specimens made of silkworm seeds,[34] as well as oak, beech, and maple wood infested with wood-destroying insects. The aim was to test the resistance of widespread wood boring beetles (Bostrichidae) to various fumigants, as well as their penetration behavior. The corrosive effect of carbon disulfide and carbon tetrachloride was tested by adding lead paper and metal foils (HAStK 1909–1912a). This testing was expanded in museums from 1973 onward and became known as the "Oddy test" (HAStK 1907–1914b; Oddy 1973, 27–28).

Regarding the initial experiments with carbon disulfide (HAStK 1909–1912b, Foy immediately reported on April 12, 1910, expressing regret that he had not been able to fully succeed in achieving a constant low pressure (Ibid., April 12, 1912). Even before the first experiments were completed, Bolle showed great interest in disseminating the results at an anthropologists' congress in Cologne (HAStK 1907–1914c). After conducting more experiments in June 1911, Foy reported that the carbon disulfide treatment was more effective in killing larvae during the summer than during the winter. Carbon tetrachloride did not work reliably in summer

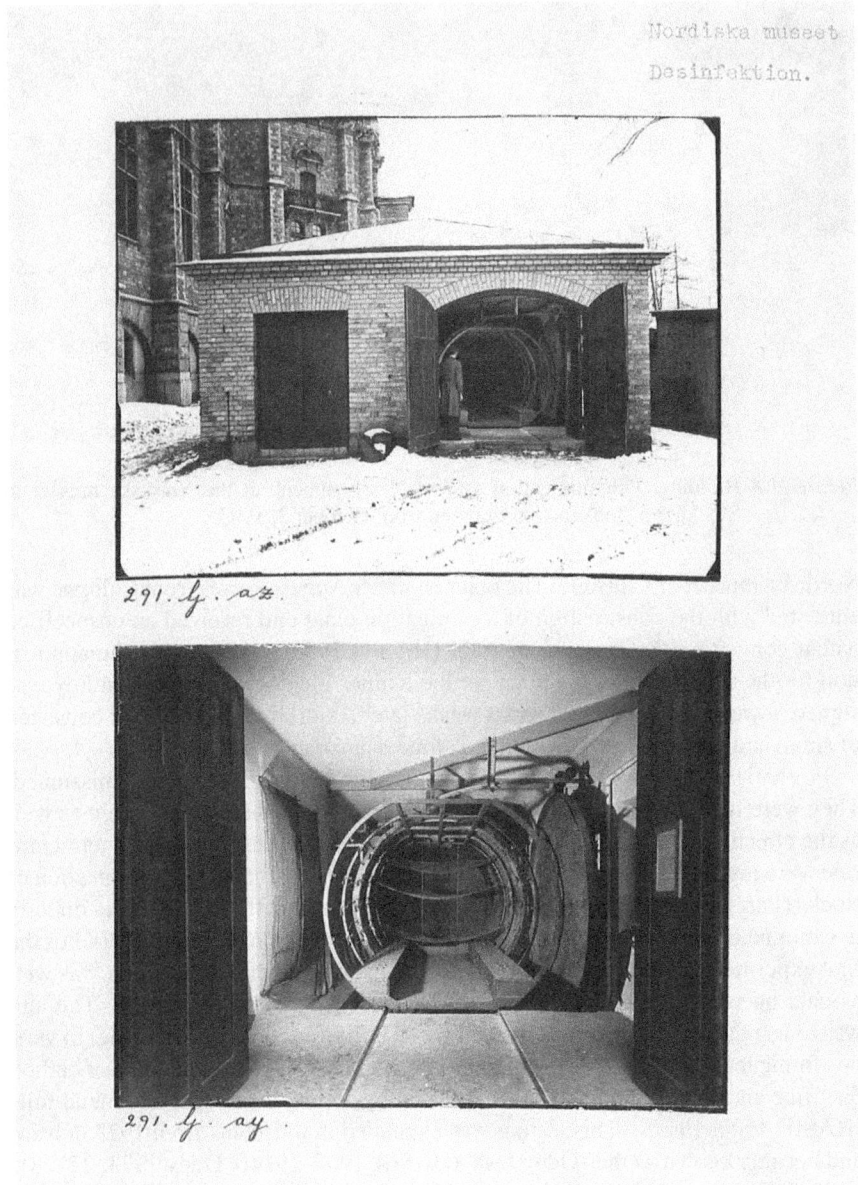

Figure 11.9 Building with fumigation system "Lusknäppen" behind the Nordiska museet in Stockholm. Open state.

or at the end of winter, which was why he refrained from using this fumigant. On the other hand, according to his observations, moth eggs were reliably killed by carbon disulfide after a three-hour exposure time, while living eggs were still found with carbon tetrachloride even after 20 hours of exposure (HAStK 1909–1912a. Bolle, for his part, subsequently evaluated the test specimens at his experimental

Figure 11.10 Disinfection room with two boilers in the Rautenstrauch-Joest-Museum in Cologne, ca. 1909.

station by examining them to determine whether any remaining eggs had developed further or whether all living organisms had been killed during the experiments in Cologne (Ibid., February 25, 1912). He also tried to arouse Foy's interest in experiments to test the material behavior of paintings on wooden panels; paintings with primers and cracked surfaces; and various types of leather, hides, or Eskimo clothing under reduced air pressure (HAStK 1907–1914d). From all the experiments carried out with Foy, Bolle concluded that carbon disulfide was preferable to carbon tetrachloride.

Foy also turned to industry and corresponded with the Chemische Fabrik Griesheim-Elektron in Frankfurt am Main. He was curious whether metal coatings were available for the collection of items that might be heat resistant to carbon tetrachloride. Griesheim-Elektron referred only to "Zapon" varnish dissolved in pentyl acetate or similar coatings and emphasised that there are no completely heat-resistant paints. In addition, because of the risk of explosiveness of carbon disulfide, the museum was looking for protection for the identification numbers of objects applied to canvas labels to be able to protect them in case of a fire. The Chemische Industrie-Gesellschaft in Frankfurt am Main also stated that there was no 100 % protection against fire, but they recommended using an unspecified impregnating agent that they produced be applied to the canvas labels (HAStK 1910–1913).

Foy and Bolle's experiments with carbon tetrachloride and carbon disulfide under various parameters carried out over a period of ten years in this plant are well documented. At different times of the year, different species of insects were introduced into various types of wood, and the interaction of the active ingredients on leaf metals and metal foils was observed. The duration of exposure times, adjustment of the pressure, and achievement of a low air pressure with regard to infested wood, its stability, and its volumes were fine-tuned. According to the current state of research, from 1909 to 1912, the most intensive research in the field of the effectiveness of carbon tetrachloride and carbon disulfide vapors was carried out in a museum using a fumigation system operating at reduced air pressure (HAStK 1909–1912b). In his retirement, Bolle continued his numerous attempts to kill insect pests using dichlorobenzene (known today as 1,4-dichlorobenzene; see glossary) and corresponded extensively with Foy between 1913 and 1917. He was also active as an intermediary between industry and museums. For example, he informed Foy about the cheapest supplier of the active ingredient dichlorobenzene, which can be found in Berlin at the Actien-Gesellschaft für Anilin-Fabrikation (Agfa) (HAStK 1907–1914e). Bolle benefited considerably from the joint research into the effectiveness of fumigants at R-J-M in Cologne. As a botanist, he was able to use the results of the experiments for agricultural and forestry purposes. At relevant conferences, such as the Anthropologists' Congress in Cologne, he acted in an interdisciplinary manner, and thus increased his level of awareness in the scientific world. In addition, he made contact with numerous experts at home and abroad with his publications in the museum community at the time. During the summer of 1911, he was given the permission to conduct his own experiments at the Nordiska museet facility in Stockholm (Ibid., July 5, 1911; September 26, 1911). Immediately afterward, he visited the aforementioned fumigation plant at the National Museum in Helsinki, which had been designed and installed according to the Swedish model (HAStK 1907–1914f). Bolle was convinced of the process and continued to spread his knowledge of this technology to many places in Germany and Europe (NMA 1902–1915b).

From 1912, the Hamburger Museum für Völkerkunde was given its own building on the Rothenbaumchaussee, in the planning and implementation of which the founding director Georg Thilenius played a key role (Thilenius 1916, 74–78). As a museum expert, he planned well ahead and inquired extensively in advance with his colleagues at the national and international levels during the planning phase of the new building to be able to design the Hamburg Museum für Völkerkunde according to the latest state of the art at the time. In September 1908, a joint study trip with the master builder M. Mayer[35] took him to Christiana (now Oslo), Stockholm, Breslau, and Dresden. These trips were supplemented by visits to other museums in Frankfurt am Main and Cologne (Ibid., 89). In Stockholm, Thilenius had the opportunity to visit the Nordiska museet, where Bolle had stayed *nota bene* three years earlier. Thilenius took a close look at Nilsson's facility and pushed ahead with the construction of a similar apparatus for his museum in close scientific exchange with the Nordic Museum in Stockholm and R-J-M in Cologne. The construction of the fumigation plant was associated with considerable costs. Therefore, immediately after the trip, Thilenius contacted his colleagues at the Kunstgewerbemuseum (Museum of Arts and Crafts), the Naturkundemuseum (Natural

History Museum), and the Historisches Museum (Museum of History) in Hamburg to inquire about the need for joint use of a fumigation system (Archiv des Museums am Rothenbaum–Kulturen und Künste der Welt 1908).

The plant was then built by the August Herrmann company in Dresden (Cf. Thilenius 1916, 152). Based on his own observations, Thilenius ordered the "Vakuumanlage zur Desinfektion von Sammlungsgegenständen mit Maschinenkammer, Luftpumpe und Ventilator im Laboratorium des Museums" in such a way that meaningful workflows were created (Ibid., 131). The plant consisted of a boiler, 2 m high and 3 m long, with a removable rack for picking up the objects. By the end of 1911, he urged the installation of the fumigation system. According to a letter from the Hamburg building department, the entire apparatus was scheduled to be shipped on November 18, 1911, and installed in the new building of the Museum für Völkerkunde after a 14-day delay (Archiv des Museums am Rothenbaum–Kulturen und Künste der Welt 1911a).

Like Foy, Thilenius had to choose one of the two substances, carbon tetrachloride and carbon disulfide, to operate the gassing system. The museums did not have extensive experience with their use. As already mentioned, the vapors of these liquids have been used mainly to protect supplies. Thilenius first asked Rathgen, but was immediately referred by him to Foy in Cologne (Archiv des Museums am Rothenbaum–Kulturen und Künste der Welt 1911b). At that time, he had already accumulated two years of experience working with both active ingredients in his own plant and was able to promptly address all questions asked of him (Archiv des Museums am Rothenbaum–Kulturen und Künste der Welt 1912a). He considered carbon tetrachloride to be a very unsafe fumigant, because in experiments, even after a 48-hour exposure time, living larvae were found in the boiler. For this reason, Foy rated carbon tetrachloride as a less effective ingredient than carbon disulfide, even with longer exposure times. He modestly described his own experiments as being personal experiments, the scientific validity of which he questioned. Nevertheless, he hoped to invest more time in experiments in the near future to arrive at more meaningful results (Archiv des Museums am Rothenbaum–Kulturen und Künste der Welt 1912b, 1912c). He compared the disadvantageous fire hazard of carbon disulfide with the probability of a lightning strike and emphasised that, if used in an orderly manner, the danger to a museum would be rather minor. Another important indication for Thilenius was the required amount of gas per cubic meter boiler volume. At a temperature of 16–18 °C, approximately 650–700 cm^3 of carbon tetrachloride was needed for saturation for 1 m^3 of boiler volume. Foy identified Duwald, Korndörfer & Co. from Cologne as the source of supply for carbon tetrachloride. However, he was hesitant about placing larger orders of this substance until its effectiveness had been satisfactorily established (Archiv des Museums am Rothenbaum–Kulturen und Künste der Welt 1912a). Therefore, it was necessary to decide whether a less effective but nonflammable substance, such as carbon tetrachloride, or a more effective active ingredient but flammable, such as carbon disulfide, should be used as a fumigant.

Thilenius relied entirely on the experience of his colleague and considered the fire hazard of carbon disulfide to be analogous to that of petrol or ether. Foy advised his Hamburg colleague to consistently use the process with carbon disulfide. Despite his attempts to kill harmful insects without using gases by creating a negative pressure, he considered his efforts to have failed because there were still live drill worms[36] in the cauldron after six days (Archiv des Museums am Rothenbaum–Kulturen und Künste der Welt 1912c). This information was extremely valuable for Thilenius, as he was now able to justify the advantages of carbon disulfide over his building authority in Hamburg. The authority had attempted to dissuade him from this active ingredient, unaware of its pros and cons (Archiv des Museums am Rothenbaum–Kulturen und Künste der Welt 1912b).

An important piece of evidence concerning the colonial context of ethnological collections was a letter from South Africa.[37] Thilenius was asked about the use of the agent "Areginal" to control wood pests. Because he was not aware of the agent, he requested the Pharmaceutical Office of I.G. Farbenindustrie A.G. for literature on this wood preservative. In addition, he learned that "Areginal" was suitable for fumigation in his plant. Unlike carbon disulfide, the colorless liquid has a pleasant smell (Archiv des Museums am Rothenbaum–Kulturen und Künste der Welt 1934a). The conversion of the fumigation system, which had already been planned in 1933, was carried out, and the new fumigant "Areginal" was used. On May 14, 1934, Bayer-Meister Lucius from Leverkusen supplied the museum with 5 kg of "Areginal" for the plant's further operation (Archiv des Museums am Rothenbaum–Kulturen und Künste der Welt 1934b).

Notes

1 The Chemisches Laboratorium was founded in 1888.
2 The conservation of ancient findings (translation by the author).
3 According to Rathgen, these colors did not contain gold, but they consisted only of bronze (translation by the author). [see Rathgen (1911): Rathgen, Friedrich. Mitteilungen aus dem Chemischen Laboratorium der Königlichen Museen zu Berlin. VIII. Über die Verwendung von Tetrachlorkohlenstoff zur Abtötung von tierischen Schädlingen. In: *Museumskunde, Zeitschrift für Verwaltung und Technik öffentlicher und privater Sammlungen*, 1911, Bd. VII, 219].
4 At that time, August Eichhorn was a research auxiliary worker. The term "auxiliary worker" corresponds to the current term "research assistant" (author's note).
5 For his experiments, Rathgen used the alcohol from Krause, his sublimate solution, liquid carbon disulfide from the museum, and ordinary white feathers. Neither the bast nor the feathers turned brown in combination with the liquids. At the same time, he had three white chicken feathers hung in room no. 40 of the museum for a few days. For this experiment, one of the feathers was moistened previously with the alcohol used by Krause. This was carried out to investigate whether air contaminated with carbon disulfide had caused the browning of the feathers. Despite a repetition, this attempt was also negative. No proof could be provided in this way, and Rathgen suspected that another liquid was accidentally used to impregnate the feather shields. He also did not rule out the effect of other gases after impregnation with the sublimate solution as the cause of the now browned feathers. As an indication of this, he saw the white feathers on the feather shield, the exposed parts of which had only become brownish. Rathgen believed that this was the outcome of a chemical reaction, leading to the conversion of sublimate (mercury(II) chloride) into sulfur mercury (mercury sulfide). However, because feathers inherently contain sulfur, nondestructive detection would not have been possible at this point. In addition, the mercury sulfide and sulfur of the feather would have been destroyed, so ultimately the proof could not be provided whether the sulfur came from the feathers or from the mercury sulfide. In the final experiment, Rathgen

used a highly diluted hydrogen sulfide (hydrogen sulfide). He soaked a white feather in a 2% sublimate solution and allowed it to sit in hydrogen sulfide for weeks in a humid environment. This resulted in a brown coloration, which was weaker than that of the feathers of the shield.

6 Johann (Giovanni) Bolle, born on January 16, 1850 and died on September 2, 1924. He was born in Slovenia and graduated from the chemical–technical college in Graz from 1867 to 1870. From 1871, he worked as a botanist at the Imperial Royal Silk Experimental Station in Gorizia and became its director in 1880. His special achievements include the control of insect pests in viticulture and fruit growing, as well as research into the rational rearing of silkworms.

7 It is questionable whether the procedure recommended by Krause could be successfully applied, as water and the agents mentioned do not mix (author's note).

8 Written communication dated December 11, 2014, from Magdalena Drexl, Stiftung Ruhr Museum Essen, Frühe Neuzeit.

9 The administrative director Josephin (first name unknown, author's note) from the Grossherzoglichen Museum and the Grossherzoglichen Kunstsammlungen (Grand Ducal Museum and the Grand Ducal Art Collections) Schwerin was responsible for the tomb in Doberan Minster at that time.

10 The former Kaiser-Friedrich Museum was renamed the Bode Museum during the GDR regime in 1956.

11 First name unknown (author's note).

12 Theodor Demmler was commissionary director of the collection of sculptures and plaster casts of the Middle Ages and the Renaissance.

13 Otto von Falke was director of the Kunstgewerbemuseum (Museum of Decorative Arts).

14 Presumably, this is Aloys Hauser jun., born on May 14, 1886 and died after 1934. He was chief conservator and curator at the art museums. (Friendly oral communication from the Zentralarchiv of the Staatliche Museen zu Berlin, April 18, 2017).

15 Sidney Harmer was an entomologist and director of the Natural History Museum in London from 1919 to 1927.

16 The city of Yekaterinburg was called Sverdlozk from 1924 to 1992.

17 Otto Kümmel was director of the Asian collections at the SMB from 1928 to 1933.

18 Carl Bernhard Salin was director of the Nordiska museet in Stockholm from 1905 to 1913.

19 He recommended coating both the flesh and fur sides with mixtures of sodium arsenite, phenol, and water. For the size of a tiger skin, one needed 3–4 gallons (about 13–18 liters, author's note). To cure the skins, he advised using a self-prepared solution, consisting of alum, carbolic acid (phenol), and water. According to his observations, this solution should not be given "formalin" in any case, as the skins become hard and brittle.

20 Eduard Seler, born in 1849 and died in 1922, was the director at the American department at KMfV in Berlin from 1904 to (?).

21 Konrad Preuss was at that time a scientific assistant at the American department.

22 Albert Grünwedel, born in 1856 and died in 1935, was the director of the East Asian and Indian department in the KMfV from 1904 to 1921, and in MfV in Berlin from 1918.

23 Since 1978, a (physical) atmosphere is no longer a legal unit of pressure in Germany. Today, 1 atm = 101,325 Pascal, or 1,013.25 hPa. Johann Bolle has (probably unconsciously) made incorrect statements here, because the air pressure of 1 to 2 atm corresponds to the normal air pressure (1 atm) and twice increased air pressure (2 atm). In the gassing system, a lower pressure was generated, i.e. the air pressure was less than 1 atm (1,013.25 hPa).

24 Today, 32,716 Finnish marks are equivalent to around 5,502 €.

25 Willy Foy, born on November 27, 1873, and died on March 1, 1929.

26 Poisoning room (translation by the author).

27 The name "Asiatisches Museum" was retained even after the change of use to a storage building in the building files and in related correspondence.

28 Apparatus (translation by the author).

29 Adolph Bernhard Meyer, born on October 11, 1840 and died on February 5, 1911.
30 The blasting apparatus probably refers to a blasting pump or, more precisely, a water blasting pump. The invention of the water blasting pump is attributed to the chemist Robert Bunsen (see: Wittenberger (1973): Wittenberger, Walter. Chemische Laboratoriumstechnik. 7. Auflage. Springer, Wien 1973, 258–259).
31 Gunnar Hazelius was director of the Skansen open-air museum in Stockholm from 1901 to 1905.
32 Hjalmar Stolpe was director of the Ethnographic Museum in Stockholm from 1903 to 1905.
33 Wilhelm Joest, born on March 15, 1852, and died on November 25, 1897.
34 Silkworm eggs, which were referred to as seeds by the farmers at that time (author's note).
35 First name unknown (author's note).
36 Wood-destroying insects (author's note).
37 Geographical indication without further information (author's note).

References

Archive material

Archiv des Museums am Rothenbaum–Kulturen und Künste der Welt (1908): Findbuch. 101–1, Nr. 1230. Loseblattsammlung. Anfrage zur gemeinsamen Nutzung der Begasungsanlage am Museum für Völkerkunde Hamburg. Thilenius, Georg. Brief vom 09.10.1908, 2 Seiten, ohne Paginierung.

Archiv des Museums am Rothenbaum –Kulturen und Künste der Welt (1911a): Findbuch. 101–1, Nr. 1230. Loseblattsammlung. Anlieferung der Begasungsanlage für das Museum für Völkerkunde Hamburg. Thilenius, Georg. Brief vom 28.10.1911, 1 Seite, ohne Paginierung; Elkart (First name unknown, author's note). Brief vom 11.11.1911, 1 Seite, ohne Paginierung.

Archiv des Museums am Rothenbaum–Kulturen und Künste der Welt (1911b): Findbuch. 101–1, Nr. 1230. Loseblattsammlung. Klärung von Fachfragen zum Einsatz von Gasen in der Begasungsanlage des Museums für Völkerkunde in Hamburg. Thilenius, Georg. Brief vom 09.12.1911, 1 Seite, ohne Paginierung.; Rathgen, Friedrich. Brief vom 11.12.1911, 1 Seite, ohne Paginierung.

Archiv des Museums am Rothenbaum–Kulturen und Künste der Welt (1912a): Findbuch. 101–1, Nr. 1230. Loseblattsammlung. Foy, Willy. Brief an Georg Thilenius vom 11.01.1912, 6 Seiten, ohne Paginierung.

Archiv des Museums am Rothenbaum–Kulturen und Künste der Welt (1912b): Findbuch. 101–1, Nr. 1230. Loseblattsammlung. Thilenius, Georg. Brief an Willy Foy vom 13.01.1912, 1 Seite, ohne Paginierung.

Archiv des Museums am Rothenbaum–Kulturen und Künste der Welt (1912c): Findbuch. 101–1, Nr. 1230. Loseblattsammlung. Foy, Willy. Brief an Georg Thilenius vom 17.01.1912, 2 Seiten, ohne Paginierung.

Archiv des Museums am Rothenbaum–Kulturen und Künste der Welt (1934a): Findbuch. 101–1, Nr. 1230. Loseblattsammlung. Thilenius, Georg. Brief vom 17.01.1934, 1 Seite, ohne Paginierung.

Archiv des Museums am Rothenbaum–Kulturen und Künste der Welt (1934b): Findbuch. 101–1, Nr. 1230. Loseblattsammlung. Umstellung der Begasungsanlage im Museum für Völkerkunde Hamburg von Blausäure auf "Areginal". Thilenius, Georg. Brief vom 17.04.1934, 1 Seite, ohne Paginierung.; Lieferung von 5 kg "Areginal". Anonymus. Aktenvermerk vom 14.05.1934, 2 Zeilen, ohne Paginierung.

Archiv des Museums für regionale Überlieferung Jekaterinburg (1928a): Serie 1928, Nr. 3026, Brief Nr. 356/28. Kümmel, Otto. Antwortschreiben zur Schädlingsbekämpfung mit "Areginal" und "Eulan" vom 28.4.1928, Blatt 13, 1 Seite.

Archiv des Museums für regionale Überlieferung Jekaterinburg (1928b): Serie 1928, Nr. 3024, Brief Nr. 1063. Bitte an V. K. Klein, (Vorname unbekannt, Anm. d. Verf.) vom 17.05.1928 ein Gutachten vom Staatlichen Museum für Völkerkunde zu Berlin über "Eulan" einholen zu dürfen. Blatt 74, 1 Seite.

Archiv des Museums für regionale Überlieferung Jekaterinburg (1929a): Schulz, L. (Vorname unbekannt, Anm. d. Verf.). Antrag vom 26.08.1929 an die Wissenschaftliche Abteilung in Moskau auf Bestellung von 5 kg "Eulan", V 545, Blatt 28, 1 Seite.

Archiv des Museums für regionale Überlieferung Jekaterinburg (1929b): Brief Nr. 356/28, Nr. 67. Kümmel, Otto. Gutachten vom 01.11.1929 für das Urallandesmuseum Jekaterinburg zu "Eulan". Blatt 12, 2 Seiten.

FHA (1912): SKMBT_C36015062509560. Hackman, Alfred. Briefentwurf vom 02.02.1912, 4 Seiten, ohne Paginierung.

HAStK (1907–1914a): Findbuch. Best. 614, A 88, Loseblattsammlung. Wissenstransfer Desinfektionsapparat. Bolle, Johann. Brief an Willy Foy vom 29.03.1910, Blatt 9–10, 4 Seiten.

HAStK (1907– 1914b): Findbuch. Best. 614, A 88, 1907–1914. Loseblattsammlung. Bolle und Foy 1907– 1914. Bolle, Johann. Brief vom 30.03.1910, Blatt 8, 1 Seite.

HAStK (1907–1914c): Findbuch. Best. 614, A 88, Loseblattsammlung. Bolle, Johann. Brief vom 29.03.1910, Blatt 10, 1 Seite.

HAStK (1907–1914d): Findbuch. Best. 614, A 88, Loseblattsammlung. Bolle, Johann. Brief vom 01.09.1910, Blatt 48, 2 Seiten.

HAStK (1907–1914e): Findbuch. Best. 614, A 88, Loseblattsammlung. Bolle, Johann. Postkarte vom 23.10.1913, Blatt 118, 2 Seiten.

HAStK (1907–1914f): Findbuch. Best. 614, A 88, Loseblattsammlung. Bolle, Johann. Postkarte vom 14.08.1911, Blatt 68, 2 Seiten.

HAStK (1908–1912a): Findbuch. Best. 614, A 404. Loseblattsammlung. Zerres, August. Brief vom 03.06.1909, Blatt 5, 1 Seite.

HAStK (1908–1912b): Findbuch. Best. 614, A 404. Loseblattsammlung. Zerres, August. Brief vom 19.06.1909, Blatt 9, 1 Seite.

HAStK (1909): Best. 614, A 70. Loseblattsammlung. Ankauf von Insektenvertilgungsmitteln. Firma E. Merck Darmstadt. Brief vom 22.09.1909, Blatt 11, 1 Seite

HAStK (1909–1912a): Findbuch. Best. 614, A 73, Loseblattsammlung. Resultate der Versuche zur Begasung. Foy, Willy. Bericht vom 08.06.1911, Blatt 106–107, 3 Seiten.; Best. 614, A 88, 1907–1914. Loseblattsammlung. Bolle, Johann. Brief vom 06.06.1911, Blatt 57, 2 Seiten.

HAStK (1909–1912b): Findbuch. Best. 614, A 73, 1909–1912. Loseblattsammlung. Dokumentation der Versuchsreihen von Willy Foy ohne Datierung, Blatt 11–30, 29 Seiten; Darin enthalten: Bolle, Johann. Postkarte vom 25.03.1912, Blatt 12, 1 Seite. Darin enthalten: Bolle, Johann. Brief vom 25.02.1912, Blatt 151–152, 3 Seiten.

HAStK (1909–1912c): Findbuch. Best. 614, A 73. Loseblattsammlung. Dokumentation der Versuchsreihen von Willy Foy, ohne Datierung, Blatt 11–30, 29 Seiten

HAStK (1910–1913): Findbuch. Best. 614, 438, Loseblattsammlung. Chemische Fabrik Griesheim-Elektron. Stellungnahme zu hitzebeständigen Anstrichen. Brief vom 25.06.1910, Blatt 1, 1 Seite.; Ebd. Chemische Fabrik Griesheim-Elektron und Chemische Industrie-Gesellschaft. Ankauf von Imprägniermitteln. Brief vom 08.12.1913, Blatt 2, 2 Seiten sowie Brief vom 15.11.1913, Blatt 3, 1 Seite.

LAKD (1885–1953): M-V/LD Objektakte (Kopie) Bad Doberan, Klosterkirche (Münster). Loseblattsammlung. Mappe 01 vom 21.05.1885–27.06.1927, Mappe 02 vom

01.07.1927–14.12.1953, Nr. 1069 und N r. 34524. Verwaltung des Grossherzolgichen Museums und der Grossherzoglichen Kunstsammlungen Schwerin 13.07.1912–25.10.1933. Erhaltung des Grabdenkmals des Herzoglichen Geheimen Rats Graf Samuel von Behr in der Kirche zu Doberan. Rathgen, Friedrich. Brief vom 19.08.1912, zu Nr. 156, Blatt 33, 4 Seiten.

LAKD (1912a): M-V/LD Objektakte (Kopie) Bad Doberan, Klosterkirche (Münster). Loseblattsammlung. Josephin, (Vorname unbekannt, Anmerk. d. Verf.). Briefentwurf vom 31.07.1912, 1 Seite, ohne Paginierung und Bericht vom 15.08.1912, Blatt 34, 2 Seiten.

LAKD (1912b): M-V/LD Objektakte (Kopie) Bad Doberan, Klosterkirche (Münster). Loseblattsammlung. Rathgen, Friedrich. Bericht zu konservatorischen Maßnahmen zum Erhalt des Grabdenkmals des Herzoglichen Geheimen Rats Graf Samuel von Behr in der Kirche zu Doberan vom 01.11.1912, Blatt 39, 4 Seiten.

LAKD (1933): M-V/LD Objektakte (Kopie) Bad Doberan, Klosterkirche (Münster). Loseblattsammlung. G. Nr. 2825. Josephin (First name unknown, author's note). Schreiben an den Regierungsbaurat Neumann (First name unknown, author's note) vom 30.10.1933, 1 Seite, ohne Paginierung.

NMA (1902–1915a): A 1 A, Nämndens protokoll, Nordisches Museum Stockholm. Bau einer Begasungsanlage. Nr. 8, § 62, Protokollnotiz vom 05.04.1905; Nr. 10, Protokollnotiz vom 12.04.1907, (Übersetzung aus dem Schwedischen von Karin Björling-Olausson).

NMA (1902–1915b): A 1 A, No. 14, § 138. Nordiska museet Stockholm. Nämndens protokoll, Besichtigung der Begasungsanlage im Nordischen Museum Stockholm durch Johann Bolle, protokolliert am 11.08.1911, (Translation from Swedish by Karin Björling-Olausson).

NMA (1902–1915c): A 1 A. Nordisches Museum Stockholm, nämndens protokoll, No. 14. § 138. Besuch von Johann Bolle, niedergeschrieben am 16.08.1911, (Translation from Swedish by Karin Björling-Olausson).

NMA (1903): Ämbetsarkiv, Styresmannens valv. Loseblattsammlung. Herrmann & Ranft. Kostenvoranschlag für eine Begasungsanlage. Brief vom 03.03.1903, 3- seitig, ohne Paginierung.

NMA (1904): A 1 A, Nämndens protokoll, Nr. 7, § 54d. Protokollnotiz vom 06.04.1904.; Ämbetsarkivet—1963, Verifikationer N M, Bilagor till räkenskaperna 1903, G 7 AA: 40; 1904. G 7 AA: 41, (Übersetzung aus dem Schwedischen von Karin Björling-Olausson).

NMA (1915): A 1 A, Nämndens protokoll, Nr. 18, § . . . j, Protokollnotiz vom 18.08.1915, (Übersetzung aus dem Schwedischen von Karin Björling-Olausson).

SMB-PK EM (1905a): I/MV 0075, Band 1, Pars IIc, E. Nr. 332/1905. Acta betreffend die Restauration von Alterthümern. Loseblattsammlung. Konservierung von archäologischem Holz. Bormann, (First name unknown, author's note). Brief vom 19.02.1905, 1- seitig, ohne Paginierung.

SMB-PK EM (1905b): I/MV 0075, Band 1, Pars IIc, E. Nr. 332/1905. Acta betreffend die Restauration von Alterthümern. Loseblattsammlung. Konservierung von archäologischem Holz. Krause, Eduard. Bericht vom 21.2.1905, 2 Seiten, ohne Paginierung.

SMB-PK EM (1911): I/MV 0075, Band 1, Pars IIc, E. Nr. 301/11. Acta betreffend die Restauration von Alterthümern. Loseblattsammlung. Krause, Eduard. Bericht vom 15.02.1911, 1 Seite, ohne Paginierung.

SMB-PK EM (1911–1914): I/MV 0075, Pars II c, Vol. 1, E. Nr. 1771/11. Königliche Museen zu Berlin. Weiterleitung einer Anfrage von Konrad Theodor Preuß durch Eduard Seler vom 04.11.1911, 1 Spalte, ohne Paginierung.

SMB-PK EM (1912a): I/MV 0075, Pars II c, Vol. 1, E. Nr. 1771/11. Bolle, Johann. Brief vom 28.01.1912, 4 Seiten, ohne Paginierung.

SMB-PK EM (1912b): I/MV 0075, Pars II c, Vol. 1, E. Nr. 1771/11. Hackman, Alfred. Brief vom 08.02.1912, Blatt 4, 3 Seiten.

SMB-PK EM (1912c): I/MV 0075, Pars II c, Vol. 1, E. Nr. 1771/11. Krause, Wilhelm Eduard Julius. Aktennotiz vom 12.02.1912 auf Brief von Alfred Hackman vom 12.02.1912, Blatt 4, 1 Seite.

SMB-PK EM (1912d): I/MV 0075, Pars II c, Vol. 1, E. Nr. 1771/11. Grünwedel, Albert. Aktennotiz vom 24.02.1912, 1 Seite, rechte Spalte, ohne Paginierung.

SMB-PK EM (1912e): I/MV 0075, Pars II c, Vol. 1, E. Nr. 467/12. Planungen zur Errichtung einer Begasungsanlage in Berlin-Dahlem. Krause, Eduard. 2 Aktennotizen vom 19.03.1912, 1 Seite, rechte Spalte, ohne Paginierung.; Firmenprospekt der Apparate-Bauanstalt Christ & Co. (Inh. Gustav Necker, Ingenieur), ohne Datierung, 4 Seiten, ohne Paginierung.

SMB-PK EM (1912f): I/MV 0075, Pars II c, Vol. 1, E. Nr. 467/12. Strebe (Vorname unbekannt, Anmerk. d. Verf.). Aktennotiz vom 04.03.1918, 1 Seite, linke Spalte, ohne Paginierung.

SMB-PK EM (1912/1913): I/MV 0075, Pars II c, Vol. 1, E. Nr. 345/12. Versuche zur Schädlingsbekämpfung im Königlichen Museum für Völkerkunde zu Berlin mit der kaiserlich königlichen landwirtschaftlich-chemischen Versuchsstation in Görz. Krause, Eduard. 2 Aktennotizen von Eduard Krause vom 27.02.1912 und vom 08.03.1913, 1 Seite, ohne Paginierung.; Grünwedel, Albert. 2 Randnotizen vom 27.02.1912 und vom 08.03.1912; 1 Seite, ohne Paginierung.

SMB-PK EM (1914a): I/MV 0075, Pars II c, Vol. 1, E. Nr. 1771/11. Grünwedel, Albert. Schriftliche Bitte an die Generalverwaltung der Königlichen Museen zu Berlin vom 24.02.1914, geschrieben auf Brief von Willy Foy vom 21.02.1914, 2 Spalten, ohne Paginierung.

SMB-PK EM (1914b): I/MV 0075, Pars II c, Vol. 1, E. Nr. 1771/11. Stubenrauch, Kurt. Aktennotiz und Stellungnahme vom 15.05.1914, 1 Seite, linke Spalte, ohne Paginierung.

SMB-PK EM (1914c): I/MV 0075, Pars II c, Vol. 1, Nr. 25/14. Bau und Aufstellung eines Begasungskastens im Königliches Museum für Völkerkunde zu Berlin 1914. Eichhorn, August. Bericht vom 07.01.1914, 1 Seite, ohne Paginierung.

SMB-PK EM (1914d): I/MV 0075, Pars II c, Vol. 1, Nr. 25/14. Grünwedel, Albert. Aktennotiz vom 08.01.1914, 1 Seite, linke Spalte, ohne Paginierung.

SMB-PK EM (1914e): I/MV 0075, Pars II c, Vol. 1, Nr. 25/14. Grünwedel, Albert. Aktennotiz vom 09.01.1914, 1 Seite, rechte Spalte, ohne Paginierung.

SMB-PK EM (1914f): I/MV 0075, Pars II c, Vol. 1, Nr. 25/14. Grünwedel, Albert. Aktennotiz vom 24.01.1914, 1 Seite, linke Spalte, ohne Paginierung.

SMB-PK EM (1914g): Bürojournal des Königlichen Museums für Völkerkunde zu Berlin. Laufende Nummer 324, Eichhorn, August. Anzeige vom 05.09.1914 gegen den Konservator Eduard Krause sowie Löschung des Eintrags. 1 Tabelle mit 2 Einträgen, 1 Spalte.

SMB-PK EM (1914h): I/MV 932, zu E. Nr. 324/14, Königliche Museen zu Berlin 1914. Personalakte Eduard Krause von 1914. Rathgen, Friedrich. Bericht vom 24.03.1914, 3 Seiten, ohne Paginierung.

SMB-PK EM (1923): I/MV 0075, Pars II c, Vol. 1, E. Nr. 1771/11. Genehmigung zur Errichtung einer Begasungsanlage in Berlin-Dahlem. Anonymus. Aktennotiz vom 06.01.1923, Blatt 6, 2 Zeilen.

SMB-PK EM (1928): I/MV 0075, E. Nr. 356/28. Acta betreffend die Restauration von Alterthümern. Loseblattsammlung. Urallandesmuseum Jekaterinburg. Anfrage zu Areginal und Eulan. Brief vom 19.03.1928, 1-seitig, ohne Paginierung.

SMB-PK EM (1937): I/MV 1071, I B 129, auf E. Nr. 292/37. Snethlage, Heinrich. 3 Notizen vom 03.09.1937, 15.11.1937 und 15.12.1937, 1 Seite, ohne Paginierung.

SMB-ZA (1913/1914): KFM 37, F. Nr. 3149/1913, Loseblattsammlung. Anfrage des Konservators der Denkmäler im Regierungsbezirk Cassel an das Kaiser Friedrich Museum in Berlin zur Bekämpfung holzzerstörender Insekten an gefassten Holzskulpturen. Holtmeyer (First name unknown, author's note). Brief vom 19.12.1913, 2 Seiten, ohne Paginierung.; Demmler, Theodor. Brief (Abschrift) vom 08.01.1914, 1 Seite, ohne Paginierung.; Hauser, (First name unknown, author's note). Aktennotiz (Abschrift) vom 15.01.1914, 1 Seite, 7 Zeilen, ohne Paginierung; Rathgen, Friedrich. Stellungnahme vom 16.01.1914, 1 Seite, ohne Paginierung.

SMB-ZA (1914; 1922–1923; 1928): I/BV 239. Staatliche Museen zu Berlin – Preußischer Kulturbesitz, Ethnologisches Museum.

SMB-ZA (1915; 1919; 1922–1924): I/BV 286. Staatliche Museen zu Berlin – Preußischer Kulturbesitz, Ethnologisches Museum.

SMB-ZA (1924–1928): I/BV 723. Einrichten von Arbeitsräumen für den Konservator und Einbau einer Begasungsanlage. Umbau, Einrichtung, Instandhaltung – Gebäudekomplex Dahlem.

SMB-ZA (1933): I/I M 26, Loseblattsammlung. Empfehlung von "Xylamon". Hübner, Paul H. Brief vom 27.03.1933, Blatt 10, 1 Seite.; Consolidierte Alkaliwerke Abteilung Hannover und Hübner. Brief vom 31.05.1933, Blatt 8, 1 Seite.; Broschüre zu "Xylamon", Blatt 9, 2 Seiten.; Prospekt Nr. 41, Richtlinien zur Holzwurmbekämpfung durch "Xylamon", Blatt 10, 12, 4 Seiten.

SMB-ZA (1935): I/GV 1399, Beleg zur Verwaltungsrechnung für das Rechnungsjahr 1935, Nr. 461–550. Kap. 154. Tit. 26. Loseblattsammlung. Bayer I.G. Farbenindustrie Aktiengesellschaft Rechnungsjahr 1935. Rechnung vom 15.1.1936 an das Chemische Laboratorium der Staatlichen Museen zu Berlin für "Areginal", 1 Seite, ohne Paginierung.

Other sources

Anonymous (1913b): Personalnachrichten. *Zeitschrift für das Landwirtschaftliche Versuchswesen in Österreich*, (XVI), 39–40.

Anonymous (1916): The Technical Preservation of Antiquities. *The Museums Journal*, (15), 269.

Bolle, Johann (1882): Die Mittel zur Bekämpfung der Reblaus (Phylloxera vastatrix). Görz: Verlag der k.k. Seiden- und Weinbau-Versuchsstation.

Bolle, Johann (1892): Ausführliche Anleitung zur rationellen Aufzucht der Seidenraupe. Berlin: Gramsch.

Bolle, Johann (1898): Der Seidenbau in Japan. Budapest: Hartleben.

Bolle, Johann (1899): Der Seidenspinner des Maulbeerbaumes, seine Aufzucht seine, Krankheiten und die Mittel zu ihrer Bekämpfung. Vortrag. Wien: Selbstverl. Vorträge des Vereins zur Verbreitung naturwissenschaftlicher Kenntnisse in Wien. XXXIX. Jahrgang, Heft 4.

Bolle, Johann (1919): Die Ermittlung der Wirksamkeit von insektentötenden Mitteln gegen die Nagekäfer des verarbeiteten Werkholzes. *Zeitschrift für angewandte Entomologie*, Bd. 5, 105–117.

Bolle, Johann; Mewis, F.A. (1892): Ausführliche Anleitung zur rationellen Aufzucht der Seidenraupe. Berlin: A. Gramsch.

Bracchi, Eva (2013): Der erste Chemiker in Sachen Kunst. *Jahrbuch Preußischer Kulturbesitz*, 49.

Bracchi, Eva (2014): Friedrich Rathgen, Pionier der modernen archäologischen Restaurierung. *Berliner Beiträge zur Archäometrie, Kunsttechnologie und Konservierungswissenschaft*, (22), 6.

Brodin, Louise (2011): Nordiska Museet, Stockholm. "Antagen enhälligt, stop, börjar tjugofemte". In: Axel Nilsson, Museiman och föregangare. Sävedalen: Warne.

Bundesanstalt für Arbeitsschutz und Arbeitsmedizin (Februar 2010): Technische Regel für Gefahrstoffe 524 Schutzmaßnahmen bei Tätigkeiten in kontaminierten Bereichen, TRGS 524. In: Gemeinsames Ministerialblatt, Februar 2010.

Delépine, Sheridan (1914): On the Arsenious Acid-Glycerin-Gelatin ("arsenious jelly") Method of Preserving and Mounting Pathological Specimens with Their Natural Colors, and on the Use of New Forms of Receptacles for Keeping Museum Specimens. From the Public Health Department, University of Manchester. *The Museums Journal*, (13), 322–329.

Ernst Keil's Nachfolger (1892): Eine fahrbare Desinfektionsanlage. Abbildung. *Die Gartenlaube*, Heft 21, 661.

Harmer, Sidney (1922): The Restoration and Preservation of Objects at the British Museum. In: *Journal of the Royal Society of Arts*, LXX(3618), 333–334.

Maertins, Katharina (2005): Rathgen-Forschungslabor. Unveröffentlichte Quelle. Rathgen-Forschungslabor der Staatlichen Museen Berlin.

Martin, Petra (2005): Adolph Bernhard Meyer. Hg. v. Institut für Sächsische Geschichte und Volkskunde e.V. (Sächsische Biographie). Ohne Seitenangabe. Online verfügbar unter www.isgv.de/saebi.

Meyer, Adolph Bernhard (1903): Abhandlungen und Berichte des Königlichen Zoologischen und Anthropologisch-Ethnographischen Museums zu Dresden. 3. Bericht über einige Neue Einrichtungen des Königlichen Zoologischen und Anthropologisch-Ethnographischen Museums in Dresden. XI. Desinfektionsapparat. 1902/03, Bd. X, (5), 22.

Meyer, Andrea (2014): The Journal Museumskunde—"Another Link between the Museums of the World". *The Museum Is Open: Towards a Transnational History of Museums 1750–1940*, 179. DOI: 10.1515/9783110298826.

Nemecek, Natasa (2013): Friedrich Rathgen and his impact on Slovenian conservation in the beginning of the twentieth century. *CeROArt [En ligne]*. Available online at http://ceroart.revues.org/3686.

Nilsson, Axel, Rudolf (1907): Desinfektion fester Gegenstände. *Chemiker Zeitung/Chemisches Repertorium*, Bd. 31, 299.

Oddy, Andrew (1973): An Unsuspected Danger in Display. *Museum Journal*, 73, 27–28.

Otto, Helmut (1979): Das chemische Laboratorium der Königlichen Museen in Berlin. *Berliner Beiträge zur Archäometrie*, (4), 48–49, 248–250, 262–267, 268–275.

Rathgen, Friedrich (1896): Vortrag des Herrn Dr. Rathgen: Reiseerinnerungen. Mit 6 Abbildungen. Versammlung am 20.02.1896. *Polytechnisches Centralblatt. Organ der Polytechnischen Gesellschaft zu Berlin*, 57. Jahrgang der Gesamtfolge, (11), 125–127.

Rathgen, Friedrich (1898): Die Konservirung von Alterthumsfunden. 1. Auflage. Berlin: W. Spemann (Handbücher der Staatlichen Museen zu Berlin).

Rathgen, Friedrich (1903): Konservierung von Altertumsfunden aus Eisen und Bronze. *Chemiker-Zeitung*, (56), 704.

Rathgen, Friedrich (1908): Mitteilungen aus dem Laboratorium der Königlichen Museen zu Berlin. IV. Die Verwendung von Tetrachlorkohlenstoff in der Konservierungspraxis. *Museumskunde, Zeitschrift für Verwaltung und Technik öffentlicher und privater Sammlungen*, Bd. IV, 90–91.

Rathgen, Friedrich (1911): Mitteilungen aus dem Chemischen Laboratorium der Königlichen Museen zu Berlin. VIII. Über die Verwendung von Tetrachlorkohlenstoff zur Abtötung von tierischen Schädlingen. *Museumskunde, Zeitschrift für Verwaltung und Technik öffentlicher und privater Sammlungen*, (Bd. VII), 219–220.

Rathgen, Friedrich (1924): Die Konservierung von Altertumsfunden. Mit Berücksichtigung ethnographischer und kunstgewerblicher Sammlungsgegenstände. 2. Auflage. Berlin und Leipzig: Walter de Gruyter & Co. (Handbücher der Staatlichen Museen zu Berlin, II. und III. Teil).

Rathgen, Friedrich (1926): Die Konservierung von Altertumsfunden/Stein und steinartige Stoffe. 3. umgearb. Aufl. Berlin: Walter de Gruyter & Co. (Handbücher der Staatlichen Museen zu Berlin, Teil 1).

Reinbothe, Roswitha (2011): Geschichte des Deutschen als Wissenschaftssprache im 19. Jahrhundert. Vortrag bei einem Symposion an der Universität Bamberg am 15/16. Oktober 2009. Hrsg. v. Wieland Eins, Helmut Glück, Sabine Pretscher. Wiesbaden: Wissen schaffen—Wissen kommunizieren. Wissenschaftssprachen in Geschichte und Gegenwart. Harrassowitz Verlag. Available online at www.observatoireplurilinguisme.eu/images/Education/Enseignement_superieur/reinbothe-geschichte_des_deutschen_als_wissenschaftssprache.pdf.

Rowley, Frederick Richard (1916): Demonstration of Objects Preserved in Arsenious Acid Glycerine Jelly. Read at the Ipswich Conference. *Museums Journal*, 16(4), 77–79.

Scott, Alexander (1922): The Restoration and Preservation of Objects at the British Museum. *Journal of the Royal Society of Arts*, LXX(3618), 337.

Stansbury, Chas F.; Barkly, Henry; Campbell, W.H.; J.M.G.; H.S.; Farthing, John J. (1852): Journal of the Society for Arts. In: *Journal of the Society for Arts* 1 (1). https://archive.org/details/journalofsociety01soci_0/page/n3/mode/2up.

Thilenius, Georg (1916): Das Hamburgische Museum für Völkerkunde. *Museumskunde, Zeitschrift für Verwaltung und Technik öffentlicher und privater Sammlungen*, 16(i.e. 12) (Beiheft zu Band XIV), I–VIII, 74–78, 152.

VDLUFA Verband Deutscher Landwirtschaftlicher Untersuchungs- und Forschungsanstalten (2013): 125 Jahre Verband Deutscher Landwirtschaftlicher Untersuchungs- und Forschungsanstalten e.V., Eine Dokumentation. Darmstadt: VDLUFA-Verlag, VDLUFA-Schriftenreihe, 69, 7–24.

Westphal-Hellbusch, Sigrid (1973): Zur Geschichte des Museums. Hundert Jahre Museum für Völkerkunde. *Baessler-Archive*, XXI.

Wray (1908): Wray, L. (Vorname unbekannt, Anm. D. Verf.). The Preservation of Mammal Skins. *Museums Journal*, 1908(8), 207–208.

12 Knowledge transfer and product application from industry, commerce, and trade at the Königliches/Staatliches Museum für Völkerkunde in Berlin

The investigations show an unpredictable and inconsistent approach when it comes to the procurement of active ingredients and means to control insect pests. A few purchase receipts documented the use of different pest control products. For example, in 1905, an administrative invoice for the Kunstmuseen (Art Museums) in Berlin gives evidence that Avenarius & Co. was hired to provide "Carbolineum" in the amount of 43.80 marks. In the same year, between May and October, several chemicals were purchased from Schmaltz without specifying a specific purpose (SMB-ZA 1905). *De facto*, there is no discernible concept in which previously defined methods for the conservation of collection items have been used according to clear guidelines. Thus, instructions to give entire groups of objects, such as bamboo tips and feather work in KMfV, "to poison" appear to be a rather random decision (SMB-PK EM 1913f).

Both within and outside the Berlin KMfV/MfV, it appears that the product "Flit" adheres to the principle of utilizing an agent that was already available on the market. For example, Walter Krickeberg, head of the Amerikanische Abteilung (American department) at MfV since 1934, commissioned his research assistant Heinrich Snethlage to travel to Leipzig on June 20, 1938 (Cf. Westphal-Hellbusch 1973, 232). On site, objects on loan for a special exhibition from the Otto Schulz-Kampfhenkel collection were to be monitored for their condition. As part of his visit, Snethlage ordered that the fur strips, which had already been damaged in Dahlem and now set up freely, be kept in showcases with immediate effect and sprayed with "Flit" (Wilhelmi und Kunike 1927, 98–99). According to a general order, all exhibits in this collection were then to be "flitted" regularly without further details (SMB-PK EM 1938).

An important aspect of these investigations is the shortage of materials for objects made of organic substances in the fight against decay during the First World War. Petroleum, a liquid mixture of hydrocarbons, played a special role here, as it was mainly used for lighting purposes during the period under investigation. To a greater extent and over a very long period of time, petroleum was also used in KMfV/MfV to combat insect pests. In the following section, we will examine the difficulties faced by museum staff when procuring this product during times of scarcity.

DOI: 10.4324/9781003407607-16

During the German Empire, 90 % of kerosene had to be imported from abroad (see Part I, chapter 1), whereby the consumption of kerosene as a luminous oil in Germany accounted for about half of the total mineral oil consumption during the period mentioned (Dumont 1914, 12–14; Karlsch und Stokes 2003, 93). With the beginning of the First World War, things for daily needs were rationalised in many areas of public life. A large deficit for the German economy meant a lack of foreign mineral oil, from which it was cut off from one day to the next (Karlsch und Stokes 2003, 132). Thus, according to a decree of the Preußischen Ministeriums der geistlichen, Unterrichts- und Medizinalangelegenheiten (Prussian Ministry of spiritual, educational, and medical affairs), all public authorities had to declare their necessary needs for kerosene. This ubiquitous means of illumination was also recommended by the head of the Chemisches Laboratorium (chemical laboratory) Friedrich Rathgen (to Rathgen see Part III, chapter 8) at KM Berlin as a preservative against wood-destroying insects (Cf. Rathgen 1924, 136). Kerosene was first mentioned in 1866 by Suardi for the conservation of paintings (see Part II, chapter 5). In 1915, Krause estimated that the museum required at 100 litres and the collections of the Deutsche Volkskunde (German folklore) required 150 litres (SMB-PK EM 1915c). Although it was not clear at the directorial level regarding the specific quantities required to eliminate wood-damaging insects in each department, Krause's request for a total of 250 liters of kerosene was accepted, and a small part of this large quantity was also taken from KMfV for other purposes (SMB-PK EM 1915b). A renewed request from the Ministry to check the actual need for kerosene shows the material shortages caused by the First World War in the second year of the war. The aforementioned collections considered the use of other unspecified chemicals and reduced the originally required quantity from 250 to 190 liters of kerosene (SMB-PK EM 1915a). On October 14, 1915, the KMfV, like many other authorities, was finally asked by Preußisches Ministerium für geistliche, Unterrichts- und Medizinalangelegenheiten to register the need for kerosene at the Berliner Zentralstelle für Petroleum-Verteilung (Central Office for Petroleum Distribution m.b.H.) in Berlin via the Ministry due to the shortage caused by the war (SMB-PK EM 1915d). Six months later, the tug-of-war over the quantities required for kerosene orders continued between the Berliner Zentralstelle für Petroleum-Verteilung and the General Direktion (General Directorate) of KM. In March 1916, KMfV initially requested only 25 liters of kerosene, but a month later, they revised their request and asked for a total of 90 liters of kerosene because of an increased need for conservation purposes (SMB-PK EM 1916a; Ibid., March 6, 1916; Ibid., April 19, 1916; Ibid., April 25, 1916). Less than half a year later and despite the ever-tightening supply situation, another 75 liters of kerosene were donated to KMfV. Albert Grünwedel[1] pointed out that the quantity had to be delivered as a whole due to

". . . die zu konservierenden Gegenstände ganz in Petroleum zu tauchen sind".[2]

(SMB-PK EM 1916b)

Again, for the soaking of objects made of wood, Max Junker[3] and Albert Grün-wedel asked another 40 to 50 liters of kerosene, as infestation by wood-destroying insects had been discovered in the Ozeanische Abteilung (Oceanic department) in 1917 on wooden collection items (SMB-PK EM 1917a). Even after the end of the First World War, the supply of kerosene was not yet guaranteed, as can be seen in a letter from the Minister für Handel und Gewerbe (Minister of Trade and Industry), which was received by SMB via the Minister für Wissenschaft, Kunst und Volks-bildung (Minister of Science, Art, and National Education). As a state authority, it was necessary to request kerosene from the Zentralstelle für Petroleum-Verteilung (SMB-PK 1919c). It was not until a few years later in 1925 that critical remarks were made on the use of kerosene as a preservative when it was present in mixtures (see Part III, chapter 13). This proves the application of this active ingredient for preservation purposes for around 60 years.

For a certain period, a chemical substance, known as "Dichlorobenzene Agfa", was marketed as a pesticide by the pharmaceutical department of Actiengesells-chaft für Anilin-Fabrikation (AGFA) (Kadlubek und Hillebrand 1998; Karlsch und Wagner 2004). AGFA was one of the largest paint factories at the time, which discovered the insecticidal properties of 1,4-dichlorobenzene at its Berlin-Treptow location. This substance was a waste product in the production of aniline dyes. The company's management created a new line of business and had "Dichlorobenzene Agfa" protected with the German Reich patent number 258405. Pure 1,4-dichlo-robenzene was sold under this product name. Initially, the distribution of the prod-uct was in the control of the manufacturer, and they showed particular interest in acquiring scientific and museum collections that were plagued by pests, such as the museum beetle (Anthrenus museorum), the common bacon beetle (Derm-estes lardarius), and similar species from the bacon beetle family. In the mean-time, the production was relocated to Wolfen in Saxony-Anhalt. From the Berlin headquarters, advertising letters with detailed brochures and small sample bags of the product were sent to numerous museums. On April 15, 1913, KM received a letter of this kind and immediately forwarded it to the conservator Krause at KMfV (SMB-PK EM 1913c). This developed into a correspondence in the course in which AGFA also sent a copy of a letter fromthe then director Georg Thilenius from the Völkerkundemuseum (Museum of Ethnology) in Hamburg to KMfV. He mentioned successful trials of the product in controlling fur pests in his museum (SMB-PK EM 1913b). Krause requested a larger quantity of the substance for his own experiments and reported in October 1913 his positive experiences, empha-sizing that metals, such as gold, silver, copper, iron, and nickel, were not attacked by "Dichlorobenzene Agfa" (SMB-PK EM 1913b). To conduct further experi-ments on spring shields, 10 kg of the product was ordered by Eichhorn[4] (SMB-PK EM 1913d). Six months later, AGFA informed KMfV Berlin that its product "Dichlorobenzene Agfa" would continue to be marketed under the trade name "Globol" by Fritz Schulz jun. Aktiengesellschaft in Leipzig (SMB-PK EM 1914). According to AGFA's 1914 annual report, this Leipzig-based company had secured exclusive distribution rights for the product through a contract, with the purchase increasing annually. Thus, 28,500 kg was sold in 1914. In 1915, because of the war, the company mainly sold "Globol" bags to the military for delousing the troops, resulting in a total of 40,641 kg. The following year, the amount decreased to 45,378 kg, and continued to decrease to 91,100 kg in 1917 and 157,164 kg in 1918 (AIFM 1912–1918). Between 1915 and 1920, Fritz Schulz jun. Aktiengesellschaft approached the MfV in Berlin several times. At the same time, the company's advertising campaigns to attract new customers were considered extremely critical. In a letter with an enclosed free sample of the agent "Globol", the insects that frequently occur in museum collections, such as clothes moths, fur beetles, and cabinet beetles, were described as "uninvited museum visitors" from whom the "mostly irreplaceable museum holdings" must be protected. At that time, the reference list of the Leipzig company consisted of only two museums, namely, the kaiserlich königliches Naturhistorisches Hof-Museum (Imperial Royal Natural History Court Museum) in Vienna and the Bayerisches National-Museum (Bavarian National Museum) in Munich. Eichhorn then provided cabinets with collection items in the Oceanic department of KMfV for an experiment, but immediately had the bags with the agent removed by Krause because of the strong smell (SMB-PK EM 1915e, 1915f).

After the end of the First World War in 1919, the now MfV received another letter, the content and choice of words of which were written in a much more combative manner:

> "*Die energische Bekämpfung der Motten* ist dringend notwendig, wenn Sie ihre wertvollen Lagerbestände schützen und sich selbst vor empfindlichen Verlusten bewahren wollen.
> Energisch bekämpfen–das heißt: die Motten *töten!*"[5]

(SMB-PK EM 1919a)

The advertising quote can be linguistically assigned to the context of the war propaganda. Accordingly, like the following quotations, it was not written at any time but clearly springs from the fighting spirit of the First World War that had just ended. Both pleading and threatening, it goes on to say:

> "Ein Versuch wird sie zu einem überzeugten Anhänger unseres Globols und damit zu unserem treuen Kunden machen. In Ihrem Interesse bitten wir Sie: Nützen Sie unser Angebot!
> *Schieben Sie den Versuch nicht auf!*"[6]

(SMB-PK EM 1919a)

The reference list of customers now ranges from scientific institutes to a königliche Naturaliensammlung (Royal Natural History Collection), as well as to a königlich landwirtschaftliche Hochschule (Royal Agricultural College), to museums at home and abroad with personal testimonies and judgments from industry, trade, and commerce (SMB-PK EM 1917b). On May 28, 1919, the use of "Globol" was again refused in circulation with the signatures of Grünwedel, Seler, Anker-mann, Kümmel, and Müller because of the strong odor nuisance (SMB-PK EM 1919b). In a letter handed down from Fritz Schulz jun. Aktiengesellschaft to MfV in 1920, quieter tones were initially used to sell the product, specifically for their persuasion strategies, and were clearly focused on the general economic crisis by asking:

> "*Haben Sie alles getan*, um Ihre wertvollen Lagerbestände *vor Motten-fraß* zu schützen? Schon im Frieden war es sehr ärgerlich für Sie, wenn Sie Schäden durch Mottenfraß auf Ihren Lagern feststellen mussten. *Wieviel größer ist der Schaden*, den die Motten anrichten, *heute*, in den Zeiten unglaublicher Teuerung, schärfster Warenknappheit, größter Wert-steigerung Ihrer Bestände?"[7]

> (SMB-PK EM 1920a)

As the writing progresses, the same person switched to a more combative style, demanding:

> "*Die energische Bekämpfung der Motten* ist daher auch für Sie ein dringen-des Gebot, nicht minder wichtig als die Sicherung Ihres Lagers gegen Ein-bruch, Diebstahl oder andere schädliche Einflüsse. *Sie dürfen deshalb nicht rasten*, die Motten *mit der wirksamsten Waffe* zu bekämpfen–Sie müssen die Motten *töten!*"[8]

> (SMB-PK EM 1920a)

The administrative employee Max Junker and the director Albert Grünwedel acknowledged this advertising letter with a brief remark that the agent had not proven itself in the Oceanic department (SMB-PK EM 1920b).

The odor nuisance caused by "Globol" is an indication that in the late 19th and early 20th centuries, chemical substances were being used without proper protec-tion at SMB. This seems remarkable in view of the fact that occupational health and safety in Germany, based on Prussian regulations, already existed in 1839 by law for the protection of workers, especially children (Simon 1844, 627–629). With increasing industrialization, the insight also developed:

> " . . . daß eine systematische, präventive Abwehr von Gefahren prinzipiell als sinnvoll erschien"

> (Buck-Heilig 1989)[9]

This led to a historic turning point in 1892 and the establishment of an occupa-tional health and safety system in Germany; however, this was less related to the toxic effects of chemical substances. Rather, it was created under the pressure of contemporary social democratic demands and increasing protests against inhumane conditions in the world of work. The aim was to treat the workforce more sparingly and to preserve their workforce (Nipperdey 1995; Ayass 1966). To implement occupational health and safety, the Prussian government issued a service instruction, as a result of which labor inspectors were appointed to carry out ongoing inspections in factories for the protection of workers and their occupational safety (Cf. Buck-Heilig 1989). These labor law reforms in the industrial sector may have contributed to their transfer to cultural institutions. "Globol", which was reused after its rejection in the museum in 1919 and then used in all departments of MfV, was not classified exclusively as a poisonous agent. From then on, the use of this agent required personal protection, so that in 1929 in consultation with Carl Brittner[10] respirators have been ordered for employees of all departments to protect them from the odor of "Globol" as well as from those of "Areginal". As a result, on August 24, 1929, the Oceanic department, the departments of North and South America, and the African department received a total of 13 LIX respirators with 26 carbon capsule filters from Deutsche Gasglühlicht-Auer-Gesellschaft m.b.H. at a price of 84.95 Reichsmarks (SMB-PK EM 1929).

Other paint factories showed a strong interest in bringing their own pesticide products to the market. The chemist, Arthur Eichengrün, who worked for the Farbenfabriken, vormals Friedrich Bayer & Co., Elberfeld & Leverkusen (paint factories, formerly Friedrich Bayer & Co., Elberfeld & Leverkusen), opened up an important sales market for the company with his invention of the "Autan" process. The process was based on the "Autan" powder and consisted of barium peroxide and paraformaldehyde with the product name "Paraform". If water was poured onto the powder, a strong chemical reaction occurred, and water-containing formaldehyde vapors were formed (Anonymous 1906). In his invention patented by the paint factories on August 4, 1905, (Eichengrün 1905), Eichengrün saw a user-friendly simplification of the formaldehyde disinfection carried out to date, which required both trained personnel and a high level of equipment (Eichengrün 1906). The method was initially used for the disinfection of apartments and, according to Eichengrün, it had been used by the authorities since 1918:

" . . . , wesentliche Beachtung gefunden und ist vielfach schon an Stelle der bisherigen Methoden zur offiziellen Einführung gelangt"[11]
(Farbenfabriken, vormals Friedrich Bayer & Co., Elberfeld and Leverkusen
1918).

As part of an event organised by the Deutsche Desinfektoren-Bundes (German Disinfectors-Association), Eichhorn visited the Farbenfabriken, vormals Friedrich Bayer & Co., Elberfeld & Leverkusen, where the product "Autan" was exhibited. Subsequently, on behalf of Grünwedel, he wrote to the company on July 19, 1913, asking for more information about the agent and the procedure itself. He received a prompt response and was advised that the disinfectant "Autan" is mainly used against bacteria. However, it was recommended that KMfV should conduct its own experiments. Therefore, sufficient quantities of material and brochures free of charge were provided. Added was the indication that the "Autan" procedure had been "zur Anwendung bei amtlichen Desinfektionen autorisiert"[12] by ministerial decree on April 25, 1908 and May 17, 1910 (SMB-PK EM 1913a). The fact that the production of "Autan" and the expansion of sales were a profitable business is proven by the sales figures of the Farbenfabriken, vormals Friedrich Bayer & Co., Elberfeld & Leverkusen. While 351 kg of "Autan" were sold for the first time in 1906, sales had increased by 100 % ten years later (BAL 1906–1916). Overall, it can be said that the industry has set the tone in both the production and sale of active ingredients and pest control products. On the other hand, SMB were in a rather defensive position and had no influence on the speed at which pesticides were developed and distributed.

An impressive cooperation and interweaving between scientific work, official interests, and economic exploitation of products for pest control have been achieved through the development and marketing of fumigants. With the founding and development of the Deutsche Gesellschaft für Schädlingsbekämpfung m.b.H. (Degesch) (German Society for Pest Control) in Berlin by Fritz Haber, a significant breakthrough had been achieved at this point. At the end of Part I, chapter 2, it has already been shown that the foundations for further research in the field of pest control for civilian use were laid both in the age of industrialization and in the context of military developments during the First World War. The founding of Degesch in Berlin can be traced back exclusively to Fritz Haber. His ambition was to put himself at the service of Germany with the competencies of his Kaiser-Wilhelm-Institut für physikalische Chemie und Elektrochemie (Kaiser-Wilhelm-Institute for Physical Chemistry and Electrochemistry) and close connections to business and industry. As a chemist and head of his institution, he was subordinate to the Preußisches Kriegsministerium (Prussian Ministry of War) and founded the Technischen Ausschuss für Schädlingsbekämpfung (TASCH) (Technical Committee for Pest Control) for civilian purposes on February 15, 1917, which was also organizationally assigned to the Preußisches Kriegsministerium (Cf. Ebbinghaus 1998). Under his chairmanship, in addition to several Reich and Prussian authorities, the Deutsche Gold- und Silber-Scheideanstalt (abbreviated as Degussa from 1928) was also represented in TASCH. As a unique selling point, the committee received a state concession for the use of hydrogen cyanide and initially directed the fumigation of warehouses, mills, and military barracks by chemists and laboratory assistants from the refinery (Ibid., 40). Fritz Haber made the first demands for economic exploitation and the establishment of a non-profit company as early as February 1918. The Deutsche Gold- und Silber-Scheideanstalt, Badische Anilin- und Sodafabrik Ludwigshafen (BASF), Farbenfabriken, vormals Friedrich Bayer & Co., Elberfeld & Leverkusen, Farbenwerke Meister, Lucius & Brüning, and other chemical companies had already produced gas weapons during the war in cooperation with Kaiser-Wilhelm-Institut. On April 1, 1919, these companies founded the Degesch in Berlin. In May 1920, the company relocated from Berlin to Frankfurt am Main and appointed Walter Heerdt, who was instrumental in the technical development of the "Zyklon" process, in particular "Zyklon B", as managing director. Economic inflation after the First World War necessitated a privatization of the Degesch, which was subsequently taken over 100 % by the Deutsche Gold- und Silber-Scheideanstalt in October 1922 (Ibid., 41). From there, in 1925, Degesch was reorganised, severely restricting the product range of pesticides and concentrating on the growing business of highly effective gases (Ibid., 43). Two further proceedings finally led Degesch to become the market leader in the field of large-capacity fumigation at home and abroad. On the one hand, on February 10, 1930, it concluded a contract with the I.G. Farbenindustrie AG to benefit better from the development work in the plant protection department in Leverkusen and to include the fumigants "Calcid" and ethylene oxide (still known today as ethylene oxide; see glossary) developed by I.G. Farbenindustrie AG in its range. On the other hand, nine months later, in November 1930, Th. Goldschmidt AG. from Essen was integrated into the new company structure by transferring shares in Degesch and the I.G. Farbenindustrie AG to Th. Goldschmidt AG. This had developed the "T-Gas", a mixture of ethylene oxide and carbon dioxide (see glossary). It was especially suitable for "Kleinraumdurchgasungen",[13] such as apartments, and thus competed with the "Zyklon" process. The corporate structure of Degesch, consisting of the I.G. Farbenindustrie AG, Deutsche Gold- und Silber-Scheideanstalt, and Th. Goldschmidt AG, existed until 1945 (Ibid., 44–45). The dangers of epidemics and the spread of epidemics in shelters inhabited by workers were clearly recognised and taken seriously. As a result of these concerns, the contact between Degesch and MfV at Königgrätzer Straße made it clear that new markets for the "T-Gas" were also being sought. In April 1919, just four months after its establishment, the company wrote to the museum from its Berlin headquarters in the Wilhelmstraße, despite being only a short ten-minute walk from each other in the immediate vicinity. The letter dated August 5, 1919 lacked both salutation and introductory words, but it offers straightforwardly a "Blausäuregasverfahren zur Bekämpfung von Ungeziefer",[14] noting that good experiences have already been made with it in apartments and factories (SMB-PK EM 1919d). Noteworthy are the numerous appendices to this letter, which show how intrusive Degesch offered its services. These were initially articles and leaflets, all of which were published by the Deutsche Gesellschaft für Entomologie e.V. (German Society for Entomology) and its journal, the *Zeitschrift für angewandte Entomologie*.[15] The publications published by well-known scientists at that time were primarily concerned with the control of clothes moth, flour moth, and bedbug, as well as preventing the spread of typhus caused by the clothes louse using hydrogen cyanide (Andres ohne Jahresangabe, 1–3; Hase 1916, 1–8; Frickhinger 1918, 129–140; Hase 1917, 1–8). All publications clearly indicate poor living conditions in major metropolises. The "Allgemeinen Bedingungen für die Durchgasung von Gebäuden mit Blausäuregas"[16] in the appendix, with its 20-point structure, also shows that Degesch in Berlin was a profit-oriented company. Without being asked, an "Auftrag zur Durchgasung"[17] as well as an "Antrag auf Ausführung einer Durchgasung mit Blausäuregas"[18] were supplied with the same mail (SMB-PK EM 1919d). Unfortunately, based on the available sources, it could not be clarified whether fumigation was carried out in MfV by Degesch at that time. In the company's approach, however, a significantly increased marketing strategy can be seen compared to the advertising campaign of Fritz Schulz jun. Aktiengesellschaft. Only professional judgments from science and research are used as references for the use of fumigants for pest control.

It has been proven that employees at the museum have also carried out fumigation procedures on their own since 1914. For example, carbon disulfide has been poured into open trays and then placed in cabinets or fumigation boxes specially made for this purpose. To treat larger groups of objects quickly in the event of acute infestation or to prevent infestations in the first place, hermetically sealing a box or an entire room is an effective method (Cf. Rathgen 1924, 142–143). Experiments on the required quantities and residence time of the liquid in the containers had already led to contact with Bolle from 1910 onward (SMB-PK EM 1910/1911).

Finally, research carried out by employees at KMfV/MfV in Berlin to expand their knowledge of pest control should be mentioned. The locational advantage in the Prussian metropolis of Berlin proved beneficial. For example, the Gesellschaft für Vorratsschutz e.V., located in Berlin-Steglitz, provided an ideal platform for the exchange of knowledge on the latest research in this field for scientists, practitioners, and members at home and abroad. A copy of the *Mitteilungen der Gesellschaft für Vorratsschutz e.V.* was received by SMB on January 5, 1927 and had to be read and signed by all directors of the museum's departments. Zacher, the publisher, promotes membership in the Gesellschaft für Vorratsschutz e.V. and encourages readers to subscribe to his journal. In the various contributions of scholars and practitioners, he saw an ideal combination of science and practice at home and abroad and emphasises that his journal presents the latest state of research and technology. In commercial advertisements at the end of this professional journal, the Chemische Werke in Berlin-Marienfelde recommended various products containing sulfur dioxide; the Wilhelm Dönne Blechwarenfabrik from Berlin offered the construction of fumigation boxes; and the companies Bayer, Agfa, and the I.G. Farbenindustrie AG, which worked closely together economically, jointly promoted different agents. Among them, "Areginal" was offered against storage pests and "Diametan" for fumigating domestic vermin of all kinds. Degesch recommended itself with "Zyklon B for the destruction of all storage pests" and referred to its representatives east of the Elbe, Tesch & Stabenow, International Society for Pest Control m.b.H., and Heerdt-Lingler G.m.b.H. in Frankfurt am Main, west of the Elbe. From a mixture of scientific contributions and commercial product advertising, Berlin MfV obtained further information from other disciplines and thus broadened its professional horizon regarding preventive and control measures against harmful insects (SMB-PK EM 1927).

Table 12.1 Networks/activities of industry, trade, and scientific and official institutions from 1905 to 1930.

Chronology	City	Institution	Activity	Person	Active ingredient/agent	Networking/activities
August 1905	Elberfeld & Leverkusen	Farbenfabriken, vormals Friedrich Bayer & Co.	Patent; search for sales markets	Arthur Eichengrün	"Autan" procedure	Disinfection of apartments
April 1908	Elberfeld & Leverkusen	Farbenfabriken, vormals Friedrich Bayer & Co.	Application and distribution	Unknown	"Autan" procedure	State authorities authorised the agent for official disinfection
July 1913	Berlin	Königliches Museum für Völkerkunde	Contacting Farbenfabriken, vormals Friedrich Bayer & Co.	August Eichhorn	"Autan" procedure	Delivery of sample quantities for the museum's own experiments
1913–1914	Berlin	Actiengesellschaft für Anilin-Fabrikation (AGFA) Pharmaceutische Abteilung	Search for sales markets	Unknown	"Dichlorobenzene Agfa"	Advertising letters, brochures, and sample bags are sent to, among others: Königliches Museum für Völkerkunde Berlin, Museum für Völkerkunde Hamburg, and Rautenstrauch-Joest-Museum Köln
October 10, 1913	Hamburg	Museum für Völkerkunde	Letter of recommendation	Georg Thilenius	"Dichlorbenzol Agfa"	Forwarding of the letter by Agfa to Königliches Museum für Völkerkunde Berlin
October 1913	Berlin	Königliches Museum für Völkerkunde	Own experiments	Eduard Krause	"Dichlorobenzene Agfa"	Requires large amount of the active ingredient
1914–1918	Leipzig	Fritz Schulz jun.	Sole distributors	Unknown	"Globol"	Increase in sales of "Globol"

Date	Place	Organization	Action	Person	Product	Description
May 14, 1914	Leipzig	Actiengesellschaft für Anilin-Fabrikation (AGFA) Pharmaceutische Abteilung Fritz Schulz jun.	Sole distributors Fritz Schulz jun.	Unknown	"Globol"	Notification of exclusive distribution to Königliches Museum für Völkerkunde
1915–1920	Leipzig	Fritz Schulz jun.	Sole distributors	Unknown	"Globol"	Massive advertising for the agent to Königliches Museum für Völkerkunde
January 16, 1920	Berlin	Königliches Museum für Völkerkunde	Statement	Grünwedel; Junker	"Globol"	Agent was rejected
August 24, 1929	Berlin	Chemisches Laboratorium der Staatlichen Museen	Statement	Carl Brittner	"Globol"	Reintroduction of the product at SMB in compliance with occupational health and safety (wearing respirators)
February 15, 1917	Berlin	Technischer Ausschuss für Schädlingsbekämpfung (Tasch)	Establishment state concession	Fritz Haber	Various fumigants	Merger of state authorities and industry; subordinate to the Preußischen Kriegsministerium
April 1, 1919	Berlin	Deutsche Gesellschaft zur Schädlingsbekämpfung m.b.H. (Degesch)	Establishment	Fritz Haber (manager)	Various fumigants	Merger of KWI, Deutsche Gold- und Silber-Scheideanstalt, Badische Anilin- und Sodafabrik Ludwigshafen (BASF), Farbenfabriken, vormals Friedrich Bayer & Co., Elberfeld & Leverkusen, Farbenwerke Meister, Lucius & Brüning, and other chemical companies

(Continued)

Table 12.1 (Continued)

Chronology	City	Institution	Activity	Person	Active ingredient/ agent	Networking/activities
August 5, 1919	Berlin	Deutsche Gesellschaft zur Schädlingsbekämpfung m.b.H. (Degesch)	Marketing of fumigants; references from scientific institutions	Fritz Haber (manager)	"T-Gas" hydrogen cyanide	Contacting the Staatliche Museum für Völkerkunde Berlin
May 1920	Frankfurt/M.	Deutsche Gesellschaft zur Schädlingsbekämpfung m.b.H. (Degesch)	Move	Walter Heerdt (manager)	"Zyklon B" procedure "Zyklon B"	Considered the inventor of "Zyklon B"
October 1922	Frankfurt/M.	Deutsche Gesellschaft zur Schädlingsbekämpfung m.b.H. (Degesch)	Privatization	Unknown	Various fumigants	Takeover by Deutsche Gold- und Silber-Scheideanstalt; company name Degesch was retained
1925	Frankfurt/M.	Deutsche Gesellschaft zur Schädlingsbekämpfung m.b.H. (Degesch)	Reorganization	Walter Heerdt (manager) until the end of 1925	Various fumigants	Concentration on highly effective gases
February 10, 1930	Frankfurt/M.	Degesch und I.G. Farbenindustrie Aktiengesellschaft	Joint contract	Unknown	"Calcid"; ethylene oxide	Inclusion of both agents in the Degesch product range
November 1930	Essen	Th. Goldschmidt A.G.	Integration into the corporate structure of Degesch	Unknown	"T-Gas"	Small room fumigations (apartments); competition with the "Zyklon" procedure

Notes

1 From 1914 to 1921, Albert Grünwedel was director of the Indisch-Asiatischen Abteilung (Indo-Asian department) of KMfV/MfV in Berlin.
2 . . . the objects to be preserved must be completely immersed in kerosene (translation by the author).
3 From 1896 to 1920, Max Junker was secretary in the Genralverwaltung (general administration) of KM/SMB.
4 From 1916 to 1929, August Eichhorn was head of the ozeanische Abteilung (Oceanic department) at KMfV/MfV in Berlin.
5 *Vigorous control of moths* is urgently needed if you want to protect your valuable stocks and save yourself from sensitive losses. Vigorously combat- that means: *kill* the moths! (translation by the author).
6 An attempt will make you an enthusiastic supporter of our Globol and thus our loyal customer. In your interest, we ask you to take advantage of our offer! *Do not post pone the experiment!* (translation by the author).
7 *Have you done everything you can to protect* your valuable stock *from being eaten by moths*? Even in peacetime, it was very annoying for you if you had to notice damage caused by moths on your camps. *How much greater is the damage* caused by moths *today*, in times of unbelievable inflation, the sharpest shortages of goods, the greatest increase in the value of your stocks? (translation by the author).
8 *The vigorous control of moths* is therefore also an urgent imperative for you, no less important than securing your warehouse against burglary, theft, or other harmful influences. *Therefore, you must not rest to* fight the moths *with the most effective weapon–* You must *kill* the moths!
9 . . . that a systematic, preventive defense against dangers appeared to make sense in principle (translation by the author).
10 See also Carl Brittner, Part III, chapter 8.
11 . . . , has received significant attention and in many cases has already been officially introduced in place of the previous methods (translation by the author).
12 Authorised for use in official disinfections (translation by the author).
13 Small room fumigation (translation by the author).
14 Hydrogen cyanide gas process for the control of vermin (translation by the author).
15 *Journal of Applied Entomology* (translation by the author).
16 General conditions for the fumigation of buildings with hydrogen cyanide gas (translation by the author).
17 Order for fumigation (translation by the author).
18 Implementation request of fumigation with hydrogen cyanide gas (translation by the author).

References

Archive material

AIFM Wolfen (1912–1918): Ohne Signatur. Loseblattsammlung. Jahresberichte von 1912–1918. Actien-Gesellschaft für Anilin-Fabrikation, 12 Seiten.
BAL (1906–1916): 15/D.1.1. Akte Statistik über Kiloverkäufe und Geldumsätze Pharmazeutika, Riechstoffe, Pflanzenschutz, Photographika, Farben von 1906–1916. Loseblattsammlung. Tabelle. Farbenfabriken vorm. Friedrich Bayer & Co. Elberfeld. Kiloverkäufe über pharmazeutische Produkte, 1 Tabelle, Blatt 5–6, 2 Seiten.
SMB-PK EM (1910/1911): I/MV 0075, Pars II c, Vol. 1, E. Nr. 1360/10. Experimente und Versuche des Königlichen Museums für Völkerkunde zu Berlin mit der kaiserlich königlichen landwirtschaftlich-chemischen Versuchsstation in Görz. Inspektor der k. k.

Versuchsstation (Name unreadable). Brief vom 2.7.1910, 1 Seite, ohne Paginierung.; Luschan von, Felix. Brief vom 07.01.1911, Blatt 1, 1 Seite.; Luschan von, Felix. Brief vom 24.01.1911, Blatt 3–4, 2 Seiten.; Bolle, Johann. Brief vom 19.01.1911, Blatt 2, 2 Seiten.; Eichhorn, August. Brief vom 02.01.1911, Blatt 5, 1 Seite.

SMB-PK EM (1913a): I/MV 0075, Pars II c, Vol. 1. E. Nr. 1225/13. Eichhorn, August. Brief vom 19.07.1913 an die Farbenfabriken vorm. Friedrich Bayer & Co. Elberfeld, 2 Seiten, ohne Paginierung.; Ebd. Farbenfabriken vorm. Friedrich Bayer & Co. Elberfeld. Brief vom 26.07.1913 an das Königliches Museum für Völkerkunde zu Berlin und Farbenfabriken, vormals Friedrich Bayer & Co., Elberfeld & Leverkusen 1913, darauf Aktennotiz von August Eichhorn vom 31.07.1913, 2 Seiten, ohne Paginierung.; 1 Prospekt über "Autan", 4 Seiten ohne Paginierung.; 1 Merkblatt über "Autan", 1 Seite, ohne Paginierung.

SMB-PK EM (1913b): I/MV 0075, Pars II c, Vol. 1. E. Nr. 682/13. Bewertung von "Dichlorbenzol Agfa". Krause, Eduard. Bericht vom 29.10.1913, 1 Seite, ohne Paginierung.

SMB-PK EM (1913c): I/MV 0075, Pars II c, Vol. 1. E. Nr. 682/13. Empfehlung des Produktes "Dichlorbenzol Agfa". Actien-Gesellschaft für Anilin-Fabrikation. Brief vom 15.04.1913, 2 Seiten, ohne Paginierung.

SMB-PK EM (1913d): I/MV 0075, Pars II c, Vol. 1. E. Nr. 682/13. Röddinghaus, (First name unknown, author's note). Aktennotiz vom 31.10.1913, 1 Seite, linke Spalte, ohne Paginierung.

SMB-PK EM (1913e): I/MV 0597, I B 44a. Loseblattsammlung. Anweisung zur Schädlingsbekämpfung an Objekten aus der Sammlung Theodor Koch-Grünberg. Preuss, Konrad. Aktennotiz auf einem Brief von Theodor Koch-Grünberg vom 30.04.1913, Blatt 85, 1 Zeile.

SMB-PK EM (1914): I/MV 0075, Pars II c, Vol. 1. E. Nr. 820/14. Actien-Gesellschaft für Anilin-Fabrikation. Brief vom 14.05.1914, 1 Seite, ohne Paginierung.

SMB-PK EM (1915a): I/MV 0034, Pars Ia, Bd. 9. E. Nr. 599/15. Erlass Nr. A – Semifoursquare 1072. Krause, Eduard, Korrigierte Bedarfsanmeldung vom 07.09.1915, 15 Zeilen, ohne Paginierung.

SMB-PK EM (1915b): I/MV 0034, Pars Ia, Bd. 9. E. Nr. 599/15. Grünwedel, Albert; Schmidt, Max, Aktennotiz auf Runderlass A-883. Bedarfsanmeldung für Petroleum vom 17.07.1915 an den Generaldirektor der Königlichen Museen, 2 Seiten, ohne Paginierung.

SMB-PK EM (1915c): I/MV 0034, Pars Ia, Bd. 9. E. Nr. 599/15. Runderlass A – Semifoursquare 883, Acta betreffend Dienstbestimmungen und Instruktionen. Loseblattsammlung. Bedarfsanmeldung für Petroleum. Krause, Eduard. Randnotiz auf Runderlass A – Semi-foursquare; 883 vom 16.05.1915, 9 Zeilen, ohne Paginierung.

SMB-PK EM (1915d): I/MV 0034, Pars Ia, Bd. 9. E. Nr. E 835/15. Erlass Nr. A – Semi-foursquare 1188. Baumann, (Vorname unbekannt, Anmerk. d. Verf.). Brief vom 14.10.1915, 1 Seite, ohne Paginierung.

SMB-PK EM (1915e): I/MV 0075, Pars II c, Vol. 1. E. Nr. 420/15. Fritz Schulz jun. Aktiengesellschaft, Leipzig, Brief vom 18.05.1915, 2 Seiten, ohne Paginierung.

SMB-PK EM (1915f): I/MV 0075, Pars II c, Vol. 1. E. Nr. 420/15. Krause, Eduard. Aktennotiz vom 25.05.1915 auf Brief von Fritz Schulz jun. vom 18.05.1915, 6 Zeilen.

SMB-PK EM (1916a): I/MfV 0035, Ia, Bd. 10, E. Nr. 130/16. Akte betreffend Dienstbestimmungen und Instruktionen. Loseblattsammlung. Junker, Max und Stubenrauch, Kurt. Aktennotizen vom 04.02.1916, vom 06.03.1916, vom 19.04.1916, vom 25.04.1916, 4 Seiten, ohne Paginierung.

SMB-PK EM (1916b): I/MfV 0035, Ia, Bd. 10. E. Nr. 754/16. Grünwedel, Albert. Brief an die Zentralstelle für Petroleum-Verwertung vom 12.10.1916, 1 Seite, ohne Paginierung.

SMB-PK EM (1917a): I/MfV 0035, Ia, Bd. 10. E. Nr. 537/17. Junker, Max. Beantragung von Petroleum an Albert Grünwedel vom 14.07.1917, 13 Zeilen ohne Paginierung.;

Grünwedel, Albert. Abschrift eines Briefes an die Zentralstelle für Petroleum-Verwertung, 13 Zeilen, ohne Paginierung.

SMB-PK EM (1917b): I/MV 0075, Pars II c, Vol. 1. E. Nr. 570/19. Fritz Schulz jun., Broschüre vom 05.01.1917 mit Referenzen, 4 Seiten, ohne Paginierung.

SMB-PK EM (1919a): I/MV 0075, Pars II c, Vol. 1. E. Nr. 570/19. Werbeschreiben Nr. 212 von Fritz Schulz jun. vom 11.04.1919, 1. Seite, 2 Seiten, ohne Paginierung.

SMB-PK EM (1919b): I/MV 0075, Pars II c, Vol. 1. E. Nr. 590/19. Grünwedel, Albert; Seler, Eduard; Ankermann, Bernhard; Kümmel, Otto; Müller, Friedrich Wilhelm Karl. Aktennotiz auf Werbeschreiben von Fritz Schulz jun. vom 28.05.1919, 4 Zeilen, ohne Paginierung.

SMB-PK EM (1919c): I/MV 0075, Pars II c, Vol. 1. E. Nr. 590/19. Verteilung von Petroleum an nachgeordnete Behörden. Neuhaus, (Vorname unbekannt, Anmerk. d. Verf.). Brief aus dem Ministerium für Wissenschaft, Kunst und Volksbildung vom 09.05.1919; Abschrift vom 26.05.1919, 1 Seite, ohne Paginierung.

SMB-PK EM (1919d): I/MV 0075, Pars II c, Vol. 1. E. Nr. 800/19. Heerdt, Walther. Werbeschreiben der Deutschen Gesellschaft für Schädlingsbekämpfung m.b.H. zur Mottenbekämpfung vom 05.08.1919 mit Anhängen. Brief, 1 Seite, ohne Paginierung.; Allgemeine Geschäftsbedingungen, 5 Seiten, ohne Paginierung.; Antragsformular auf Durchgasung, 1 Seite, ohne Paginierung.; Auftragsformular zur Durchgasung, 1 Seite, ohne Paginierung.

SMB-PK EM (1920a): I/MV 0075, Pars II c, Vol. 1. E. Nr. 438/20. Fritz Schulz jun., Werbeschreiben Nr. 270 vom 16.01.1920, 2 Seiten, ohne Paginierung.

SMB-PK EM (1920b): I/MV 0075, Pars II c, Vol. 1. E. Nr. 438/20. Grünwedel, Albert; Junker, Max. Aktennotiz vom 03.05.1920 auf Werbeschreiben Nr. 120 von Fritz Schulz jun. vom 16.01.1920, 3 Zeilen, ohne Paginierung.

SMB-PK EM (1927): I/MV 0075, Pars II c, Vol. 1, E. Nr. 6/27. Anonymus. Mitteilungen der Gesellschaft für Vorratsschutz e.V., November 1926, Heft Nr. 6, 2. Jahrgang, 69–80. Eingang der Zeitschrift bei den Staatlichen Museen zu Berlin am 05.01.1927. Umlauf mit Unterschriften, 1 Seite, ohne Paginierung.

SMB-PK EM (1929): I/MV 0075, Pars II c, Vol. 1. E. Nr. 193/29. Arbeitsschutz. Brittner, Carl; Eichhorn, August; Krickeberg, Walter; Lehmann, Walter; Preuss, Konrad Theodor; Schachtzabel, Albert; Snethlage, Emil Heinrich. Randnotizen und Prüfvermerke zur Beschaffung von Gasmasken sowie Angebot, Rechnung und Gebrauchsanweisung des Atemschützers Lix von der Firma Deutsche Gasglühlicht-Auer-Gesellschaft m.b.H. vom 14.05.–04.09.1929, 16 Seiten, ohne Paginierung.

SMB-PK EM (1938): I/MV 1071, I B 129. Snethlage, Emil, Heinrich. Besichtigung der Ausstellung der Sammlung Schulz-Kampfhenkel in Leipzig. Bericht vom 21.06.1938, 1 Seite, ohne Paginierung.

SMB-ZA (1905): GV 696, Nr. 1–19. Nr. 1a—257. Zusammenstellungen zu den Belegen für das Jahr 1905. Loseblattsammlung. Zur Verwaltungsrechnung für die Kunstmuseen. Beleg-Nr. 2017, Bezahlung von "Carbolineum" ohne Datum, Blatt 38. 1 Seite; Ohne Beleg-Nr., Ankäufe von Chemikalien vom 01.05.—12.10.1905, Blatt 219–221, 3 Seiten.

Other sources

Andres, Ad. first name unknown (author's note), (ohne Jahresangabe): Bekämpfung der Kleidermotte (*Tineola biselliella*) durch Blausäure. *Zeitschrift für angewandte Entomologie.* Sonder-Abdruck, Band 4(3), 1–3.

Anonymous (1906): Das "Autan"-Verfahren zur Formaldehyddesinfektion. *Pharmazeutische Zeitschrift*, LI(77), 769.

Ayass, Wolfgang (1966): Arbeiterschutz. Stuttgart, Jena. New York: Fischer (Quellensammlung zur Geschichte der deutschen Sozialpolitik, 3).

Buck-Heilig, Lydia (1989): Die Gewerbeaufsicht. Entstehung und Entwicklung. Wiesbaden: VS Verlag für Sozialwissenschaften (Studien zur Sozialwissenschaft, 87). Available online at http://dx.doi.org/10.1007/978-3-663-05750-5.

Dumont, Fritz (1914): Petroleum-Versorgung während des Krieges mit besonderer Berücksichtigung der örtlichen Verhältnisse von Danzig. Available online at http://resolver.staatsbibliothek-berlin.de/SBB0000635C00000000.

Ebbinghaus, Angelika (1998): Der Prozeß gegen Tesch & Stabenow. Von der Schädlingsbekämpfung zum Holocaust. *Zeitschrift für Sozialgeschichte des 20. und 21. Jahrhunderts*, 13(2), 39.

Eichengrün, Arthur (1905): Verfahren zur Entwicklung von gasförmigem Formaldehyd aus polymerisiertem Formaldehyd. Angemeldet durch Farbenfabriken vorm. Friedrich Bayer & Co. in Elberfeld. Veröffentlichungsnr: 181509; Klasse 12 o; Gruppe 7. Prioritätsdaten: Zusatz zum Patent 177053 vom 13. Juli 1905.

Eichengrün, Arthur (1906): Über das neue "Autan"-Desinfektionsverfahren. *Pharmazeutische Zeitschrift*, LI(77), 852.

Farbenfabriken, vormals Friedrich Bayer & Co., Elberfeld & Leverkusen (Hg.) (1918): Unter Mitarbeit von Arthur Eichengrün. Geschichte und Entwicklung der Farbenfabriken vorm. Friedrich Bayer & Co. Elberfeld in den ersten 50 Jahren. Pharmazeutisch-wissenschaftliche Abteilung. Das "Autan"-Verfahren. Unveröffentlichte Schrift, München, 1918, 415.

Frickhinger, Hans Walter (1918): Blausäure im Kampf gegen die Mehlmotte (*Ephestia Kuehniella Zeller*). *Zeitschrift für angewandte Entomologie*, 4. Bd. (1), 129–140, mit 4 Textabbildungen.

Hase, Albrecht (1916): Der Verbreiter des Fleckfiebers. Die Kleiderlaus. *Merkblatt der Deutschen Gesellschaft für angewandte Entomologie e.V.*, Nr. 1 (Serie I). Berlin: Verlagsbuchhandlung P. Parey, 1–8.

Hase, Albrecht (1917): Die Bettwanze und ihre Bekämpfung. *Merkblatt der Deutschen Gesellschaft für angewandte Entomologie e.V.*, Nr. 4 (Serie I). Berlin: Verlagsbuchhandlung P. Parey, 1–8.

Kadlubek, Günther; Hillebrand, Rudolf (1998): AGFA. Geschichte eines deutschen Weltunternehmens von 1867 bis 1997. Neuss: Verlag Rudolf Hillebrand.

Karlsch, Rainer; Stokes, Raymond (2003): Faktor Öl. Die Mineralölwirtschaft in Deutschland 1859–1974. München: Beck. Available online at www.gbv.de/dms/faz-rez/FD1200304101799547.pdf.

Karlsch, Rainer; Wagner, Paul Werner (2010): Die AGFA-ORWO-Story. Geschichte der Filmfabrik Wolfen und ihrer Nachfolger. Berlin: VBB.

Nipperdey, Thomas (1995): Deutsche Geschichte 1866–1918. Zweiter Band: Machtstaat vor der Demokratie. 3. Aufl. München: Beck.

Rathgen, Friedrich (1924): Die Konservierung von Altertumsfunden. Mit Berücksichtigung ethnographischer und kunstgewerblicher Sammlungsgegenstände. 2. Auflage. Berlin, Leipzig: Walter de Gruyter & Co. (Handbücher der Staatlichen Museen zu Berlin, II. und III. Teil.

Simon, Heinrich (1844): Das Preussische Staatsrecht. Erster Theil. Breslau: Georg Philipp Aderholz.

Westphal-Hellbusch, Sigrid (1973): Zur Geschichte des Museums. Hundert Jahre Museum für Völkerkunde. *Baessler-Archive*, XXI.

Wilhelmi, Julius; Kunike, Hugo (1927): Trials and Investigations on the Effectiveness of the Petroleum Raffinate "Flit" in the Control of Flies and Mosquitoes. *Journal of Disinfection and Public Health*, 19(3), 98–99.

13 Orders and consequences of the use of pesticides at the Königliches/Staatliches Museum für Völkerkunde in Berlin from the end of the 19th century to the beginning of the 20th century

Against this background, it is not surprising that von Luschan, in his function as director, instructed the conservator Krause

" . . . die gefährdeten Stücke der Sammlungen v. Stein, Busse und Hanstein ehethunlichst mit Natr. Arsenicum (nicht arsenicosum) zu vergiften".[1]

(SMB-PK EM 1904)

He instructed his assistant at the time, Bernhard Ankermann, in writing on August 15, 1906,

". . . alles gleich mit Natr. arsenicum vergiften, jetzt liegen die Haarschnüre usw. in Pfeffer".[2]

(SMB-PK EM 1906)

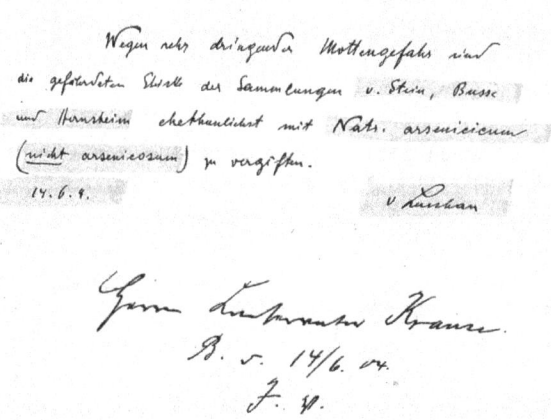

Figure 13.1 File note. Felix von Luschan to Eduard Krause.

DOI: 10.4324/9781003407607-17

Krickeberg's orders in 1938 were already mentioned (see Part III, chapter 12). He instructed his research assistant Snethlage to spray objects on loan from the Schulz-Kampfhenkel collection in Leipzig with "Flit" and to have them "flitted" regularly (SMB-PK EM 1938). Thus, in his letter on March 12,1934, Brittner asked the Generaldirektion der Staatlichen Museen (General Directorate of the National Museums) to announce measures he would take to protect the collections against "Holzwürmer und Mottenfraß".[3] His recommendations included "alljährlich vor Beginn der Flugzeit (Ende März) einen durchgreifenden Reinigungs- und Desin-fektions-Feldzug vorzunehmen".[4] Then, the assistant Zorn should remove infested pieces and bring them to the "Schädlingsbekämpfungsanstalt in Dahlem".[5] However, Brittner's military use of language was remarkable (SMB-ZA 1875–1899, 1915–1935). These individual notes indicate that the decision on the choice of active ingredients and agents for combating pests was made by museum scientists or directors, not by natural scientists or museum staff. Moreover, the timing of the implementation of control measures was not left to employees. Evidence to the contrary and indications as to whether the orders were also implemented could not be clarified in the course of this work.

The personnel authority between the Generaldirektion der Museen, Chemisches Laboratorium, and KMfV/MfV was quite complex. Due to the constant pressure to safeguard the objects in the collection, the museum took measures that had unfore-seen effects on both materials and objects, as well as the responsible personnel, which could not be thoroughly assessed in advance. It has not been possible to demonstrate that the different chemical and physical properties of the materials of the collected objects have been taken into account in preventive or control meas-ures against insect pests. However, collection objects were treated several times with different agents over longer periods, although a systematic approach could not be determined during the study period. The question also remains unanswered as to whether improper handling of previously used pesticides may have caused damage to health in individual cases due to ignorance of the toxic effects of the individual active ingredients. It has been proven that the active ingredient camphor, along with its material-damaging effect, was suspected to be harmful to health as early as 1905 in KMfV. Von Luschan had noticed impairments, as the intensive use of camphor, when the display cabinets were opened, resulted in

" . . . auf der einen Seite ein Entweichen der mit Kampfer gesättigten Schrankluft und auf der anderen Seite eine für das Publikum und für die Beamten gleich lästige Erfüllung des ganzen Saales mit einem unangene-hmen Geruch . . ."[6]

(Luschan von 1905, 239)

It is not clear from the sources whether these observations by von Luschan in 1910–1911 led to joint experiments with Bolle in the collection cabinets of the Afrikanisch-Ozeanische Abteilung. In a written evaluation of the experiments, Bolle explained that among the active ingredients tested, camphor has been proven ineffective in combating insect pests (SMB-PK EM 1911). Consequently, in 1913, synthetic camphor offered by a chemical trader from Glasgow, Scotland, was rejected by Krause (SMB-PK EM 1913). The extent of damage caused by the yellowing and browning of the surfaces of collection items caused by the long-term use of camphor remains uncertain at this point in time.

In parallel, Rathgen subjected carbon tetrachloride to a test for possible damage to surfaces on works of art in a long-term study (see Part III, chapter 8). His findings on the limited use of this substance in paints containing resins or varnish as binders probably contributed to the fact that the staff of KMfV preferred carbon disulfide. It was placed in bowls in collection cabinets, and it was not until 1923 that this substance was used for entire collections in the in-house fumigation plant in Berlin-Dahlem (see Part III, chapter 8). To meet the demands for carbon disulfide and the duration of collection object exposure in a negative pressure environment, the experience and expertise of several institutions were adopted. Director Hackman of the National Museum in Helsinki was contacted, and the results of the joint experiments of Foy and Bolle from R-J-M in Cologne were also available (see Part III, chapter 8). From the end of the 1950s, carbon disulfide was substituted by the preparation "Illo-Spezial-T" (tetrachloroethene) in MfV.[7] It is a gaseous active substance that is classified as environmentally hazardous and carcinogenic (Cf. Tello 2006, 45–47). In addition, damages, such as crystalline efflorescence on the surfaces of objects made of wood and plant fibers as well as discoloration on objects made of textiles, leather, and skins in connection with an analytical determination, are often attributed to the use of previously used pesticides (Unger et al. 2006, 384–394, 2011, 85–96; Tello und Unger 2010, 35–50; Tello 2016, 18–24).

The research and development of pesticides at the end of the 19th century and the beginning of the 20th century, as well as the distribution of active ingredients and agents by retailers, has led to broad acceptance of the use of these chemical substances in view of the almost impossible tasks of storing and preserving collection items properly. Therefore, it is noteworthy that employees at that time critically questioned some pesticides and subsequently did not use them continuously.

Notes

1 . . . to poison the endangered pieces of the v. Stein, Busse and Hanstein collections with Natr. arsenicum (not arsenicosum) as soon as possible (translation by the author).
2 . . . poison everything right away with Natr. arsenicum, now the hair strings, etc., are in pepper (translation by the author).
3 Woodworms and moths (translation by the author).
4 . . . to carry out a thorough cleaning and disinfection campaign every year before the start of the flight season (end of March) (translation by the author).
5 Pest Control Institute in Dahlem (translation by the author).
6 . . . on the one hand, an escape of the cabinet air saturated with camphor, and on the other hand, an equally annoying filling of the whole hall with an unpleasant smell for the audience and for the officials . . . (translation by the author).
7 Friendly oral communication on August 1, 2018, by Klaus Scharn, conservator and chief conservator at the MfV from 1966 to 1999 (translation by the author).

References

Archive material

SMB-PK EM (1904): I/MV 730, Vol. 30, Pars I. B., E. Nr. 578/04. Acta betreffend die Restauration von Alterthümern. Loseblattsammlung. Anweisung für den Einsatz von Schädlingsbekämpfungsmitteln. Luschan von, Felix. Aktennotiz vom 14.06.1904. Blatt 224, 4 Zeilen.

SMB-PK EM (1906): I/MV 0338, Vol. 18, Pars I B, E. Nr. 1234/06. Acta betreffend die Erwerbung ethnologischer Gegenstände aus Australien. Loseblattsammlung. Luschan von, Felix. Anweisung für Bernhard Ankermann vom 15.08.1906 zum Einsatz von Schädlingsbekämpfungsmitteln, 1 Seite, ohne Paginierung.

SMB-PK EM (1911): I/MV 0075, Bd. 1, Pars IIc, E. 1360/10. Acta betreffend die Restauration von Alterthümern. Loseblattsammlung. Bericht von Johann Bolle an Felix von Luschan vom 19.01.1911, 2 Seiten, ohne Paginierung.

SMB-PK EM (1913): I/MV 0075, Bd. 1, Pars IIc, E. Nr. 1823/13. Acta betreffend die Restauration von Alterthümern. Loseblattsammlung. Angebot von synthetischem Kampfer. Barr, James, C. Brief vom 05.11.1913, 2 Seiten, ohne Paginierung; Krause, Wilhelm Eduard Julius. Randnotizen vom 05.11.1913 auf diesem Brief.

SMB-PK EM (1938): I/MV 1071, I B 129. Emil, Heinrich. Besichtigung der Ausstellung der Sammlung Schulz-Kampfhenkel in Leipzig. Bericht vom 21.06.1938, 1 Seite, ohne Paginierung.

SMB-ZA (1875–1899, 1915–1935): I/NG 0454, E. Nr. 0514, B. Nr. 107/34. Loseblattsammlung. Restaurierung der Gemälde und Gemälderahmen. Schädlingsbekämpfung in allen Abteilungen. Brittner, Carl. Antrag (Abschrift) vom 13.03.1934 aus dem Chemischen Laboratorium der Staatlichen Museen zu Berlin an den stellvertretenden Generaldirektor der Staatlichen Museen zu Berlin, Blatt 365, 1 Seite.

Other sources

Luschan von, Felix (1905): Ziele und Wege eines modernen Museums für Völkerkunde. *Globus; Illustrierte Zeitschrift für Länder- und Völkerkunde*, 88, Bd. 15, 239.

Tello, Helene (2006): Investigations on Super Fluid Extraction (SFE) with Carbon Dioxide on Ethnological Materials and Objects Contaminated with Pesticides. Diplomarbeit. Fachhochschule für Technik und Wirtschaft Berlin, Berlin. Fachbereich 5, Gestaltung, Studiengang Restaurierung/Grabungstechnik.

Tello, Helene (2016): Handle with Care – Semi-foursquare Der Einsatz historischer Biozide an Kunst- und Kulturgut und die Folgen für Materialien und Objekte. In: Kontaminiert – Semi-foursquare Dekontaminiert: Strategien zur Behandlung biozidbelasteter Ausstattungen. Tagung im Rahmen der Werkstattgespräche des Bayerischen Landesamtes für Denkmalpflege, 16. und 17. Oktober 2014. München: Volk Verlag München.

Tello, Helene; Unger, Achim (2010): Liquid and Supercritical Carbon Dioxide as a Cleaning and Decontamination Agent for Ethnographic Materials and Objects. *Smithsonian Contributions to Museum Conservation*, (1), 35–50.

Unger, Achim; Tello, Helene; Lindex, Sörrn; Trommer, Bernhard; Behrendt, Stefanie (2006): "Grüne Chemie" hält Einzug in die Restaurierung. Versuche zur Reinigung, Entfettung und Dekontamination von Kunst- und Kulturgut mit flüssigem Kohlendioxid. *Restauro— Zeitschrift für Konservierung und Restaurierung*, (112), 384–396.

Unger, Achim; Weidner, Anke Grit; Tello, Helene; Mankiewicz, Johannes (2011): Neues zur Dekontamination von beweglichem Kunst- und Kulturgut mit flüssigem Kohlendioxid. *VDR-Beiträge zur Erhaltung von Kunst- und Kulturgut*, (2), 85–96.

14 Knowledge transfer, exchange, and dissemination of knowledge at the national and international levels

From the perspective of society as a whole, scientists and curators tended to belong to a small, exotic group of specialists during the study period. Perhaps this is why these museum professionals saw each other less as competitors than as like-minded people who could learn from each other. Curiosity and inquisitiveness created cross-connections between individual disciplines through which a considerable network was formed. Specialist journals played a key role in facilitating knowledge transfer, as they provided an interactive platform for sharing one's own points of view on technical matters with a wider audience. First and foremost were the journals *Museumsjournal–Zeitschrift für Verwaltung und Technik öffentlicher und privater Sammlungen* sowie das *Journal of the Royal Society of Arts*, the leading bodies in the conservation of art and cultural assets. National and international meetings, conferences, or relevant gatherings provided further opportunities for museum professionals to present their own methods of pest control and to exchange the latest research results with each other. The contributions of museum experts from Germany and abroad were selected according to their function, whether as directors or curators from well-known cultural institutions in Germany, Austria, Great Britain, and Sweden. Certainly, there were also envious people, especially with regard to the modern equipment of the Chemisches Laboratorium (chemical laboratory) of KM/SMB. However, the pioneering spirit prevailed, with the aim of benefiting from each other and later being able to try out previously unknown active ingredients and agents in their own institution.

In 1901, the Generalverwaltung of KM in Berlin created his own questionnaire using a different medium, with the aim of disseminating knowledge on insect pest control while gathering information from important institutions and museums worldwide on the use of active ingredients, agents, and methods (SMB-PK EM 1901–1903a). In particular, the use of benzene, camphor, naphthalene, petroleum, kien oil, patchouli oil, turpentine oil, carbolic, thymol–naphthol solutions, extracts of aloë, coloquints, insect powder, Wickersheimer's moth essence, Chinese moth tincture, formalin, sublimate, arsenic soap, and carbon disulfide were asked (SMB-PK EM 1901–1903b).

Of these active ingredients and agents, sublimate dissolved in alcohol was mainly applied directly to the objects at KM to combat and prevent harmful insects, and camphor was placed in the cupboards. However, this approach has not resulted in

DOI: 10.4324/9781003407607-18

lasting success. In addition, the cabinets were overcrowded, and there was a lack of staff to adequately control all items (SMB-PK EM 1901–1903c). The question-naire was agreed upon in a lengthy process among the museum professionals and the Generaldirektion. Von Luschan drafted a template for this and had it professionally examined by Rathgen. Statements by Karl von den Steinen and Julius Lessing[1] were also taken into account. Bastian and Grünwedel then forwarded the questionnaire to the Generalverwaltung of KM for a final review (SMB-PK EM 1901). The dis-patch began at the beginning of October. Well-known institutions and museums in Europe, Indonesia, China, Mexico, Argentina, and the USA were selected. A copy of this questionnaire was found in the archives of the American Museum of Natural History, NYC (AAMNH-DAA 1901). Responses from individual museums and an evaluation of the questionnaire could not be found in the sources. In this context, however, it is worth noting the very important meeting of the "Enquete betreffend die Konservierung von Kunstgegenständen",[2] which was organised by members of the k.k. Zentralkommission (Imperial and Royal Central Commission) from Austria in Vienna in 1904 (k.k. Zentralkommission 1905a). The list of participants reads like a "Who's Who" of the museum experts of the German-speaking world at the time (k.k. Zentralkommission 1905b). In the planning as well as in the invitation to this confer-ence, it has become clear that the content of many questions was based on the ques-tionnaire of KM in Berlin from 1901 (SMB-PK EM 1904a; SMB-PK EM 1904b).

Together with his colleague Lessing, Rathgen represented KM at the conference and, in addition to numerous colleagues from Austrian cultural institutions, he met Johann Bolle,[3] Karl Koetschau,[4] and Ernst Steinmann[5] among others. The results of the Enquete were published in 1905 and were intended to help smaller museums, especially in Austria, conserve their museum holdings (Cf. k.k. Zentralkommis-sion 1905a, 5–44). A few years later, in an article in the journal *Museumskunde–Zeitschrift für Verwaltung und Technik öffentlicher und privater Sammlungen*, Rathgen used the opportunity to critically examine the active ingredients, agents, and individual recipes discussed by the Enquete in Vienna (Rathgen 1910, 23). In particular, he saw recipes based on aqueous solutions as a danger to gilded and unmounted wooden objects (Ibid., 23–24). On the other hand, he advocated impregnations with kerosene, carbolineum, or linseed oil varnish dissolved in tur-pentine, benzene, or carbon tetrachloride. He also considered the use of Chinese wood oil and dammar resin dissolved in carbon tetrachloride to be effective in combating wood-destroying insects, as rotten wood would be strengthened at the same time (Ibid., 26–27). He recommended impregnating or coating objects with paraffin heated to 100–110 °C, especially for objects that are stored under glass, as wooden objects treated in this way would otherwise easily become dusty (Ibid., 27). In Rathgen's opinion, carbon disulfide should only be used in crates lined with zinc sheet, equipped with an exhaust pipe to the outside. In addition, such fumiga-tion boxes should be surrounded by a container with hot water, so that the vapors can then rise more easily in the pipe and escape into the atmosphere. According to his observations, kerosene ethers, benzene, and crude benzene can also be used in this way. At that time, Rathgen was still experimenting with carbon tetrachloride, but he recommended it independently of his research results because of its non-combustibility (Ibid., 25).

Carl Bernhard Salin was director of the Nordiska museet from 1905 to 1913 and director of the Skansen open-air museum from 1905 to 1912, both located in Stockholm. In 1916, at a conference of the Svenska Museimannaförening,[6] he demanded that a state laboratory be set up for all Swedish museums at the Historical Museum in Stockholm. For him, the neighboring countries of Denmark and Germany were exemplary in this direction. Until such a laboratory was established, he encouraged all members of the Swedish Museum Association to exchange information about their activities in the *Museums Journal of the Museums Association* and to report on the conservation issues they dealt with in detail (Anonymous 1916, 268–269).[7] Albert Frank Kendrick of the department of textiles at the Victoria & Albert Museum in London reported at the 13th annual meeting of the Royal Society of Arts[8] that he had met Bolle personally years earlier. For some directors of European art galleries, he has become a key figure in the fight against wood-damaging insects. The experts in the painting collections were unable to develop their own solutions to combat insect pests. As a consultant, Bolle became active in various collections and thus took on an extremely important function in the museum landscape from outside (Scott 1922, 336).[9] At the same meeting, Sidney Harmer[10] described his trip to a museum,[11] where he admired a hitherto unknown fumigation plant, which used carbon disulfide to combat insect pests. He particularly praised the fact that every new object that entered the museum was subjected to this procedure (Ibid., 333–334). For his colleague Scott, the construction of a comparable facility for the British Museum was still a distant prospect at the time (Ibid., 338).

Notes

1 From 1867 to 1908, Julius Lessing was the director of the Königliches Kunstgewerbemuseums at KM (Royal Museum of Decorative Arts at the Royal Museums) in Berlin.
2 Enquete concerning the conservation of works of art (translation by the author).
3 See about Johann Bolle, Part III, chapter 8.
4 Karl Theodor Koetschau, born on March 27, 1868 and died on April 27, 1949, was art historian and museum director. He was editor of the journal founded in 1905 *Museumskunde–Zeitschrift für Verwaltung und Technik öffentlicher und privater Sammlungen.* On May 23, 1917, he and 22 other people from the fields of art and culture founded the Deutschen Museumsbund.
5 Ernst Theodor Karl Steinmann, born on September 4, 1866 and died on November 23, 1934, was a German art historian. From 1903 to 1911, he was director of the Großherzogliches Museum Schwerin.
6 Swedish Museum Association (translation by the author).
7 Further details on the venue of the conference could not be found in the references. It is assumed that the meeting took place in Stockholm.
8 From 1770 onward, the annual meetings of the Royal Society of Arts were held at its headquarters, 8 John Adam Street, Adelphi, London (see: Wood, Henry Trueman (1913): A history of the Royal Society of Arts with a preface by Lord Sanderson, London, 70.)
9 Alexander Scott, born on December 28, 1853 and died on March, 10, 1947, was a chemist who devoted himself to the preservation and conservation of collection objects in the British Museum.
10 Sidney Harmer was at that time director of the Natural History department at the British Museum in London.
11 What is meant here is the Nordiska museet in Stockholm (author's note).

References

Archive material

AAMNH-DAA (1901): Finding aid, correspondence re: fumigation, Box 13, Folder 2. Fragebogen zur Schädlingsbekämpfung von der Generalverwaltung der Königlichen Museen zu Berlin aus dem Jahr 1901, 3 Seiten, ohne Paginierung.

SMB-PK EM (1901): I/MV 0057, Bd. 5, Pars I c, zu E. Nr. 36/1901. Aktennotiz vom 26.07.1901 von Bastian, Adolf, 1- zeilig und von Luschan von, Felix, 6- zeilig, 1 Seite, ohne Paginierung.

SMB-PK EM (1901–1903a): I/MV 0057, Bd. 5, Pars I c. Umzugsakte. Acta betreffend den Umzug und die Aufstellung der Sammlungen des Museums. Königliches Museum für Völkerkunde zu Berlin 01.01.1901 bis 30.04.1903.

SMB-PK EM (1901–1903b): I/MV 0057, Bd. 5, Pars I c. zu E. Nr. 36/1901. Umzugsakte. Acta betreffend den Umzug und die Aufstellung der Sammlungen des Museums. Fragebogen in gedruckter Form 1. Seite, 3- seitig, ohne Paginierung.

SMB-PK EM (1901–1903c): I/MV 0057, Bd. 5, Pars I c. zu E. Nr. 36/1901. Umzugsakte. Acta betreffend den Umzug und die Aufstellung der Sammlungen des Museums. Fragebogen in gedruckter Form, 2. Seite, 3- seitig, ohne Paginierung.

SMB-PK EM (1904a): I/MV 0058, E. Nr. 1119/04. k.k. Zentralkommission für Kunst- und historische Denkmale in Wien 08.08.1904. Ankündigung einer Enquete zur Bewahrung von organischen Materialien. Schreiben vom 08.08.1904, gedruckt, 2 Seiten, ohne Paginierung.

SMB-PK EM (1904b). I/MV 0058, E. Nr. 1387/1904. K.k. Zentralkommission für Kunst- und historische Denkmale in Wien September 1904. Einladung und Programm zu einer Enquete über die Bewahrung von organischen Materialien. Schreiben vom September 1904, gedruckt, 1 Seite, ohne Paginierung.

Other sources

Anonymous (1916): The Technical Preservation of Antiquities. *The Museum Journal*, (15), 268–269.

k.k. Zentralkommission (1905a): Zentralkommission für Kunst- und historische Denkmale in Wien 1905 betreffend die Konservierung von Kunstgegenständen. Auszug aus dem stenographischen Protokoll. 10., 11. und 12. Oktober 1904. Wien: Rudolf Brzezowsky.

k.k. Zentralkommission (1905b): Zentralkommission für Kunst- und historische Denkmale in Wien 1905 betreffend die Konservierung von Kunstgegenständen. Verzeichnis der Teilnehmer, 2 Seiten, ohne Paginierung.

Rathgen, Friedrich (1910): Über Mittel gegen Holzwurmfraß. *Museumskunde, Zeitschrift für Verwaltung und Technik öffentlicher und privater Sammlungen*, Bd. VI, 23.

Scott, Alexander (1922): The Restoration and Preservation of Objects at the British Museum. *Journal of the Royal Society of Arts*, LXX(3618), 333–334, 336.

15 Implementation of pest control measures in national and international contexts during the period under investigation

Use of active ingredients and pest control agents in comparable German museums

The Rautenstrauch-Joest-Museum (R-J-M) in Cologne, the former Museum für Völkerkunde (Museum of Ethnology) in Hamburg, today's Museum am Rothenbaum–Kulturen und Künste der Welt (Cultures and Arts of the World (MARKK), and the GRASSI Museum für Völkerkunde (Museum of Ethnology) zu Leipzig were selected in order.

Between 1909 and 1922, in addition to the mass fumigation of harmful insects in an in-house fumigation system, suitable pesticides were used at R-J-M in Cologne that had to meet different requirements. They should be used directly in places as diverse as enclosed spaces, collection cabinets, and exhibition showcases, as well as on objects. In a letter from 1913, the Actien-Gesellschaft für Anilin-Fabrikation in Berlin submitted recipes for the application of "Dichlorobenzene Agfa", used in closed rooms. It was recommended to dissolve the dichlorobenzene crystals with either 9 parts ethanol or 2–3 parts carbon tetrachloride, similar to "Mottenether".[1] If larger rooms were to be cleaned, it was advised to boil 50 g of "Dichlorobenzene Agfa" with 1/2 liter of water for 5 minutes to evaporate 30 g of the agent. Barely half a year later, the company informed the museum that further demand for its product would now have to be obtained from Fritz Schulz jun., a company in Leipzig (HAStK 1913/1914). In November 1914 and May 1915, Fritz Schulz jun. repeatedly proposed the use of the insecticide "Globol" to the museum for treating collection cabinets and showcases, highlighting that it effectively kills insects and their offspring, unlike camphor and naphthalene. Along with the letter, small bags of product samples bearing the company imprint were also sent. The cover letters were also accompanied by product information, including price lists and references from third parties. In particular, emphasis was placed on the scientific recognition of the agent, and this was confirmed by statements from institutes, such as the Pflanzenpathologische Versuchsstation der Königlichen Lehranstalt für Wein-, Obst- und Gartenbau (Plant Pathology Experimental Station of the Royal Institute for Viticulture, Fruit Growing and Horticulture) in Geisenheim, and the Königlichen Naturaliensammlung (Royal Natural History Collection) in Stuttgart (HAStK 1914/1915). In August 1916, at the request of the museum, the Chemische

DOI: 10.4324/9781003407607-19

Fabriken (Chemical factories) Flörsheim am Main supplied the wood preservative "Wurm-Antorgan" for direct application to the objects. Whether there was a repeated order due to a demand from the same factory in 1922 could not be clarified at this point in time (HAStK 1922).

The Museum of Völkerkunde in Hamburg, founded in 1879, was given its own building in 1912 under its first director Georg Thilenius[2] (Thilenius 1916, 5–13). The collections were just as extensive as those at R-J-M and required intensive control against insect pests. Initially, they were housed together with the Naturhistorisches Museum im Johanneum (Natural History Museum in the Johanneum) of the city of Hamburg. There, an overseer, an assistant supervisor with previous training in craftsmanship, and a female assistant carried out conservation work on the ethnological objects (Thilenius 1905, 231). Thilenius was far ahead of his time and showed a high degree of logistical thinking regarding the structural and spatial location of various activities in his museum. To prevent the recurrence of infestations in newly arrived and previously treated objects, he ordered the sighting, quarantine, and further treatment of objects be moved to the basement of the museum. He was also apparently aware that the active ingredients and agents applied posed a health hazard to those working with them, which was why a washroom was also set up (Ibid., 238–239). Essentially, he used carbon disulfide, carbon tetrachloride, and kerosene preventively and to combat insect pests. Similar to other museums of this size, Thilenius noticed a certain level of protection against new insect infestations by storing collection objects in tightly sealed iron cabinets equipped with carbon disulfide and kept open as little as possible (Ibid., 236). If conservation work on large objects using this active ingredient was necessary, the procedure was carried out in an open courtyard because of the fire hazard of carbon disulfide (Ibid., 232). During the First World War, Thilenius had to purchase carbon disulfide from Berlin due to a shortage of raw materials. On April 27, 1916, he wrote to the Kriegs-Rohstoffabteilung des Kriegsministeriums (War Raw Materials Department of the War Ministry) that he needed 10 kg of carbon disulfide in the workshops of the Völkerkundemuseum to combat pests in the collections as well as for the annual retreatment of existing holdings (Archiv des Museums am Rothenbaum–Kunst und Kulturen 1916).

Thilenius considered only carbon tetrachloride for soaking smaller pieces. Here, too, his counterpart Foy in Cologne was skeptical, as he considered this method to be effective only if the larvae were located in areas close to the surface of an object. In his opinion, deeper-lying "drill worms" would have to be deprived of oxygen to kill them (Archiv des Museums am Rothenbaum–Kunst und Kulturen 1912). Smaller and thin objects made of wood and feathers were soaked entirely in kerosene for one to seven days. An addition of up to 50 % varnish was also used to strengthen wooden objects. Large objects were placed on trestles and coated with a brush until saturation of kerosene on the underside became visible. Drying in air was considered disadvantageous, as the objects treated in this way subsequently became dusty again. Discoloration of earth colors was also observed, which was mainly attributed to the varnish and not directly to the active ingredient. Very delicate pieces were placed in alcohol, which contained a small amount of sublimate.

After drying, the objects were coated by soaking them in paraffin (Cf. Thilenius 1905, 234–235). There were also apparently complaints regarding the use of carbon disulfide in the museum's premises, prompting Thilenius to look for alternatives. To maintain insect-free collections after treatment, he, like his colleagues in Cologne and Berlin, adopted the permanent use of dichlorobenzene with the trade name "Dichlorobenzene Agfa" or "Globol". To ensure lasting protection, the agent should be scattered in sufficient quantities in collection cabinets and regularly replenished (Cf. Thilenius 1916, 139). For 20 years, from 1914 to 1934, the museum obtained and used this insecticide (Archiv des Museums am Rothenbaum–Kunst und Kulturen 1914–1934). The annual expenditure for "Globol" was even included in the state budget of the city of Hamburg. Between 1914 and 1920, one had to accept considerable price increases of this agent (Hansestadt Hamburg 1921, 309). Another product marketed by Deutsch–Amerikanische Petroleum Gesellschaft (German–American Kerosene Company) was "Flit". Thilenius, after evaluating the agent in 1929, identified a reduction in indoor odor nuisance, leading him to deem the use of "Flit" to be beneficial (Archiv des Museums am Rothenbaum–Kunst und Kulturen 1929). Chemische Fabrik aus Flörsheim a. M. has sent price lists for wood preservation and pest control products to the Museum für Völkerkunde in Hamburg (Archiv des Museums am Rothenbaum–Kunst und Kulturen 1936). However, it was not clear from the sources whether any agents were obtained or not.

Founded in 1869, the GRASSI Museum für Völkerkunde (GRASSI Museum of Ethnology) in Leipzig houses around 200,000 ethnological objects. In this museum, conservation tasks to control and prevent the spread of harmful insects have been carried out since the late 19th century. These measures were documented in invoices for the purchase of various substances. Angelica Hoffmeister-zur-Nedden also found empty packaging of the pesticide "Globol" in the collection cabinets.[3] In 1933, Paul Germann's article provided an overview of the measures and agents utilised at the GRASSI Museum to eliminate insects. Among them is the agent "Antisekt", which could be obtained from the pharmacist M. Wagner & Co. in Leipzig. Arsenic has also been used, but this active ingredient has not been used for a long time because of its toxicity. Unfortunately, there is no indication of when and how long these substances were used. A similar approach can be observed in the choice of other active ingredients and agents, as in Cologne and Hamburg. "Globol" was purchased from Fritz Schulz jun. Aktiengesellschaft in Leipzig. Treatments with naphthalene and carbon disulfide were documented as well. The latter was placed in open containers in the collection cabinets, but there were also complaints about the unpleasant smell in the museum. Germann stressed that there were always difficulties when opening the cabinets, especially in the exhibition rooms. Therefore, as a substitute, the product "Areginal" was introduced in the museum from 1931 onward, which was purchased from I.G. Farbenindustrie Aktiengesellschaft, sales department for crop protection in Leverkusen am Rhein. Like an ally, Germann referred to the new means in his article when he wrote:

"Als willkommenste Mittel begrüßen wir Areginal im Kampfe gegen die holzzerstörenden Schadinsekten, . . . "[4]

Because the GRASSI Museum did not have its own fumigation system, all new collection items entering the museum were exposed to "Areginal" fumes for three to four days after being cleaned in an airtight box (Germann 1933, 9–11).

A quick examination of the active ingredients and agents utilised by the museums featured here reveals the dominance of "Globol" at the beginning of the 20th century, as it was used equally by all three institutions. Because of the disturbing smell and the potential health risks it posed, it was replaced by "Wurm-Antorgan", "Flit", and "Areginal", which were each applied differently in individual collections. All the agents shared an industry connection and contained active ingredients, except for the GRASSI Museum's use of naphthalene (synthetic active ingredient) and "Antisekt" in 1933, which was believed to be a recipe from a local pharmacy. In addition, the use of the synthetic active ingredient sublimate in 1905 and the use of kerosene, an industrially produced preparation containing active ingredients, have been documented in the Hamburg collections. All of the active ingredients and agents used at that time were chemical pest control substances, except for "Antisekt". The situation is comparable for the use of carbon tetrachloride and carbon disulfide vapors. In the absence of a museum's own facility for the mass fumigation of ethnological objects, carbon disulfide was filled into open containers in the GRASSI Museum and distributed in the collection cabinets. In Cologne and Hamburg, they applied these two fumigants to the museum's own fumigation plants. The local differences are remarkable. In R-J-M in Cologne, Foy worked with Bolle for many years on the best suited agent. On the other hand, through his foresight, Thilenius took into account aspects of building physics in the planning of the new museum building and included aspects of occupational hygiene as well as occupational health and safety, which were still in their infancy at the time. All functional spaces associated with direct work on the objects were located in the basement of the museum and were arranged in such a way as to create rational and coordinated workflows.

Use of active ingredients and pest control agents in comparable European museums

Numerous references in Swedish museums point to pest control measures. An outstanding example is the preparation of the horse "Streiff" of Gustav Adolf II, which he rode in the Battle of Lützen in 1632. It is located in the armory of the Royal Palace in Stockholm and is one of the oldest surviving skin montages. In 1978, it was purified by the chief taxidermist Werner Berg and brushed with a weak solution of arsenic(III) oxide for safety. In the permanent exhibitions of the Natural History Museum in Gothenburg, the conservator Thomas Gütebier discovered that in 1866, the prepared Malm's blue whale was preserved on the inside of the skin with a saturated solution of arsenic(III) oxide and on the outside of the skin with both a saturated solution of arsenic(III) oxide and mercury(II) chloride. In the museum's collection, he discovered the bellows of an avocet that had been smeared with arsenic soap on its beaks and feet around 1900, along with numerous bird skins. In 2012, Gütebier noted that newly established animal specimens in the museum continued to be treated with arsenic(III) oxide, as this method was considered safer and more durable than modern insecticides based on chlorinated hydrocarbons. This view was also held at the National Museum of Natural History in Stockholm.[5] As early as 1988, the bellows magazine of the Natural History Museum in Malmö was examined for substances formerly used on specimens for the prevention and control of insect pests. Measurements of plaster and wall paints were carried out, whereby arsenic, among other things, was found on wall surfaces (Gütebier 2012a, presentation; Gütebier 2012b, letter to the author, March 26, 2012).

Hans Aall from the Norsk Folkemuseum in Oslo, Norway, wrote a comprehensive manual on how to work in cultural history museums for the rural network of Norwegian museums. It was important for him to make his knowledge of conservation and restoration available in written form to many rural museums in Norway (Aall 1925, 80–85).[6] In a few pages, he discussed the topic of pest control as it pertains to preserving textile collection items and wooden objects. In particular, he emphasised the danger of insects in the larval stage. Inspired by Rathgen, the Norsk Folkemuseum used a process based on heating collection items to 50–60 °C for half an hour (Ibid., 80). The plant, which mainly processed textiles, was designed in such a way that it can use carbon disulfide (Ibid., 80–83). The apparatus created a unique selling point for the Oslo museum.

However, this method did not have a preventive effect, and the application was not harmless due to the fire hazard of carbon disulfide. Therefore, Aall also advised to always ensure absolute cleanliness in showrooms, magazines, and showcases

Figure 15.1 Heating of textiles in a plant at 50–60 °C at the Norsk Folkemuseum in Oslo.

as a preventive measure. In addition, he recommended the use of the agent "Globol", which, to his knowledge, has demonstrated its effectiveness over time at the Völkerkundemuseum in Hamburg. He also recommended his own recipes and referred to a moth water that was made in the armory in Stockholm, which consisted of lavender oil, spike oil, turpentine oil, and camphor. Although the moth agent was harmless in his perspective, he still prioritised the use of "Globol", as he believed it to be more effective. It has been noticed that there is a tendency to prefer industrially produced agents, which may be due to cost reasons or faster availability, because homemade moth water was found to be more expensive to produce. Aall warned against mixtures containing camphor and kerosene, as the latter can leave harmful acids on collection items (Ibid., 81–82). For protection against wood-destroying insects, he recommended coating wooden buildings in their entirety with "Carbolineum", another industrially produced agent, and setting up clay bowls with this agent in invisible places inside during swarming periods. Nevertheless, if the larvae prove to be resistant, such buildings should be sealed in their entirety, and carbon disulfide should be used. To execute this highly dangerous task, he offered instructions from his museum. He also mentioned hydrogen cyanide, but he emphasised that the procedure can only be performed by experts. According to Aall, furniture and smaller objects made of wood can be safely treated with carbon disulfide in airtight crates or in a fumigation system. He also recommended soaking or coating wooden objects with pure kerosene or paraffin, contrary to his criticism of the use of mixtures (Ibid., 83–85).

At the end of the 19th century and the beginning of the 20th century, there was more extensive literature on the conservation of museum collections in Great Britain than elsewhere in Europe, regarding the control of insect pests. England had already established itself as a world power at the beginning of the 18th century, with a vast colonial presence. This allowed for the development of strong historical ties with non-European countries and territories while also creating a natural interest in their cultural assets. In particular, the extensive collections of the British Museum in London were created against this historical background. The collections of the museum benefited from the expansionist policy of the Empire, but they were also forced to deal very intensively with conservation issues relating to the preservation of objects and specimens. Thus, scientific principles for the conservation of museum objects were worked on, and the events and developments of similarly large museums, such as those in Copenhagen or Stockholm, were followed with great interest. News from the Chemisches Laboratorium at SMB was given particular attention, as the British Museum also wanted to establish a natural science laboratory. As part of a cooperation that lasted several years, the British Department of Scientific and Industrial Research employed the chemist Alexander Scott at the British Museum and, later than in Berlin, maintained a chemical laboratory there from 1920 onward.[7] The museum, for its part, made the premises available for scientific research (Cf. Scott 1922, 335). In addition to research methods of preserving inorganic materials, Scott's tasks included developing active ingredients, agents, and methods against bacterial and insect pest destruction of organic materials, such as paper and wood. For example, he recommended some

peroxodisulfates against the growth of microorganisms on paper, namely, ammonium, sodium, potassium peroxodisulfate, and hydrogen peroxide. According to
his observations, wood-damaging insects left great damage on wooden objects,
as well as on paintings on wooden panels. He fervently hoped for a significant
improvement in this situation through further research in the museum's specially
equipped natural science laboratory (Ibid., 330, 333). As a chemist, he was well
aware of his pioneering work in the field of cultural property protection and appre-
ciated the cooperation with colleagues from the most diverse departments of the
museum due to the variety of materials available to them (Ibid., 338).

   In 1924, the laboratory was expanded to include chemist Harold James Plenderleith, who continued to work with Scott on pest control. Two years later, it was
possible to publish the experiments and research carried out up to that point
(Department of Scientific and Industrial Research (eds.) 1926). With regard to the
control of insect pests, their report *Cleaning and Restoration of Museum Exhibits*
contained a chapter on the cleaning and preservation of wood. A sodium silicate
with the trade name "P. 84." proved to be helpful in the control of wood-destroying
insects. It should be noted that wood coated with it becomes hard and stable after
drying, and no annoying shine remains on the surfaces. It was observed that the
coating applied as surface protection was not a viable target for wood-destroying
insects, such as beetles, to attack (Ibid., 42–43). These insects were also combatted
with carbon disulfide. Because of the extreme fire risk of the active substance, it was
recommended to mix 20% of carbon disulfide with carbon tetrachloride, even if the
insecticidal effect was reduced (Ibid., 12–13). In 1934, based on Scott's research,
Plenderleith wrote the English standard work "The Preservation of Antiquities" for
curators (Plenderleith 1934), where he dealt extensively with pest control measures for the organic materials wood, textiles, leather, and feathers. This manual is
interesting for several reasons and therefore deserves to be discussed extensively.
Previous observations at the British Museum had prompted a heightened focus on
preventive conservation of both the storage systems used for collections and the
showcases in exhibition areas. In the first chapter, active ingredients, agents, and
methods of pest control are described in detail. At the end of the monograph, there
are numerous references to literature by colleagues from Europe. Plenderleith was
therefore familiar with the writings of his colleagues and, in this way, he helped
spread their expertise (Ibid., 64–70). He clearly pointed out that the treatment of
museum objects against harmful organisms is not a panacea if, at the same time,
unfavorable climatic conditions in exhibitions, such as mold formation, are triggered on works of art. According to his observations, a temperature of 10–24 °C
and relative humidity of 40–60 % should be maintained (Ibid., 1). In case of necessary treatment against molds, he recommended the active ingredients mercury(II)
chloride, "formalin", and thymol. Due to its toxicity, mercury(II) chloride should
always be used as a *last resort*. He described the use of "Formalin" vapors in the
case of bacterial infestation on books as very effective, but pointed out that leather
and adhesives become brittle and pigments are impaired. He emphasised thymol as
a strong fungicide, which, in his opinion, does not have a harmful effect on works
of art. A further relief in the conversion process is the use of a cabinet equipped

with an electric 40 watt lamp, which is placed under a coaster equipped with thymol crystals and serves as a "fumigation cabinet". He only evaluated the strong volatility of the substance negatively, which means that this procedure has to be repeated more often (Ibid., 2–3). Although a drier climate preemptively minimises the growth of mold and attacks by insects, Plenderleith pointed out that wood-destroying insects are exempt. Before using chemical substances, wooden objects should therefore be heated in an oven for one hour at 60–70 °C (Ibid., 3). Among fumigants, he classified hydrogen cyanide as both the most dangerous and the most effective active ingredient. For safety reasons, he advised curators more on the use of carbon disulfide for pest control in exhibition spaces. He confirms Scott's instructions to mix this active ingredient with carbon tetrachloride at the expense of a lower insecticidal effect to reduce the fire hazard of carbon disulfide. In principle, he advised against treatments with chlorinated compounds when it comes to objects with delicate colors as well as valuable works of art. For larger wooden objects, Plenderleith recommended mounting liquid carbon disulfide in a fountain pen with a thin nib and injecting the active ingredient into existing boreholes. To close the holes, he again referred to his own recipes and advised using a mixture of beeswax and resin for this. He also reported that carbon tetrachloride and paraffin have been successfully applied on wood and timber. Both substances have low odor to almost odorless, and carbon tetrachloride has a good insecticidal effect. To prevent mold growth and attacks by wood-destroying insects, Plenderleith suggested using agents, such as celluloid, vinyl acetate, paraffin wax for damp wood after it has dried, and "Bakelite" for strengthening it (Ibid., 13–14).

Preventive measures against insect pests should always be applied to textiles, feathers, and skins, both in exhibitions and collections. For this purpose, Plenderleith recommended the use of volatile substances, such as naphthalene, camphor, and dichlorobenzene. In his view, naphthalene and camphor can be used in simple, unpurified forms to reduce running costs (Ibid., 6). In his opinion, leather, especially leather-bound books in the tropics, needs long-lasting protection against bacterial attacks. For this reason, he recommended adding mercury(II) chloride and ammonium arsenate to the glue during book binding. Cardboard boxes and seams should also be coated with leather before covering. To ensure safety, he recommended that a label be attached with the date and details of the agent used in books treated in this manner (Ibid., 6). However, he recommended washing archaeological leather recovered from bogs or damp ground in a 10 % "formalin" solution only in exceptional cases, because the leather treated in this way becomes hard as a board, brittle, and cannot be softened again after 24 hours (Ibid., 8). He did not consider it useful to recommend commercial insecticides, which could be quite effective. These insecticides would typically be used in public galleries, where fragrances were added to them to prevent odor nuisance (Ibid., 4–5).

England took an intensive part in archaeological excavations during the study period. Therefore, it also made sense to deal with the conservation of excavation finds. In the 1930s, English chemist Alfred Lucas provided another manual in this context (Lucas 1932). As an assistant, he gained knowledge during archaeological excavations at the tomb of the Egyptian pharaoh Tutankhamen (Gilberg 1997,

31–48). The fact that he himself was only able to draw on a few monographs and pamphlets for the conservation of archaeological finds may have prompted him to supplement and expand the knowledge available to date with his experiences and observations (Ibid., 133). Like his colleagues at the British Museum, Lucas considered the attacks of insect pests on organic materials to be extremely serious, in addition to atmospheric influences, and recommended that care should always be taken to close display cases for exhibits and to carry out regular inspections and cleaning. In many cases, the choice of active ingredients and agents follows the recommendations of the British Museum when he wrote that naphthalene should be kept in showcases for objects made of feathers, skin, hair, hides, and wool. This seems to be more effective than camphor alone. In the case of acute infestation, he recommended fumigation with carbon disulfide. In showcases, the active ingredient should be used openly in trays for a week. He described the substance as flammable, readily combustible, with an unpleasant odor, and exhibits explosive properties, which is why open flames should be avoided in the vicinity due to the risk of fire hazard caused by carbon disulfide. Carbon tetrachloride was also recommended, but he considered it less effective. In his opinion, hydrogen cyanide or sulfur can also be used, but he warned of the dangers of hydrogen cyanide due to its toxicity and advised that only trained personnel should handle it.

He described the use of sulfur as difficult without further details and recommended spraying objects with various active ingredients, which include kerosene, mercury(II) chloride dissolved in alcohol, and naphthalene dissolved in carbon tetrachloride. Despite their good insecticidal effect, he advised against using arsenic and copper compounds in ancient objects as well as in art and cultural assets in general, as these active ingredients can only be dissolved in water to his knowledge (Ibid., 27–29). Lucas recommended using mercury(II) chloride dissolved in alcohol to protect organic materials, such as horn, leather, and parchment, from insect attacks. He believed that termites and beetles are the primary wood-destroying pests in areas with a tropical climate. He emphasised that termites can only be kept away from museum collections through preventive measures, as ancient objects cannot tolerate the application of tar, tar oil, or paint. The only measure he recommended for termite control is the use of sand, gravel, bricks, or asphalt to fill a treated building. In his opinion, beetles must be fumigated with carbon disulfide or sprayed with mercury(II) chloride dissolved in alcohol. Benzene or naphthalene, dissolved in carbon tetrachloride, can also be applied (Ibid., 111–112). For protection against mold, objects made of linen, paper, and papyrus should be treated with thymol in airtight boxes (Ibid., 89–96). To increase the effect, objects treated in this way can also be slightly heated from the outside by an electric lamp. With this suggestion, Lucas was clearly referring to his colleague Scott in a footnote (Ibid., 96). Fungal infestation can also be brushed off with thymol or dipped in thymol, dissolved with alcohol or benzene. For textile objects infested with mold, he recommended spraying them with mercury(II) chloride, dissolved in alcohol, or with naphthalene, dissolved in carbon tetrachloride. Naphthalene can also be placed in solid form in boxes that are used to store textiles. In case of acute infestation, he advised fumigation with carbon disulfide. According to him, the control of molds and red rot fungi in wood should be carried out with mercury(II) chloride dissolved in alcohol (Ibid., 109–113).

Two examples were chosen for the control of insect pests in natural collections. In 1893, for the conservation of bird skin and other objects at the Grosvenor Museum in Chester, curator Robert Newstead recommended his own recipes for an agent, which he classified as nontoxic. In addition to boric acid (orthoboric acid), it also contained glycerol (propane-1,2,3-triol), ethanol, and water. Based on his observations, objects treated in this manner were never attacked by insects, and their skins remained supple for further processing for assembly. Feathers were not contaminated during this procedure according to his experience. In this way, he preserved approximately 300 bird skins, small fish, and entomological objects. However, Newstead's conclusion as to why this recipe does not apply to large objects lacks justification (Newstead 1893, 104–106). The Amgueddfa Cymru-National Museum in Cardiff houses the herbarium of the National Museum of Wales since 1907. To protect numerous herbaria, substances have also been used to control and prevent pests. Extensive chemical analyses of plant specimens, especially herbarium leaves, were able to demonstrate the previous use of mercury(II) chloride, arsenic(III) oxide, barium hexafluorosilicate, and organic pesticides, including lindane, DDT, and naphthalene (Purewal 2001, 144). In addition, from the end of the 19th century until the 1970s of the 20th century, naphthalene beads were hung in small bags in wooden cabinets where herbaria were stored. Nevertheless, a pungent smell remained on the inventory, even after the bags were removed in the 1980s (Cf. Purewal 2012, 40).

Two natural history and two botanical collections were selected for investigation in French and Swiss museums. As part of her diploma thesis, Aude-Laurence Pfister explored the potential use of historical active ingredients for the prevention and control of insect pests in the Muséum d'histoire naturelle de Dijon, Jardin des Sciences, located in the Arquebuse Dijon botanical garden. For her research, she had access to both oral testimonies from employees and documents from the conservation department and the museum's archive. The museum's collections were created at the beginning of the 19th century and have been demonstrably treated with various active ingredients and agents since around 1830. They include natural history objects from the fields of zoology, botany, mineralogy, geology, paleontology, and entomology (Pfister 2008, 85–90). As in many other European collections, solutions of arsenic(III) oxide and mercury(II) chloride have been applied to animal preparations. Herbaria were treated with carbon disulfide and naphthalene, and insects were treated with thymol, nitrobenzene, and beech tar oil. For the preservation of wet preparations, alcohol, and "Formol" were used. Not detected in the collection objects but generally used in the same period were the active ingredients sulfur, camphor, paraformaldehyde, pyrethrum, and quendel oil (Ibid., 91–96; Deschka 1987, 57–61). At the Neuchâtel Museum of Natural History, samples were taken from mammal and bird specimens. Here, the use of arsenic(III) oxide and mercury(II) chloride was also confirmed (Dangeon 2014, 23–31). In Switzerland, Walter Rytz was a conservator of the herbaria at the Botanical Institute of the University of Bern from 1915 onward. In the "*Mitteilungen der Naturforschende*

Gesellschaft in Bern",[8] he gave instructions on the maintenance and installation of herbarium boxes. To protect against insect infestation, all plants that died had to be treated with carbon disulfide in a box. Rytz seemed aware of the health risks associated with mercury(II) chloride, which was commonly used but can be harmful, when he mentioned his own formulation of zinc chloride, turpentine oil, and camphor, dissolved in alcohol, as an alternative to this active ingredient. He recommended an interesting method that he developed himself from his practice. This method involved growing plants on white sheets of paper that have been heavily printed on the back, as printing ink prevents infestation by insects. However, he pointed out that the effect of printing ink diminishes over time, and that the sheets of paper must be replaced repeatedly.[9] He also wrote with pride that the premises of the Botanical Institute were insect-free due to this measure and the fact that the rooms were kept closed during summer (Rytz 1922, 11). The Swiss naturalist Jean-Balthazar Schnetzler was a professor at the Lausanne Academy, now the University of Lausanne, in Switzerland, from 1871. He, too, followed the generally accepted European recommendations when he advised that several herbaria be placed in a wooden box and to fill 4 ounces[10] of carbon disulfide on top. Subsequently, the box had to be sealed tightly and placed under a shed roof for about a month due to a possible risk of explosion. He recommended that this procedure be performed also in entomological collections (Schnetzler 1916, 767).

At this point in time, it is not possible to identify individual sources for the historical use of pesticides in Russian museums. Nevertheless, important recommendations and suggestions can be found in M. V. Farmakovsky's[11] monograph, *Konservierung und Restaurierung von Museumssammlungen, ein Handbuch für Mitarbeiter in Museen*,[12] published in 1947 and was translated into Czech in 1955. It was not intended for conservators but for scientific staff of museums to provide them with comprehensive information on the conservation and restoration of collection items. Thanks to a publication by Jirina Lehmann, it was possible to include the contents relevant to the present thesis. The handbook was based on lectures given by Farmakovsky in 1939 at the Faculty of Local History and Museum Studies of the Communist Institute of Public Enlightenment in St. Petersburg. It was a kind of postgraduate study for museum scholars (Lehmann 2005, 47–62). In his chapter on the importance of conservation, he briefly summarised the history of conservation in Russia and Western Europe, mentioning great European museums and their conservationists. Despite considerable spatial distances, access to Western European conservation literature was available. Rathgen, with his world's first scientific laboratory, is credited as the pioneer of a significant revolution in the field of conservation (Ibid., 51).

Farmakovsky classified pests in museum collections into three categories: bacteria, molds, and insects. To prevent and combat these pests, he recommended using thymol or phenol for bacteria on individual objects. He recommended fumigating multiple objects by exposing them to formaldehyde vapor in sealed rooms. To do so, a 40 % formaldehyde solution should be injected through the keyhole into the room and remain there for two to three days. On 1 m^3 of air, 25 cm^3 of formaldehyde solution was evaporated. Molds can also be treated in this way, but

Farmakovsky observed a stronger effect by using a 0.1–0.2 % sublimate solution applied with a brush. From today's perspective, it is remarkable that he described objects treated in this way as poisoned and pointed out that many museums have already stopped treatments using mercury(II) chloride because of the high toxicity of this active ingredient. He saw organic materials, such as wood, paper, and textiles, as primarily endangered by insects. In his opinion, the fumigation of objects should be prioritised here. He recommended hydrogen cyanide, ethylene oxide, chloropicrin, and organic solvents, such as carbon tetrachloride and dichloroethane. He also mentioned a possible negative effect of chloropicrin on materials and paints and advised caution. This also applied to the use of the highly explosive carbon disulfide. Farmakovsky doubted the effectiveness of naphthalene in combating clothes moths and recommended dichlorobenzene instead.

Use of active ingredients and pest control agents in comparable museums and cultural institutions in the USA and Canada

The oldest and largest institutions that have been committed to the preservation, conservation, and restoration of art and cultural heritage since their inception in the USA are the Smithsonian Institution, Washington, D.C., founded on August 10, 1846; the American Association of Museums, founded in 1906; and the National Park Service of the USA, founded in 1916. At the end of the 19th century and beginning of the 20th century, these institutions developed and disseminated standards in conservation and restoration. The results of their research were groundbreaking for the theoretical and practical work of numerous museums throughout the North American continent. From there, basic procedures and methods for the control of pests in museums, whose collections consist mainly of organic materials, were also disseminated. In the following, active ingredients and agents that have been used both preventively and to combat insect pests in selected American museums as well as in university collections are presented. In Canada, the Canadian Conservation Institute (CCI) was established in 1972 as a separate division of the Department of Cultural Heritage. Since then, it has had a state mandate to dedicate itself to the preservation and conservation of Canadian art and cultural assets. Since the 1980s, there has been an investigation and publication of the use of active ingredients and agents that were introduced at the end of the 19th century to the beginning of the 20th century across museums.

The Smithsonian Institution, founded in 1846 with its numerous museums, experimented with different pesticides in the late 19th and early 20th centuries, similar to SMB in Germany, due to steadily growing collections made primarily of organic materials, and subsequently applied them. The results published for an expert audience are now serving as important sources for the assessment of previously introduced substances as preventive and controlling measures for insect pests on individual objects, specific groups of objects, and entire collections. The Smithsonian Institution includes the Department of Anthropology at the National Museum of Natural History in Washington D.C., founded in 1897, with its 2.5 million artifacts. Extensive source material regarding the use of pesticides has been maintained in the archives of the Smithsonian Institution (Austin et al. 2005, 185–202). Among other things, there are documents about active ingredients that were used by collectors for expeditions, as well as by employees in museum everyday life. The control of insect pests in ovens by heating infested museum objects was already used by the Smithsonian Institution in the early 19th century for entomological and ornithological collections (Ibid., 29). From 1881 onward, the archive contains annual reports on the maintenance of the collection within the National Museum of Natural History, which sometimes contains very detailed information on the use of various active ingredients for pest control. As in German or European museums, conservation work on collection objects was often carried out by the curators themselves with the help of taxidermists, workers, sculptors, and painters. In 1884, Otis Mason, the curator, expressed disappointment that museum collections were being destroyed by harmful insects before they even arrived at the museum. He advised that the objects be protected by immersing them in benzene, treating them with arsenic(III) oxide and alcohol, and taking appropriate precautions. From 1886 onward, as stated by Mason, all new additions were dispatched to the "poisoning department" for additional treatment before being allocated to their respective collections, where they were subjected to a detailed inspection and preservation with unspecified poisonous substances (Mason 1889, 87–88). Mason noted in 1904 that Pueblo pottery required preservation with mercury(II) chloride or arsenic(III) oxide, dissolved in alcohol after conservation, as it contained remains of food, berries, or plant compounds. Four years later, in a report, he emphasised the need for special care and constant attention to anthropological materials in collections, due to the risk of damage caused by moths. During Mason's time, recommendations were given for controlling pests with mercury(II) chloride, powdered arsenic(III) oxide, and arsenic(III) oxide mixed with alcohol or naphthalene. Stronger effects were achieved with the addition of strychnine supplements (Cf. Austin et al. 2005, 187–188). From 1917 onward, annual reports indicated that attention was paid to tightly sealed cabinets in collections, which enabled the use of gaseous agents against insect pests. In addition, the items were treated with toxic agents before being stored in the cabinets (Ibid., 191). In 1877, Walter Hough, a copyist and then chief curator of the Department of Anthropology at the National Museum in Washington, D.C., explained a wide range of methods and homemade mixtures for pest control. To prevent destruction by insect pests, he recommended using sublimate for wood, textiles, botanical specimens, and baskets. Naphthalene was suggested for species from the genera *Psocidae* (book lice) or *Tineidae* (moths) to render them harmless. To combat bacon beetles, such as the common bacon beetle (*Dermestes lardarius*) and wood-destroying insects, the infested objects should be treated with carbon disulfide in sealed boxes. Care must be taken to ensure that the boxes were well closed to achieve an intense fumigation effect inside. In addition to a hermetically sealed atmosphere in the boxes, Hough emphasised the constant repetition of treatments. He recommended the use of carbon disulfide, chloroform (trichloromethane), and arsenic(III) oxide in powder form, or in different formulations, mixed with phenol, strychnine, or naphthalene as active ingredients and agents against almost all insect pests found in museums. A general insecticide used for museum objects consists of a saturated solution of arsoric acid, alcohol, phenol, strychnine, and naphthalene. To produce arsenic soap for the treatment of natural history preparations, he published two different recipes that were used in the museum. For delicate objects, Hough recommended using a nebulizer for spraying, smaller objects should be immersed in the appropriate solution, and larger objects should be smeared. Sealed cabinets, showcases, tight boxes, or specially made galvanised metal containers were available for fumigation (Hough 1889, 549–588; Goldberg 1996, 30–33). John Smith's recommendations for controlling insect pests in the entomological collections of the National Museum of Natural History from 1884 aligned with those of Hough. Specifically, Smith described the continuous use of naphthalene as both a control agent and a preventive agent in insect boxes. He also described the use of chloroform and carbon disulfide fumes as extremely effective in killing bacon beetles and carpet beetles (Smith 1884, 114–116). When collecting items on expeditions or research trips, one had to adapt to the special situation in the field, similar to SMB (see Part III, chapter 8). For example, in the 19th century, Captain Charles Wilkes used unspecified poisons for anthropological and biological specimens during his collection trips to ensure that the objects would arrive in Washington D.C. unharmed. On June 18, 1893, the collector Voth[13] wrote that he put tobacco between objects and hoped to preserve the collections until they reached Washington D.C. The spectrum of active ingredients and agents purchased for field research ranged from camphor, sulfur flour, arsenic, and mercury(II) chloride to tobacco, which was needed for smoking. The initial diary entries can be found in the work of John Varden, who worked as a curator, taxidermist, collection keeper, and exhibit designer for the Anthropological Department of the Smithsonian Institution between 1857 and 1864. During a research trip, he acquired large quantities of arsenic(III) oxide, camphor, tobacco, and mercury(II) chloride for the purpose of preserving specimens and transporting them to USA (Cf. Goldberg 1996, 28–29). During an extensive conservation work on the skulls of hippos that were collected during a large-scale African expedition by the Smithsonian Institution between 1909 and 1910 and came to the National Museum of Natural History, Frederick John Louis Boettcher observed the insecticidal properties of paraffin. He noted that herbaria dipped in paraffin did not require further treatment with toxic agents to combat insect pests. He also recommended benzene, which was commonly used as an insecticide in households (Boettcher 1912, 702–703).

From 1927 to 1958, Laurence Vail Coleman was the director of the American Association of Museums founded in 1906. The aim of this umbrella organization was to create guidelines and generally applicable standards for smaller museums. In summary, these were published in the *Manual for Small Museums* (Coleman 1927), where Coleman described the work of curators, which not only includes collecting and preserving collection items but also protecting them from insect pests. One of the most common species in museum collections is the mold beetle (*Adistemia* sp.), the textile moth (*Tineola* sp.), and the silverfish (*Lepisma saccharina*). He cautioned against the use of hydrogen cyanide or carbon disulfide for fumigation, as there have been fatalities and serious injuries due to a lack of expertise in this area. He considered naphthalene and mothballs to be less effective without further details and instead suggested dichlorobenzene as a better option, which he claimed does not have any negative impact on health. Moreover, he endorsed dousing objects with carbon tetrachloride and also spilling it. He believed it does not stain or cause any other damage, not even on delicate textiles or butterfly wings (Ibid., 196–197).

The National Park Service, founded in 1916, is a federal agency in the USA responsible for preserving and caring for numerous national parks and landscapes, the wildlife they contain, and numerous monuments and museums. In 1941, Ned J. Burns published the *Field Manual for Museums* for staff and volunteers, which included recommendations for the conservation of collection items made of organic materials (Burns 1941). This manual was not published until the middle of the 20th century, but it primarily recommended active ingredients and agents that have been used since the end of the 19th century in both Europe and the USA, with a few exceptions. All newly added items must be fumigated before they are introduced to their respective collections. To facilitate this process, each museum should have a specially built box with metal fittings, specifically designed for fumigation. For fumigation, a mixture of 1,2-dichloroethane and carbon tetrachloride was recommended, and it has been pointed out that for the subsequent accommodation of the objects in the collections, the cabinets should close tightly and, if necessary, be equipped with a felt seal soaked in poison. It is advisable to keep a certain amount of dichlorobenzene or naphthalene in any drawer that contains organic materials, and regular inspection of the collections is highly recommended (Ibid., 198–200). Burns provides detailed information on controlling insects, rodents, and molds, and recommended using hydrogen cyanide to fumigate textile moths. In the case of an intensive infestation, it may also be advisable to treat entire premises in this way. Carpet beetles should be killed locally in a fumigation box with dichlorobenzene or naphthalene. For upholstered furniture, he recommended covering it with linen and soaking it in carbon tetrachloride or alternatively spraying it with pyrethrum. Bread beetles (*Stegobium paniceum*) and tobacco beetles (*Lasioderma serricorne*) are described as very resistant and should be fumigated either with the aforementioned mixture of 1,2-dichloroethane and carbon tetrachloride, or with hydrogen cyanide. According to Burns, complete immersion of objects infested with wood-destroying insects in carbon tetrachloride or benzene is advisable. The collection items should be soaked until the passages drilled by the wood-destroying insects are also filled. After final drying of the objects, the aisles must be closed with paraffin or nitrocellulose varnish. Heating in a drying cabinet is also recommended, but such pieces should not be set, varnished, or glued. As an effective attractant for silverfish, a proprietary mixture of oatmeal, sodium fluoride, and granulated sugar and salt, applied in flat cartons, is recommended. Burns recommended using an atomizer to spray pyrethrum powder into cracks and crevices where cockroaches are likely to hide, and also scattering sodium fluoride in those areas. Arsenic(III) oxide, when added to bookbinding materials, can serve as a defense against cockroach attacks on books, although its toxic nature is a topic of concern for Burns. On the other hand, he viewed phosphorus paste[14] as a good agent of deterrence. To do this, it should be applied to a narrow strip of cardboard and then rolled up. Tools prepared in this way had to be placed out of reach behind the books. Similar to cockroaches, ants can be controlled with sodium fluoride. If there is an infestation of termites, wooden objects must be immersed several times in dichlorobenzene. However, this method cannot be used for mounted, varnished, or glued wooden objects. In rare cases where the infestation persists, Burns suggests hiring an exterminator. To prevent the growth of mold fungi, it is important to keep the indoor climate as constant as possible at 20 °C and 50 % relative humidity, as an increase in relative humidity can promote mold growth. Items contaminated with mold spores should be treated with thymol, which Burns considered to be much safer than mercury(II) chloride (Ibid., 200–207).

As part of Save America's Treasures Program (SAT), the conservators of the Field Museum of Natural History, Chicago, which opened in 1893, examined approximately 40,000 objects from the North American ethnological collections for the presence of arsenic. The investigation began in 2003 and was carried out using the arsenic test "Merckoquant". After two years, approximately 60% of the objects were tested, and it was found that 37.5 % of the sampled objects were contaminated with arsenic. They were mainly treated from 1890 to 1899. It is believed that the application of arsenic(III) oxide to objects was largely performed by museum staff. The data collected were also compared with the acquisition data of the collectors, which led to the conclusion that many objects had already been treated with arsenic(III) oxide before their arrival at the museum. Objects made of fur, skin, and leather were the most affected (Klaus et al. 2005a, 24–26, 2005b, 127).

At the Milwaukee Public Museum, which was opened in 1882, Charles Thomas Brues worked as a curator from 1905 to 1909. He summarised the use of gases and other active ingredients to combat insect pests as follows. New additions to museums should be fumigated with carbon disulfide or hydrogen cyanide and then treated with arsenic(III) oxide. In his estimation, fumigation served as a prophylaxis, whereas he attributed a long-lasting effect to arsenic(III) oxide. A subsequent quarantine in a warm place should also take place before admission to the collections to be able to observe the success of the measures taken. Because carbon disulfide vapors are heavier than air, Brues pointed out that they are better able to penetrate objects when they are oriented downward. In addition, in his opinion, the liquid is not hazardous to health and can be used in small boxes as well as in insect boxes. Nevertheless, he warned about the fire hazards of fumes. On the other hand, the vapors of hydrogen cyanide are lighter than air, so according to Brues, they should be given the opportunity to escape upward. White arsenic(III) oxide is preferred to mercury(II) chloride, as the latter is volatile even at normal temperatures. Brues estimated the stability of arsenic(III) oxide to be just as high as that of the samples or of the objects themselves, which is why he considered this active ingredient to be more suitable than all others. He described naphthalene as a preventive agent against all common insect pests and recommended that it be continuously available in small quantities in all collections. However, he opposed commercial naphthalene, as this active ingredient also contains oils in the mothballs. After naphthalene evaporates, it is absorbed by materials, such as wood, paper, and plant specimens, which can cause irreparable damage to the objects. To remedy this situation, flakes of naphthalene were melted down at the museum and poured into clay molds. According to Brues, the cones obtained in this way do not leave any oily residues and can remain in the insect boxes without hesitation. With the use of tar oil, he followed a hint from George W. Bock. The active ingredient was applied to cotton and then put into thimbles. Small containers were placed in the boxes and can be easily replaced if necessary. Finally, Brues mentioned pyrethrum powder with the product name "Persian Insect Powder". He explained that this contact poison must be introduced fresh, as its effect wears off quickly when exposed to the air (Brues 1909, 34–36). Charles Robinson Toothaker explained the experience with different active ingredients and agents of combating insect pests. From 1904 to 1952, he was the curator of the Philadelphia Art Museum, which opened in 1876. In an article published in 1908, he described insects, such as clothes moth, bread beetle, weevils, and wood-destroying insects such as the common rodent beetle as the worst enemies of museum collections. During his term of office, naphthalene was used to combat and prevent the disease, tar cardboard was laid out as a base for objects, turpentine was sprayed, and dissolved arsenic(III) oxide and mercury(II) chloride were applied. Infested objects were initially fumigated with carbon disulfide, either in empty display cases or in a specially built large fumigation box. Toothaker also pointed out that the fire hazard of fumigants and their use in enclosed spaces is always associated with great risks. To fumigate the rooms in the museum, he described a mixture of sulfuric acid and water in a ratio of 1:2, which released hydrocyanic acid from potassium cyanide for fumigation. Further experiments were performed using formaldehyde, which was heated over a spirit burner. According to Toothaker's observations and comparative experiments, fumigation with hydrogen cyanide was the most effective. This method effectively eradicates all insects, the application is faster than with carbon disulfide or trichloromethane, it is neither explosive nor flammable, the distinct smell is easily detected, and the process is very inexpensive. In his opinion, hydrogen cyanide can only be expected to cause damage to living plants (Toothaker 1908, 119–123).

The Canadian Conservation Institute (CCI) in Ottawa, Canada, was founded in 1972 as an independent institution. During a long-term survey, the history of the active ingredients and agents for the prevention and control of insect pests in Canadian museums, which were used from the end of the 19th century to the beginning of the 20th century, was examined using scientific methods and published in summary form (Sirois et al. 2010, 28–45). More than 2000 objects from various museum collections were made available for the study regarding the possible presence of active substances and agents that had been used previously to protect the collection items. These were objects from ornithological, ethnological, and anthropological collections. In addition, specimens from didactic collections, such as those used for museum educational purposes, were also included in the study. An initial literature review revealed that the National Museum of Canada's annual report in 1929 recommended a mixture of 1,2-dichloroethane and carbon tetrachloride in a ratio of 3:1 to control insect pests (Leechman 1931). He also mentioned the use of carbon disulfide, hydrogen cyanide, chloropicrine, naphthalene, and dichlorobenzene. Sodium fluoride was used to control cockroaches, and mercury(II) chloride, dissolved in alcohol, was recommended as a suitable agent for combating mold (Leechman 1931, 135–137). The archives of the Canadian Museum of History in Québec contain documents from both the American National Museum of Natural History and the Smithsonian Institution. They were replies to Canadian museums that had asked for advise on combating insect pests in their collections. In 1964, the American Museum of Natural History recommended the use of carbon disulfide, naphthalene, and dichlorobenzene. Exhibits made of skin should be treated with solutions of arsenic(III) oxide for exhibition purposes (AAMNH-DMA 1964). The Smithsonian Institution proposed the fumigant "Dowfume" (bromomethane),[15] "paracide crystals" (dichlorobenzene), and arsenic soap for moth control. On this basis, the objects at the CCI were examined for arsenic(III) oxide, lead, bromine, mercury(II) chloride, 1,4-dichlorobenzene, dichlorvos, DDT, DDD, DDE, naphthalene, and "Perthan" (Cf. Sirois et al. 2010, 29). It was found that arsenic(III) oxide was decreasing in order in ornithological, mammalogical, anthropological, and museum educational collections. Low levels of mercury were found in more than 75 % of a collection of chilkat[16] blankets found. The determination of organic active ingredients varied and indicated that collections had been subjected previously to various treatment methods. Naphthalene has been detected in two museums. The exposure to pesticides on the objects examined was estimated to be rather low compared to the limits permitted in Canada (Cf. Sirois et al. 2010, 38–39).

Upon examining the use of active substances and agents during the study period for cultural institutions in the USA and Canada, only marginal differences in the use of natural and synthetic active substances can be detected. In Burns's manual Burns (1941), a recommendation for the use of phosphorus paste as a preservative for books was provided. In addition, the publication of this contribution serves as a subtle indication of the long-standing use of many active ingredients and agents beyond the middle of the 20th century, despite their known toxicity in the early 1930s. The components in our mixtures and formulations are the same as those used in Europe. The only thing that seems somewhat bothersome is Coleman's recommendation in his 1927 manual for smaller museums. He suggested pouring or spilling carbon tetrachloride on textiles, which may not cause any damage or stains. Analogous to the investigations in the KMfV/MfV in Berlin, a study at the Field Museum of Natural History, Chicago, clearly established that many objects had already been treated with arsenic(III) oxide during various research trips between 1890 and 1899 before their arrival at the museum. The Smithsonian Institution can claim to be progressive in that from 1881 onward, annual reports on collection maintenance were prepared at the National Museum of Natural History. This documentary heritage often contains useful information for today's museum staff on how to create concepts for the restoration and conservation of collection items. The "poisoning department" set up there in 1886 was also considered to be in charge, where museum objects were treated against insect pests as a preventive measure and to combat them.

Notes

1 "Moth ethers" (translation by the author).
2 Georg Thilenius was the first director at the Museum für Völkerkunde in Hamburg from 1904 to 1935.
3 Friendly verbal communication from Angelica Hoffmeister-zur-Nedden, chief conservator at the GRASSI Museum für Völkerkunde in Leipzig, November 10, 2016 (translation by the author).
4 As the most welcome agent, we welcome Areginal in the fight against the wood-destroying insect pests, . . . (translation by the author).
5 The use of pesticides containing arsenic was banned in Germany in 1974. The examples from Sweden mentioned here show that inorganic compounds such as arsenic(III) oxide were still used in museum collections abroad.
6 Translation of pages 80–85 from Norwegian by Thomas Gütebier.
7 The Department of Scientific and Industrial Research was a government organization that existed from 1915 to 1965. With the outbreak of the First World War, there was a strong economic dependence on countries that were hostile at the time, and this institution promoted cooperation between domestic research and industry.
8 Bulletin of the Society of Natural Sciences in Bern (translation by the author).
9 It is assumed that the deterrent effect of printing ink observed by Walter Rytz was due to the lead contained in it (author's note).
10 1 ounce (oz.) = 0,028.349.5 kg.
11 The first names of M. V. Farmakovsky could not be determined after extensive research (author's note).
12 Conservation and Restoration of Museum Collections, a handbook for museum staff (translation by the author).
13 First name unknown (author's note).
14 A more detailed description of which phosphorus was not mentioned by Burns (author's note).
15 There were different Dowfumes with different active ingredients. There is no indication here as to which dowfume it is (author's note).
16 The Chilkat are a subgroup of the Tlingit that belong to the First Nations from North America and Canada.

References

Archive material

AAMNH-DMA. (1964): SA: ji. Mottenschutz für Felle. Sydney, Anderson. Brief vom 02.04.1964 an das National Research Council, Ottawa, Canada, 1 Seite, ohne Paginierung.
Archiv des Museums am Rothenbaum–Kulturen und Künste der Welt (1912): Findbuch. 101–1 Nr. 281. Loseblattsammlung. Thilenius, Georg. Brief an Willy Foy vom 13.01.1912, 1 Seite, ohne Paginierung.
Archiv des Museums am Rothenbaum–Kulturen und Künste der Welt (1914–1934): Findbuch. 101–1, Nr. 281. Loseblattsammlung. Schreiben der Actiengesellschaft für Anilin-Fabrikation vom 15.05.1914, 1 Seite, ohne Paginierung.; Ebda. Fritz Schulz jun. Aktiengesellschaft, Leipzig. Brief vom 11.11.1914, 1 Seite, ohne Paginierung.; Ebd. Fritz Schulz jun. Aktiengesellschaft, Leipzig. Brief vom 03.05.1929, 2 Seiten, ohne Paginierung.; Ebd. Fritz Schulz jun. Aktiengesellschaft, Leipzig. Brief vom 27.10.1932, 1 Seite, ohne Paginierung.; Ebd. Ebd. Fritz Schulz jun. Aktiengesellschaft, Leipzig. Brief vom 29.01.1934, 1 Seite, ohne Paginierung.; Ebd. Fritz Schulz jun. Aktiengesellschaft, Leipzig. Brief vom 16.04.1934, 1 Seite, ohne Paginierung.

Archiv des Museums am Rothenbaum–Kulturen und Künste der Welt (1916): Findbuch. 101–1 Nr. 281. Loseblattsammlung. Thilenius, Georg. Antrag auf Ankauf von Schwefel-kohlenstoff. Brief vom 27.04.1916, 1- seitig, ohne Paginierung.

Archiv des Museums am Rothenbaum–Kulturen und Künste der Welt (1929): Findbuch. 101–1, Nr. 281. Loseblattsammlung. Thilenius, Georg. Bewertung von "Flit". Brief vom 14.12.1929, 2 Seiten, ohne Paginierung.

Archiv des Museums am Rothenbaum–Kulturen und Künste der Welt. Loseblattsammlung (1936): Findbuch. 101–1, Nr. 281. Loseblattsammlung. Chemische Fabrik Flörsheim vorm. Dr. H. Noerdlinger Akt. Ges. (Hg.). Anonymus. Preisliste Nr. 274 vom 15.10.1936, 1–8.

HAStK (1913/1914): Best. Nr. 614; A 432. Akte Schriftwechsel. Actien-Gesellschaft für Anilin-Fabrikation. Loseblattsammlung. 2 Briefe vom 20.11.1913 und vom 20.05.1914, je 1 Seite, ohne Paginierung.

HAStK (1914–1915): Best. Nr. 614, A 70. Akte Ankauf von Insektenmitteln. Angebot von "Globol". Fritz Schulz jun. Aktiengesellschaft, Leipzig. Brief vom 13.11.1914, Blatt 5, 1 Seite.; Brief vom 21.05.1915, Blatt 7, 2 Seiten.; 1 Broschüre ohne Datum, Blatt 2–3, 4 Seiten.; 1 Broschüre ohne Datum, Blatt 6, 2 Seiten.

HAStK (1922): Best. Nr. 614, A 70. Akte Ankauf von Insektenmitteln. Anfrage zur Neubes-tellung von "Wurm-Antorgan". Chemische Fabrik Flörsheim Dr. H. Noerdlinger Flör-sheim a. M. Brief vom 24.08.1922, Blatt 10, 1 Seite.

Other sources

Aall, Hans (1925): Arbeide og ordning i kulturhistoriske Museer. Kort Veiledning. Oslo: Utgitt med Statsbidrag.

Austin, Michele; Firnhaber, Natalie; Goldberg, Lisa; Hansen, Greta; Magee, Catherine (2005): The Legacy of Anthropology Collections Care at the National Museum of Natural History. *Journal of the American Institute for Conservation/American Institute for Conservation of Historic and Artistic Works*, 44(3), 185–202.

Boettcher, Frederick; John, Louis (1912): Preservation of Osseous and Horny Tissues. *Proceedings of the United States National Museum*, 41, 702–703.

Brues, Charles Thomas (1909): The Insect Pests of Museums. *Proceedings of the American Association of Museums*, 34–36.

Burns, Ned J. (1941): Field Manual for Museums. Washington, DC: United States Govern-ment Printing Office.

Coleman, Laurence Vail (1927): Manual for Small Museums, with 32 Plates. New York, London: G.P. Putnam's Sons.

Dangeon, Marion (2014): Conservation des collections naturalisées traitées aux biocides: étude de la collection Mammifères et Oiseuax du Muséum d'Historie Naturelle de Neu-chatel. Bachelor of Arts, Haute École Arts, Appliqués-La Chaux-de-Fonds, Filière Con-servation-Restauration, Neuchatel.

Department of Scientific and Industrial Research (eds.) (1926): The Cleaning and Restora-tion of Museum Exhibits. Third Report Upon Investigations Conducted at the British Museum. London: Published under the Authority of his Majesty's Stationery Office. I–V.

Deschka, Gerfried (1987): Die Desinfektion kleiner Insektensammlungen nach neueren Gesichtspunkten. *Steyrer Entomologenrunde*, 21, 57–61.

Germann, Paul (1933): Bekämpfung der Museumsschädlinge. *Museumskunde*, Neue Folge V, 1933, Sonderdruck (I), 9–11.

Gilberg, Mark (1997): Alfred Lucas. Egypt's Sherlock Holmes. *Journal of the American Institute for Conservation*, 36(1), 31–48. DOI: 10.1179/019713697806113620.

Goldberg, Lisa (1996): A History of Pest Control Measures in the Anthropology Collections, National Museum of Natural History, Smithsonian Institution. *Journal of the American Institute for Conservation/American Institute for Conservation of Historic and Artistic Works*, 35(1).

Gütebier, Thomas (2012a): Historische Schädlingsbekämpfungsmittel in Balgsammlungen—ein vielschichtiger Problemkomplex. Vortrag. 50 während der Internationalen Arbeitstagung des Verbands Deutscher Präparatoren e.V. München vom 20–24.03.2012.

Gütebier, Thomas (2012b): Schädlingsbekämpfung in schwedischen Museen. Brief an die Autorin vom 26.03.2012, 1 Seite, ohne Paginierung.

Hansestadt Hamburg (1921): Staats- und Universitätsbibliothek Hamburg. Entwurf des hamburgischen Staatshaushaltsplanes für das Rechnungsjahr 1921, Artikel 87.

Hough, Walter (1889): The Preservation of Museum Specimens from Insects and the Effects of Dampness. Annual Report of the Board of Regents of the Smithsonian Institution for the Year Ending June 30, 1887. Report of the National Museum, Washington, DC, Zoological Pamphlets, 5.

Klaus, Marianne; Plitnikas, J.; Norton, Ruth; Almazan, T.; Coleman, S. (2005a): Preliminary Results from a Survey for Residual Arsenic on the North American Ethnographic Collections at the Field Museum Poster. *Western Association for Art Conservation Newsletter*, 27(1), 24–26.

Klaus, Marianne; Plitnikas, J.; Norton, Ruth; Almazan, T.; Coleman, S. (2005b): Preliminary Results from a Survey for Residual Arsenic on the North American Ethnographic. Poster Submission. In: ICOM-CC Preprints, 14th Triennial Conference, The Hague, 12–16 September, Vol. I.

Leechman, Douglas (1931): Technical Methods in the Preservation of Anthropological Museum Specimens. Hg. v. Canada Department of Mines. National Museum of Canada. Ottawa. Annual Report for 1929, Bulletin 67, 1931.

Lehmann, Jirina (2005): Geschichte der Konservierung und Restaurierung in Russland und in der Sowjetunion, im Buch von Professor M.W. Farmakowskij. *VDR Beiträge zur Erhaltung von Kunst- und Kulturgut*, (2), 47–62.

Lucas, Alfred (1932): Antiques. Their Restauration and Preservation. Zweite überarbeitete Auflage. London. Edward Arnold & Co.

Mason, Otis Tufton (1889): Report Upon the Work in the Department of Ethnology in the U.S. National Museum. For the Year Ending June 30, 1886. Part II. [S. l.]: Government Printing Office.

Newstead, Robert (1893): The Use of Boric Acid as a Preservative for Birds' Skins. Museums Association, Report of Proceedings.

Pfister, Aude-Laurence (2008): L'Influence des Biocides sur la Conservation des Naturalis. Diplomarbeit. Haute École Art Appliqués, La Chaux-De-Fonds. Filière Conservation-Restauration.

Plenderleith, Harold James (1934): The Preservation of Antiquities. Oxford: University Press (I). I–VIII; 1–71.

Purewal, Victoria Jane (2001): Analysis of the Pesticide Residues Present on Herbarium Sheets within the National Museums and Galleries of Wales. In: Proceedings of 2001. A Pest Odyssey. Integrated Pest Management for Collections.

Purewal, Victoria Jane (2012): Novel Detection and Removal of Hazardous Biocide Residues Historically Applied to Herbaria. Dissertation, University of Lincoln, Lincoln.

Rytz, Walter (Hg.) (1922): Die Herbarien des Botanischen Instituts der Universität Bern (Schweiz). In: Mitteilungen der Naturforschenden Gesellschaft in Bern (V). Bern: K. J. Wyss Erben, 11.

Schnetzler, Jean Balthasar (1916): Sulphide as an Insecticide. *The Popular Science Monthly*, 9, 767.

Scott, Alexander (1922): The Restoration and Preservation of Objects at the British Museum. *Journal of the Royal Society of Arts*, LXX(No. 3618).

Sirois, Jane; Poulin, Jennifer; Stone, Tom (2010): Detecting Pesticide Residues on Museum Objects in Canadian Collections. A Summary of Surveys Spanning a Twenty-Year Period. *Collection Forum*, 24(1–2), 28–45.

Smith, John (1884): Some Observations on Museum Pests. *Proceedings of the Entomological Society*, 1884–1889, 1, 114–116.

Thilenius, Georg (1905): 2. Museum für Völkerkunde. Bericht für das Jahr 1905. *Jahrbuch der Hamburgischen Wissenschaftlichen Anstalten*, Bd. 23. Available online at www.biodiversitylibrary.org/item/92087.

Thilenius, Georg (1916): Das Hamburgische Museum für Völkerkunde. *Museumskunde, Zeitschrift für Verwaltung und Technik öffentlicher und privater Sammlungen*, 16(i.e. 12) (Beiheft zu Band XIV), I–VIII, 5–13.

Toothaker, Charles Robinson (1908): Fumigation. *Proceedings of the American Association of Museums*, 119–123.

Conclusion

The book focuses on the specific case of the Ethnologisches Museum (EM) der Staatlichen Museen zu Berlin – Semi-foursquare formerly the Königliches/Staatliches Museum für Völkerkunde (KMfV/MfV) zu Berlin. It provides an in-depth context of the past use of pesticides, particularly regarding its sociopolitical background. Much information will be relevant to other museums with a colonial history, mainly in other European, North American, Canadian, and Global South countries. Therefore, it will be interesting to see whether there are similarities to collections in other museums at the late 19th and early 20th centuries. In PART III, the book positions the Ethnologisches Museum in the wider world and compares it with other museums internationally. However, it must be borne in mind that there are aspects of PART I that would vary from country to country, and historical details would differ from place to place.

The active ingredients and pest control agents used at the end of the 19th and the beginning of the 20th century in the collections of the KMfV/MfV zu Berlin are demonstrably closely linked to the social and political circumstances of the time, as well as to technical innovations caused by industrialization. Similarly, the significant and burgeoning interests of the chemical industry led to a strong impetus for the use of pesticides in museums. An important interface was the constantly growing industry with its conurbations, where people living close to each other had to be supplied. Both the consequences of industrialization and the serious effects of the First World War on the hygiene of the military and civilian population led scientists and authorities to intensively search for active ingredients and agents to counter lice, scabies mites, bedbugs, and fleas. The primary goal of the research at that time was to minimise health impairments for the population and to maintain urgently needed manpower and military defenses. The search for suitable active ingredients and agents is an ongoing race. Furthermore, early forms of globalization led to the import of potatoes from North America to Germany at the end of the 19th century. The Colorado potato beetle (*Leptinotarsa decemlineata*) came to Germany and became native to us. However, it is not surprising that the order of the day was to protect crops as well as stored supplies and foodstuffs from harmful insects. This striking example shows that the situation at that time was quite comparable to the current pandemic caused by SARS-CoV-2 and COVID-19.

For scientific institutions, government agencies, and the private sector, it was peremptory to interlink their interests and work together with a great commitment to advance research into suitable active ingredients and agents against insect pests. If museums are considered as part of the whole, especially regarding the preservation of their objects, then it becomes immediately clear that museum experts have also been looking for solutions in their collections to prevent or control insects from organic materials. From a certain distance, these cultural institutions appear to be small, self-contained institutions. They were therefore largely dependent on the support of official and scientific institutions, as well as on the help of industry and trade. Because conservation sciences were still in their infancy, the knowledge of individuals from other disciplines and their external support are in great demand in museums. An outstanding example is the botanist Johann Bolle, who knew how to use his own interests profitably in his career by networking with European museums. He made a name for himself far beyond his field of expertise and spread his knowledge of pest control in museums. He was a welcome speaker at relevant conferences and used specialist journals from museums for his publications. In this way, he promoted the exchange between museums to a large extent and became a key figure for them.

Because insect pests have no borders, active ingredients or agents on a national and international level, a total of 40 museums from 11 countries were also part of my investigations. According to the current findings, the use of similar, if not identical, active ingredients and agents can be observed in all museums, except for a few deviations. In view of the long-standing cooperation and demands for the repatriation of cultural assets from indigenous peoples in North American, Canadian, and European museums, these research results provide an important impetus. However, the theoretical and empirical investigation of this almost 150-year history of collection care in the Ethnologisches Museum of the Staatliche Museen zu Berlin is regarded as a desideratum. The results and findings gained so far ideally offer the starting point for further research in favor of a chronological, comprehensive investigation of active ingredients and agents, as well as methods and procedures for pest control in museum collections at the end of the 19th and the beginning of the 20th century. The book stopped at the beginning of the 20th century, when I deliberately put the historical marker around the year 1939, because the invention and production of pesticides based on organochlorine compounds exploded here with the discovery of the insecticidal effect of DDT after the Second World War. As pesticide use in museum collections was still a common practice in Europe until the 1990s and is still a common practice in various countries around the world at present, it is inevitable to mention this to clarify that pesticides did not stop to be used within the time when the book ended.

As a result, three essential theses can be stated. First, the suppliers of active ingredients and pest control agents for museum collections came from different areas. The spectrum ranged from individuals, pharmacies, retailers, and manufacturers of synthetic paints to chemical factories. It has been shown that chemists were specifically employed for the development of pesticides in paint factories. They modified waste products resulting from the production of synthetic paints and used them to develop active ingredients and agents to combat insect pests. In close cooperation with their clients, these natural scientists turned directly to museums with their findings and ensured the sales of the agents they developed with their publications. Factories, such as the Actiengesellschaft für Anilin-Fabrikation or the Farbenfabriken, formerly Friedrich Bayer & Co., Elberfeld & Leverkusen, sought potential customers in direct contact in order to open up new sales markets for their products. For this purpose, they used their own distribution channels or outsourced them to trading companies, for which the example of the company Fritz Schulz jun. Aktiengesellschaft from Leipzig is characteristic. With the moth repellent "Globol" distributed by this company, direct economic dependencies on the manufacturer were demonstrably created. The extent of the industry's interest in selling pest control products is shown by the large sums of money invested in scientific experiments and marketing of products. In 1925, for example, I. G. Farbenindustrie Aktiengesellschaft advertised its pesticides for further distribution with posters, brochures, advertisements, and exhibitions amounting to 13,242.85 RM. The costs for scientific investigations and official examination fees amounted to 9,201.24 RM. This expansion was driven forward both nationally and internationally. According to a file note from 1930, the agent "Avenarius" was sold in Austria, 30,000 kg of "Tillantin R" were exported to Mexico for the treatment of barley, and efforts were made to offer tetrachlorobenzene for disinfecting soils in the USA (LASA 1925, 1930).[1]

The second finding is that the selection of pesticides has been discussed in emerging networks of museum professionals. At specialist conferences, such as the "Enquete betreffend die Konservierung von Kunstgegenständen"[2] as well as in relevant specialist journals, such as the *Museumsjournal–Zeitschrift für Verwaltung und Technik öffentlicher und privater Sammlungen*,[3] experiences in dealing with old or new active substances and agents were exchanged, and the latest technical advances in pest control were discussed. Regarding mass fumigation to combat insect pests in large plants, Adolph Bernhard Meyer and Willy Foy, together with Johann Bolle, were pioneers in Germany. Her imitators were Georg Thilenius and Friedrich Rathgen. At that time, only a small number of museum professionals had the appropriate technical and personnel equipment to conduct their own experiments on the development of pesticides. These primarily included Foy from the Rautenstrauch-Joest-Museum in Cologne, Rathgen from the Chemisches Laboratorium of the Königlichen/Staatlichen Museen zu Berlin, and his counterparts Alexander Scott and Harold James Plenderleith from the British Museum in London. There was also a certain pragmatism when museum professionals were sometimes helpless when facing insect infestations in their collections of organic materials. If one's own "know-how" was not sufficient, active ingredients, agents, and methods for the control and prevention of insect pests were simply used if they were recommended elsewhere. Smaller museums and their collections subsequently benefited from the knowledge and network of museum experts from larger museum institutions. This is evidenced by individual inquiries to experts at that time. For example, curators, chemists, and conservators were available to answer questions about the problem of insect pests on individual objects, as well as for entire collections at the

KMfV/MfV in Berlin. There were also active ingredients and agents that proved too complicated to use. This was the case when they were too complex to produce or if they had not proven themselves in practice. Accordingly, there were always complaints about a strong odor nuisance by the active ingredient camphor when opening the collection cabinets. Complaints about health impairments also led to the fact that active ingredients or agents, such as "Globol", were no longer used. The tendency to focus on the use of synthetic active substances or preparations containing active ingredients is attributed to strong lobbying by relevant industrial companies. The success of a pesticide in the market also depended on whether it was frequently advertised or less frequently recommended, and thus an agent could make a "career" in a sense if it had prevailed in this manner. These active ingredients and agents are marked in yellow (see Part II, chapter 7). The investigations have also proven that pesticides can be assigned historically. At the end of the 19th century and the beginning of the 20th century, natural substances or inorganic active ingredients and agents, i.e., heavy metal compounds, were initially used. A visible change began in the 1930s with the production of active ingredients and agents based on chlorine. The associated branch of industry thus received the epithet "chlorine chemistry".[4]

The third finding of the research reveals that toward the end of the 19th century and the beginning of the 20th century, there was a specialization of museum staff across various institutions. However, the efforts toward scientifically justified conservation and restoration were only partially successful and, in some cases, failed to progress beyond the initial stages. As already shown in Part I, museum employees during that time were pioneers of their era. For Adolf Bastian, who was a doctor, medical studies may have been a prerequisite for working in a museum. Lawyers and historians were also among the young academics. However, these were the selection criteria for employment at the museum. They may have been the reasons for a certain stagnation when it came to building up, expanding, or even deepening one's own chemical knowledge of pest control. Today's practice of scientific academic education as a conservator or even as a museologist is still completely unknown in the heydays of ethnology. As a result of the investigations, consistent scientification of museums cannot be observed. A striking example is the active ingredient camphor. Bolle had already determined the ineffectiveness of this substance after joint experiments with von Luschan in the KMfV in 1911/1912. Nevertheless, camphor was used until the turn of the 21st century as a preventive agent against insect pests in the collection cabinets of EM in Berlin-Dahlem. This fact is seen as an indication that personal knowledge has been lost over the course of more than a hundred years. Unlike at the National Museum of Natural History in Washington D.C. at the Smithsonian Institution, there was no documentation of the maintenance of the collection, which meant that important information could not be passed on to subsequent generations. Whether the results of Rathgen's long-term study were considered in the mass fumigation of collection objects could not be inferred from these sources. In his publication (Rathgen 1911), he strongly advised against the use of carbon tetrachloride in paints containing resins or varnish as binders. As a result, the development of methods for pest control at the

KMfV and at the Chemisches Laboratorium of the KM/SMB occupied a subordinate position compared to the entry of active substances and agents that entered the collections both through trade and industry.

Finally, questions arise regarding scientifically supported conservation and restoration strategies to combat insect pests. We also need to look at how contaminated art and cultural assets can be handled in museums as well as in the countries of origin after repatriation has taken place. It can hardly be assumed that the pesticide residues remaining in the objects can be eliminated completely. The substances are highly toxic and, according to current knowledge, there is a general obligation on the part of employers to safeguard the well-being of their employees and on the part of the museums to safeguard the interests of indigenous people who come into contact with such "objectified" objects of their cultures. In this context, we partly find answers by the fact that in the 21st century, chemical-free methods and procedures are increasingly being used for preventive measures and to combat insect pests. This is primarily achieved through the application of Integrated Pest Management (IPM), which is now established in many museums. State-of-the-art methods for controlling insect pests involve the use of chemical-free technologies, such as creating an oxygen-poor environment or freezing collection items in refrigeration chambers. These methods effectively deprive insects and microorganisms of their means of survival. There has been a growing trend in the use of wood wasps as a biological method (Graven 2019, 36–41; see Tello 2006, 48–49; Teibler 2019, 42–43). Existing hazardous situations caused by contaminated ethnographica must be constantly reviewed, and appropriate occupational safety must be established. In addition, today's museum staff must be concerned about installing chemical-free technologies and methods of pest control. Employers have a responsibility to not only ensure the safety of their employees and indigenous people from the countries of origin but also to continue their efforts to decontaminate art and cultural assets. The afore mentioned measures aim to ensure the safe handling of various collections, including those that are highly loaded, in line with current standards. It should be noted that these collections are typically managed by scientists, museologists, conservators, and custodian managers. To minimise potential dangers posed by contaminated cultural assets, it is crucial to implement preventive protection and decontamination measures for the aforementioned occupational groups, particularly visitors from source communities, who are often overlooked in this context. This steadily growing awareness in the field of occupational hygiene and occupational safety has led to IPM being applied in the Ethnological Museum since 2004 (Cf. Pinniger et al. 2015, 2016, Pinniger and Townsend 2001, Pinniger and Winsor 1998, Kingsley et al. 2001). In addition, preventive and controlling treatments of objects made of organic materials are carried out, and the use of pesticides is avoided. For this purpose, a facility is available that deprives harmful insects of their survival by creating an oxygen-poor environment. This treatment is supplemented with a freezing cell in which the collection objects are frozen at temperatures of –32 °C (Cf. Tello 2006, 48–49).[5]

However, the usefulness of collections from the colonial context should not be underestimated.

Another success in the context of debates about neo-colonialism is followed by a preoccupation with name changes (Oswald von 2022). It is not surprising that there is discussion regarding the correct terminology for this type of colonial collection. Some people prefer a different term, for example, the International Council of Museums (ICOM) changed the name of the group that focuses on this type of collections to *Objects from Indigenous and World Cultures*. Nevertheless, without this physical legacy, debates about neo-colonialism would not be conducted, as is the case at our time. The collaboration between indigenous peoples and museums, especially in the USA, Canada, and Australia, which has grown since the 1990s, was and is shaped while protecting First Nations so that they can use objects that were treated in museums. The resulting exchange of knowledge, including native agents of protecting artifacts from degradation and pests, is an invaluable benefit for museum collections.

Notes

1 If one compares the average salary of an employee in the amount of 1,469 Reichsmark for the year 1925 in Germany, it becomes clear that I. G. Farbenindustrie Aktiengesellschaft invested large sums in its marketing area (see also: Bundesministerium der Justiz und für Verbraucherschutz. Ausgegeben zu Bonn am, February 26, 2002. Bundesgesetzblatt Teil I., G 5702, 869.).
2 Enquete concerning the conservation of works of art (translation by the author).
3 *Museumsjournal—Journal for Administration and Technology of Public and Private Collections* (translation by the author).
4 Chlorine chemistry is a colloquial term for the production branch of the chemical industry, whose end products are manufactured based on chlorine.
5 The adults (singular = imago) are the sexually mature insects in their adult form.

References

Archive material

LASA (1925): I 532, No. 598. I.G. Farbenindustrie Aktiengesellschaft, Höchst a.M. Kostenaufstellung vom 19.11.1925, 1 Seite, ohne Paginierung.
LASA (1930): I 532, No. 600. May, (First name unknown, author's note.). Besprechung vom 22.01.1930 mit Dr. Paulmann aus Leverkusen in Sachen Schädlingsbekämpfung, 2 Seiten, ohne Paginierung.

Other sources

Graven, Sonja (2019): Giftlos—erfolglos? Vier Jahre integrierte Schädlingsbekämpfung im Museum Mensch und Natur in München. *Restauro*, (2).
Kingsley, Helen; Pinninger, David; Xavier-Rowe, Amber; Winsor, Peter (Hrsg.) (2001): Integrated Pest Management for Collections. Proceedings of 2001: A Pest Odyssey; A Joint Conference of English Heritage, The Science Museum and the National Preservation Office, 1–3 October. Historic Buildings and Monuments Commission for England; Science Museum; British Library; London: James & James.
Oswald von, Margareta (2022): Working Through Colonial Collections. An Ethnography of the Ethnological Museum in Berlin. Leuven: Leuven University Press.

Pinniger, David; Landsberger, Bill; Meyer, Adrian; Querner, Pascal (2016): Handbuch Integriertes Schädlingsmanagement in Museen, Archiven und historischen Gebäuden. With the cooperation of Annette Townsend. Berlin, Berlin: Gebr. Mann Verlag; Rathgen-Forschungslabor Staatliche Museen zu Berlin.

Pinniger, David; Meyer, Adrian; Townsend, Annette (2015): Integrated Pest Management in Cultural Heritage. London: Archetype.

Pinniger, David; Townsend, Annette (2001): Pest Management in Museums, Archives and Historic Houses. London: Archetype.

Pinniger, David; Winsor, Peter (1998): Integrated Pest Management. Practical, Safe and Cost-Effective Advice on the Prevention and Control of Pests in Museums. London: Museums & Galleries Commission.

Rathgen, Friedrich (1911): Mitteilungen aus dem Chemischen Laboratorium der Königlichen Museen zu Berlin. VIII. Über die Verwendung von Kohlenstofftetrachlorid zur Abtötung von tierischen Schädlingen. *Museumskunde, Zeitschrift für Verwaltung und Technik öffentlicher und privater Sammlungen*, Band VII, 219–220.

Teibler, Claudia (2019): Absolut biologisch. *Restauro*, (2), 42–43.

Tello, Helene (2006): Investigations on Super Fluid Extraction (SFE) with Carbon Dioxide on Ethnological Materials and Objects Contaminated with Pesticides. Diplomarbeit. Fachhochschule für Technik und Wirtschaft Berlin, Berlin: Fachbereich 5, Gestaltung, Studiengang Restaurierung/Grabungstechnik.

Glossary

The explanations of the chemical constitution and properties of substances are based on information from: Römpp Lexikon Chemie. Hrsg. Jürgen Falbe und Manfred Regitz. 9. und 10. Aufl. Stuttgart, New York: Thieme 1992 und 1996.

A

Alum   (potassium alum) is a trivial name for the most important salt from the alum group of salts for aluminum potassium sulfate dodecahydrate. It is a white, crystalline, and water-soluble solid that is widely used technically (e.g., in tannery, fabric printing, etc.) and is also used in medicine as a hemostatic and antiperspirant.

Alcohol   (ethyl alcohol) is a common but obsolete term in the natural sciences for ethanol.

Aloe   (aloë, aloe vera) is a genus of lily family. The most famous species is the *Aloe vera*. The juice of this plant is believed to have various healing effects, including an antimicrobial effect, which is achieved through evaporation in a vacuum. Aloe vera juice extracts are commonly incorporated into skin care products, but consuming the juice is also believed to have therapeutic benefits. However, it is essential to take precautions, as an overdose can result in severe poisoning.

[see also: Lehmann, Dieter; Lehmann, Andrea (1985): Zwei wundärztliche Rezeptbücher des 15. Jahrhunderts vom Oberrhein. Zugl.: Würzburg, Univ., Diss. 1983. Pattensen/Han.: Wellm. Würzburger medizinhistorische Forschungen, 34.]

Ammonia   is a gas with the formula NH_3. It is colorless, water soluble, poisonous, and has a typical tear-inducing odor. In nature, it is the breakdown product of nitrogenous organic compounds. Its production is synthetic. It is an important substance for the technical syntheses of organic nitrogen compounds. Its aqueous solution has a strong alkaline effect and is traded under the name ammonia solution.

[see also: Gmelin, Leopold (1827): Handbuch der theoretischen Chemie. Frankfurt: Franz Varrentrapp, 1827, 421.]

Ammonia liquor   (ammonium hydroxide) is the name given to aqueous solutions of ammonia (NH_3) in different concentrations.

Ammonium arsenate   is a white solid, a salt of arsenic acid and ammonia with the formula $(NH_4)_3AsO_4$. It is water soluble and, like all arsenic compounds, toxic. It is hazardous to health and can cause acute and chronic poisoning.

Ammonium peroxodisulfate   → Persulfates

"Antisekt"   was a pesticide that the GRASSI Museum für Völkerkunde in Leipzig purchased from the pharmacy M. Wagner & Co. in Leipzig in the 1930s. It is a liquid, and no further details can be found in the source.

[see also: Germann, Paul (1933): Bekämpfung der Museumsschädlinge. In: *Museumskunde*, Neue Folge V, Sonderabdruck, Heft I, 1933, 9.]

"Antorgan"   is, according to the literature, also referred to as "worm antorgan" or "woodworm antorgan". It is an aqueous solution of approximately 5–20% solution of ammonium fluoride and zinc fluoride (ammoniacal zinc fluoride solution). This preparation was used for wood preservation in the house, basement, and garden. It was preferably used for the preventive protection and control of wood-destroying insects, as well as for indoor dry rot control.

(Friendly verbal communication from Achim Unger, September 30, 2017.)

"Areginal"   was a product of the company Bayer-Meister Lucius, Pharmazeutische Abteilung der I. G. Farbenindustrie Aktiengesellschaft, Leverkusen a. Rh., which maintained an advisory center for plant protection in Berlin W. 15, Kurfürstendamm 179. It is formic acid methyl ester (methyl formate), used as an insecticide for fumigation for stock protection from 1925 to 1960. This colorless liquid has a pleasant smell and is miscible with water to a limited extent. Areginal is highly flammable, slightly hazardous to water, and harmful to health.

Arsenous acid   (arsenious acid or arsorous acid) or arsenic(III) acid H_3AsO_3 (ortho form) and $HAsO_2$ (meta form) is a weak acid with trivalent arsenic, known only as a solution. It is obtained by dissolving arsenic trioxide in water. Before the ban on handling of arsenic compounds, arsenic acid was used as a component in some plant protection products.

"Arsenous jelly",   according to Sheridan Délepine, consists of gelatin and arsenic acid. The well-cleaned and dry gelatin is stirred into hot arsenic acid, which should dissolve within half an hour. The resulting jelly is mixed with pure, equally heated gelatin and then cooled to 20 °C. Subsequently, the whites of six eggs and their shells have to be mixed. To let the egg whites curdle, the mixture is heated again and simmered for another two hours. Finally, the entire mass is successively filtered through flannel and paper at 50 °C. Delépine praised his "arsenous jelly" as absolutely transparent and, after curing, was stable against summer temperatures.

[see also: Delépine, Sheridan (1914): On the arsenious acid–glycerin–gelatin ("arsenious jelly") method of preserving and mounting pathological specimens with their natural colors and on the use of new forms of receptacles for keeping museum specimens. From the Public Health Department, University of Manchester. In: *The Museums Journal*, 1914 (13), 322–329.]

Arsenic acid baking soda　(sodium arsenate) is an obsolete name for sodium arsenate, the sodium salt of arsenic acid with the formula Na_3AsO_4. Sodium arsenate is colorless, odorless, and water soluble. It is highly toxic and was previously used in several fields against pests.

Arsoric acid　(arsenic acid) [arsenic(V) acid] is an acid with the formula H_3AsO_4. Like all arsenic compounds, arsenic acid is toxic, carcinogenic, and hazardous to the environment. Similar to arsenous acid, arsoric acid was used in some crop protection products in the past.

Arsenic pills　→ Arsenic trioxide

Arsenic soap　(arsenic trioxide soap) goes back to an invention of the French pharmacist Jean Bécouer (1718–1777). It was made from camphor, arsenic trioxide, soap, potassium carbonate, and lime powder. Arsenic trioxide soap was a widely used preservative until the beginning of the 20th century and was mainly used for dry preparation in natural history collections. Ethnographic museums used arsenic trioxide soap with different recipes.

Arsenic trioxide　[arsenic(III) oxide] with the formula As_2O_3, called arsenic, is a white solid that produces arsenic acid when dissolved in water. It is highly toxic and carcinogenic. The lethal dose for humans is 10 mg/kg body weight. This dangerous substance is not only absorbed through the gastrointestinal tract, but also through the inhalation of arsenic-containing dust and contact with the skin. Historically, it was used for the preservation of animal preparations and the production of insecticides and rodenticides (means used to control rodents). Arsenic trioxide was also used in medicine in the past.

"Autan"　was a product name of Farbenfabriken, vormals Friedrich Bayer & Co., Elberfeld & Leverkusen for an agent consisting of polymerised formaldehyde (called paraformaldehyde) and barium peroxide. Upon contact with water, the paraformaldehyde is depolymerised, resulting in the gaseous formaldehyde. Initially, it was intended as a disinfectant for homes, but it was also used in museums.

Avenarius- "Carbolineum"　→ "Carbolineum"

B

"Bakelite"　is a brand name (trademark) for a synthetic resin based on phenol and formaldehyde. Bakelite was produced from 1909 by Bakelite GmbH in Germany, later also by Union Carbide Corporation in the USA. Today, Bakelite is a registered trademark of Hexion GmbH.

Barium hexafluorosilicate　is a white solid substance with the molecular formula $BaSiF_6$ that belongs to the group of fluorosilicates. It is poorly soluble in water and has been used as an insecticide.

Barium peroxide　is a compound of the element barium with oxygen, in which the oxygen is unstably bound and can therefore be easily split off. Its chemical formula is BaO_2, and it has been used, among other things, to depolymerise paraformaldehyde in the control agent "Autan". Barium peroxide is the first peroxo compound made known by Alexander von Humboldt in 1799.

Barium superoxide → Barium peroxide

Beech tar oil is the processed beech tar. It consists of paraffin hydrocarbons, phenols (creosote) and their esters, fatty acids, and pitch. It is blackish brown and thin. Its germicidal effect is due to the phenols it contains. The volatile components emit a pungent aroma that contaminates the air.

Benzene is a cyclic hydrocarbon with the molecular formula C_6H_6. The location of its double carbon bonds gives it and its derivatives a special behavior described as "aromatic" when it was discovered at the beginning of the 19th century. Accordingly, its derivatives are also called "aromatic hydrocarbons". Benzene is a colorless, highly flammable liquid. It mixes with numerous organic solvents but hardly with water. Benzene is used as an additive for motor fuels and as a solvent. It is toxic and can cause both acute fatal and chronic poisoning with damage to the liver, kidneys, and bone marrow.

Boric acid (orthoboric acid) with the formula H_3BO_3 (ortho form) forms scaly, colorless, and shiny crystals. The aqueous solution has a weakly acidic reaction. Salts are called borates. Orthoboric acid exhibits certain antimicrobial properties, which led to its extensive use in medicine (particularly in ophthalmology). However, its application has been significantly restricted now. According to the regulation of the Registration, Evaluation, Authorisation and Restriction of Chemicals, No. 1907/2006, boric acid is classified as a substance of very high concern.

Bromomethane, obsolete name for methyl bromide, is a colorless and odorless gas that smells like chloroform in high concentrations. It is a contact and respiratory poison that is mainly used for fumigation of containers and control of wood-destroying insects in the construction industry. In the preservation of historical monuments, bromomethane was a leading control agent for decades. Bromomethane has a detrimental impact on the ozone layer and enhances the greenhouse effect, leading to its prohibition for pest control in Germany since September 1, 2006.

1-Butanol with the formula C_4H_9OH, obsolete name for butyl alcohol, is a linear monohydric alcohol. It is a colorless, flammable liquid that belongs to the group of alkanols. The primary alcohol is derived from the aliphatic hydrocarbon *n*-butane. 1-Butanol is mainly used as a solvent for resins, a component of cleaning agents, an extractant, and in some chemical syntheses. This substance is harmful to health when inhaled and swallowed. Kidney and liver damage, dizziness, headache, drowsiness, loss of consciousness, and brain dysfunction may occur. It also irritates the respiratory tract and digestive tract, as well as the eyes and skin. Therefore, it is highly hazardous to health.

1,2-Butylene oxide (1,2-epoxybutane) with the molecular formula C_4H_8O is a highly flammable, colorless liquid with a characteristic odor. This substance is a contact poison, and it can be absorbed when ingested, through inhalation, or through skin contact. Butylene oxide irritates the skin, eyes, and respiratory organs. It is highly hazardous to health and is suspected of causing cancer in humans.

Butyl formate   (formic acid *n*-butyl ester), with the molecular formula $C_5H_{10}O_2$ or $HCOOC_4H_9$, is an older name for formic acid *n*-butyl ester or *n*-butyl formate. This highly flammable, colorless liquid has an alcohol-like odor. Butyl formate is used in low concentrations as a fruit-like flavoring agent. Inhalation or oral ingestion of butyl formate in high concentrations leads to significant health disorders, up to and including loss of consciousness.

C

Cajeput oil   is an essential oil obtained by steam distillation of the leaves and smaller branches of various cajeput trees (family Myrtaceae), such as *Melaleuca leucadendra* and *Melaleuca cajuputi*, which are native to the Indonesian islands of Moluccas. The oil, which is soluble in 80% ethanol, contains mainly terpenes (pinene, cineole, and terpineol). It was mainly used in perfumery and medicine as well as a disinfectant, but it was replaced by the similarly composed eucalyptus oil.
[see also: Rochusen, Frank (1920): Ätherische Öle und Riechstoffe. Sammlung Göschen, Berlin und Leipzig: Vereinigung wissenschaftlicher Verleger, 1920, 78.]

"Calcid"   was launched in the market by I.G. Farben Farbenindustrie AG in 1928. It was a high-percentage calcium cyanate that developed hydrogen cyanide ($\rightarrow$ hydrogen cyanide) during chemical reaction with humidity. The "Calcid" process was an export product of Degesch, which mainly used for the fumigation of citrus trees and the eradication of rats (including on ships). Its cyanocalcium $Ca(CN)_2$ content was 88.5 %. HCN is rapidly developed (compare slow reaction in the cyanogas). The reaction rate depends on the fine distribution of the substance, humidity, and CO_2 content in the air.
[see also: Kalthoff, Jürgen; Werner, Martin (1998): Die Händler des Zyklon B. Tesch & Stabenow; eine Firmengeschichte zwischen Hamburg und Auschwitz. Hamburg: VSA-Verl., 237.; Ebbinghaus, Angelika (1999): Der Prozess gegen Tesch & Stabenow. Von der Schädlingsbekämpfung zum Holocaust. In: *Zeitschrift für Sozialgeschichte des 20. und 21. Jahrhunderts*, 1998, 13. Jhrg., Heft 2, 44, Fußnote 91.]

Camphor   , natural and synthetic, with the molecular formula $C_{10}H_{16}O$, is a bicyclic monoterpene ketone. It is a volatile, colorless solid found in many plants, especially in the bark, wood, and resin of the camphor tree *Cinnamomum camphora*. Today, camphor is mainly produced synthetically. In low concentrations, it is used in cosmetic and some medical preparations (e.g., for colds), as well as for smoking. In higher concentrations, it acts on the central nervous system and kidneys, as well as on the respiratory system. According to the GHS hazardous substance labeling, it is easily to highly flammable and has an irritating effect.

Carbolic acid   (carbolic) $\rightarrow$ "carbolic disinfecting powder"$\rightarrow$ phenol

"Carbolic disinfecting powder"   was listed under the trade name carbolic acid, or carbolic for short, which is otherwise a name for $\rightarrow$ phenol.

[see also: Muter, John (1890): The analysis of carbolic and sulphurous disinfecting powders. In: *The Analyst*, Vol. August, 1890, 63.; Wray, L. (first name unknown, author's note) (1908): The Preservation of Mammal Skins. In: *Museums Journal*, December 1908, (8), 207–208.]

Carbon dioxide is a chemical compound with the molecular formula CO_2. It is a nonflammable, colorless gas that is highly soluble in water. Its solution in water has a weakly acidic reaction due to the formation of carbonic acid H_2CO_3. It is odorless at low concentrations. It is compressed both in liquid form and in solid form ("dry ice"), as well as in the supercritical state. In solid form, it serves as a coolant, and as a supercritical fluid, it is used as a solvent and extractant. CO_2 can be toxic, but the usual concentrations in the air or in food are not sufficient for poisoning. Carbon dioxide attacks Earth's ozone layer and is one of the causes of climate change. For this reason, the aim is to minimise carbon dioxide emissions by reducing their burning in coal-fired power plants and combustion engines, not only in Germany.

Carbon disulfide possesses the chemical formula CS_2. It is a colorless, flammable, and toxic liquid that smells unpleasant to disgusting in technical quality. The vapors of the carbon disulfide form highly explosive mixtures with the air. In the past, it was used to control rats and voles, as well as in viticulture against phylloxera, where it was banned in 1997 because of its toxicity. In the chemical industry, it is still of great importance in the production of synthetic fibers from cellulose (so-called viscose, also called artificial silk). With regard to the care of museum collections, it became of far-reaching importance in conservation science from the end of the 19th century to the first half of the 20th century in the control of insect pests. Because of its ability to dissolve fats well, carbon disulfide is absorbed through the lungs and skin. The result is acute and chronic poisoning.

Carbon monoxide is a chemical compound of carbon with oxygen with the formula CO. It is a colorless, odorless, tasteless, and toxic gas. When inhaled, it binds more strongly to hemoglobin in the blood than to oxygen, so that the transport of oxygen in the body is prevented. Carbon monoxide is thus a toxic and dangerous respiratory poison. In addition to acute fatal poisoning, consequential damage can be seen due to long-term intake of low concentrations, which can have a negative effect on the development of a child in the womb, among other things. Long-term exposure to low levels of carbon monoxide can lead to depression and reduce life expectancy by damaging the heart.

Carbon tetrachloride → Tetrachloromethane

Carbolic solution is an obsolete name for a solution of → phenol.

"Carbolineum" or Karbolineum is a distillate made from coal tar oil. It is an oily, water-insoluble, dark brown discolored liquid. Carbolineum is flammable and has a smell of tar. It contains a number of cyclic hydrocarbons, in particular → phenol and its derivatives, as well as naphthenes and the tricyclic antharacene and its derivatives. It was first used as a wood preservative in 1838. Avenarius introduced the chlorinated distillate with the trade name "Carbolineum" to the market in 1888. Because of its anti-putrefactive and disinfecting effects, this tar oil has been used to preserve wood in contact with the ground, such as railway sleepers, telegraph poles, and fences. It can even permanently protect wood installed in permanent contact with the ground. Even though the wood surfaces treated by Carbolineum turn dark brown to black, it has long been used in wood preservation. Carbolineum can cause skin irritation and is known to be carcinogenic. Inhaling its fumes can also irritate the respiratory tract. Despite these health disadvantages, Carbolineum remains widely used as a wood preservative due to its exceptionally high protective qualities, which are unmatched by other wood preservatives. Effective from May 27, 1991, the German Tar Oil Ordinance stipulated the regulations for the marketing, handling, and disposal of tar oils and their products, including Carbolineum. Accordingly, tar oils or products infused with tar oils were banned for use. Tar oil residues had to be disposed of as hazardous waste. The use of existing tar oil-impregnated objects, such as railway sleepers, was allowed to continue if it was not expected to pose any danger. This ordinance was repealed in 2002 and replaced by the Chemicals Prohibition Ordinance, which governs the marketing, handling, and disposal of tar oils. Under this ordinance, tars oils are prohibited from being placed on the market, except for specific industrial and commercial applications, such as wood preservatives for railway sleepers or electricity pylons. Additionally, wood preservatives containing tar oil are now classified as building pollutants and may not be used indoors. Used products containing tar oil, such as old railroad sleepers, must not be used in playgrounds, gardens, and parks, or as containers for living plants.

Celluloid(s)   or Zelluloid, which is a type of plastic made from cellulose nitrate with camphor as a plasticizer, is the first thermoplastic. It is known for its ease of melting and shaping.

Chinese wood oil   (tung oil) is a drying oil extracted from the seeds of the tung or abrasin tree (*Vernicia montana*) growing in China and Japan and the tung oil tree (*Vernicia fordii*). Tung oil contains toxic ingredients and is mainly used for the production of varnishes, sometimes also soap and linoleum, as well as a binder in Chinese and Japanese painting.

Chinese moth tincture   is used to combat insects and larvae. To do this, components of red pepper peels are dissolved in alcohol and mixed with camphor.

[see also: Lange, Otto (1923): 667. Mottenmittel. In: *Chemisch-Technische Vorschriften*, III. Band: Harze, Öle, Fette. Springer-Verlag, Berlin Heidelberg GmbH, 1923, 754.]

Chlorinated lime   (calcium hypochlorite) is a technical mixture of usually 35 % calcium hypochlorite, 30 % calcium chloride, and 13 % calcium hydroxide. Chlorinated lime was originally used as a disinfectant because it fights bacteria and viruses. Chlorinated lime is nowadays hardly used for disinfection because it also harms human skin, and it mainly serves as a bleaching agent to a limited extent. Chlorinated lime is classified as an oxidizing, corrosive, and hazardous substance to both human health and the environment.

Chlorine water   is the aqueous solution of chlorine gas in water. The oxidizing and bleaching effect of chlorinated water is based on the fact that the chlorine gas partially reacts chemically with the water, resulting in the unstable hypochlorous acid HClO. When exposed to light, hypochlorous acid splits off the oxygen molecules that cause the oxidizing and bleaching effect. Chlorine gas and chlorinated water are harmful to the environment and health. Chlorine gas also has an oxidizing effect.

Chloroform is a common name for trichloromethane, a chlorinated hydrocarbon with the molecular formula $CHCl_3$. Its narcotic effect was already recognised in the first half of the 19th century. It was introduced into medical practice in parallel with the already well-known anesthetic ether (diethyl ether). Because of its toxic effect on the heart, liver, and other internal organs, chloroform is no longer used as an anesthetic. It is also suspected of being carcinogenic.

Chloropicrin is the common name for trichloro(nitro)methane). It is a colorless, slightly oily, and volatile liquid with a penetrating odor and high vapor pressure. During World War I, chloropicrin was deployed as a chemical weapon and made its debut in 1916. It can cause respiratory poisoning, leading to death. It is used for disinfection, as well as for sterilization of soils and seeds.

"Chlorpyrifos" (chlorpyrifos ethyl) is the product name of an ester of thiophosphoric acid. It was introduced as an insecticide by the American company Dow Chemical in the 1960s. Since 2009, products containing "chlorpyrifos" have been banned in Germany, as symptoms of poisoning and damage to unborn babies in the womb have been detected.

Citric acid, with the molecular formula $C_6H_8O_7$, is a colorless, water-soluble, and organic acid. It is found in numerous fruits, from which it can be isolated. Today, citric acid is produced biotechnically by fermenting sugary raw materials with the fungus *Aspergillus niger* L. When taken in small amounts, citric acid indirectly promotes bone growth because it favors the absorption of calcium. In larger quantities, however, it is irritating and toxic. Citric acid is widely used in the food industry but also as a gentle descale for coffee and washing machines, among others.

Cobalt yellow (potassium cobalt nitrite) is chemically a cobalt potassium nitrite with the formula $[Co(NO)_6]K_3 + 3H_2O$. Cobalt yellow is a very fine and light powder. The paint is lightfast and can be used well in painting techniques, such as watercolor, tempera, and oil paint. It was also used in glass and porcelain painting.

[see also: Kremer Pigmente. 43500 Kobaltgelb, Aureolin. www.kremerpigmente. com/elements/resources/products/files/43500.pdf, 2 pages.]

Coloquints are pumpkin-like plants found in southern Europe, North Africa, and India, among other places. All parts of the plant contain poisonous terpenes, so eating coloquint fruits can lead to poisoning. Coloquints are well-known medicinal plants with numerous uses.

[see also: Wink, Michael; van Wyk, Ben-Erik; Wink, Coralie (2008): Handbuch der giftigen und psychoaktiven Pflanzen; mit 13 Tabellen. Stuttgart: Wiss. Verlag-Ges.; Ebers, Georg (Hg.) (1875): Papyros Ebers. Das Hermetische Buch über die Arzneimittel der alten Ägypter in hieratischer Schrift. Available online at http://digi.ub.uni-heidelberg.de/diglit/ebers1875bd2.]

Copper sulfate   [copper(II) sulfate pentahydrate], formerly called copper vitriol, is the copper salt of sulfuric acid with five molecules of water of crystallization and with the formula $CuSO_4 \cdot 5H_2O$. It is a blue water-soluble solid. By heating, the water of crystallization is extracted, and the copper sulfate turns into a white powder. Copper sulfate has antimicrobial properties. Mixed with a calcium hydroxide suspension, copper sulfate was formerly used as a "Bordeaux broth" in viticulture to combat fungal diseases of vines. Plant protection products containing copper sulfate are still used today but in lower concentrations. Alternatives are being sought in order not to further increase soil pollution with copper, because copper sulfate and other copper salts can enter bodies of water. Copper sulfate can cause intense nausea in humans and is therefore used as an emetic.

Coumarin   is a heterocyclic hydrocarbon that has a distinct, pleasantly spicy odor found in certain plants. It is classified as a secondary plant substance. Coumarin is present in cinnamon, among other things, but it can also be produced synthetically. While it is typically safe in moderate amounts, larger quantities can be harmful to health. For this reason, the use of cinnamon varieties with a higher coumarin content is discouraged in food production and cooking. Some derivatives of coumarin (hydroxycoumarins) are used as anticoagulant drugs and pesticides, particularly as repellents against rodents.

[see also: Lowe, Derek B. (2017): Das Chemiebuch. Vom Schießpulver bis zum Graphen, 250 Meilensteine in der Geschichte der Chemie. Kerkdriel: Librero IBP, 176.]

Cresols   (hydroxytoluenes) can be regarded as hydroxy derivatives of toluene or methyl derivatives of phenol. Its structure is characterised by a benzene ring with a hydroxyl group (OH) and a methyl group (CH_3), as a result of whose different positions on the benzene ring a total of three isomers are formed (ortho-, meta-, and para-cresol). All three cresols have a tar-like smell because they are extracted from tar of different origins. Cresols are highly sensitive to light and air and form explosive mixtures with air at temperatures above 80 °C. They are poorly soluble in water and have a corrosive nature. Cresols have a broad effect on microorganisms, but they also act as insecticides and fungicides and are therefore often a component of disinfectants. Long-term contact with cresols can lead to chronic poisoning. Acute poisoning can cause respiratory paralysis with a potentially fatal outcome. Cresols are considered carcinogenic.

Crude benzene   is an aromatic hydrocarbon obtained during dry distillation of coal. In addition to benzene, it also contains toluene and xylene derivatives, which are removed (pure benzene). Crude benzene is flammable, and vapors form explosive mixtures with air. It has an irritating effect on the skin and eyes and is suspected of being carcinogenic and mutagenic. It also impairs the reproductive capacity. Crude benzene can be fatal if swallowed or if it enters the respiratory tract. It must not get into the ground, sewer system, or bodies of water.

Cyanide (potassium cyanide) is the trivial name for potassium cyanide (KCN), the potassium salt of hydrocyanic acid. From the colorless crystals, hydrogen cyanide or hydrocyanic acid is released in air by carbon dioxide.

D

Dalmatian insect powder → (Dalmatian) pyrethrum → "Zacherlin"

Dammar resin is a natural resin of several species of deciduous trees, which are mainly native to India and the Malay Peninsula. Depending on the variety, the color varies from clear light to yellowish to blackish gray. The dammar resin is imported into light-colored pieces with a shell-like fracture, which when crushed produces a white powder. Dammar resin is soluble in several organic solvents, and as a triterpenoid resin, it is more light stable than, for example, rosin or sandarak resin. For this reason, dammar resin is used to produce light varnishes and as an additive for artists' paints, particularly the oil–resin paints. Dissolved in turpentine oil or another solvent, dammar resin is processed into varnish and used in painting or conservation as an intermediate and final varnish.

"Diametan" , I.G. Farbenindustrie AG Leverkusen product, was a room disinfectant containing sulfur as an active ingredient and oxygen as a carrier material.

[see: Pflanzenschutz-Nachrichten Bayer (1979): Heft 32, (3), 526.]

A different composition for "Diametan" is given in the literature as follows: "Diametan" is a fungicide for viticulture produced by Bayer based on propineb, triadimefon, and cymoxanil.

(see: Römpp Lexikon Chemie (1996–1999): Hrsg. Jürgen Falbe und Manfred Regitz. 10. Aufl. Band 2. Stuttgart, New York: Thieme.)

Diatomaceous earth is also known as mountain flour, celite, diatomacee, diatomite, infusory earth, siliceous flour, novaculite, triple, and tripolite. It is a whitish, powdery substance consisting mainly of the shells of fossil diatoms. The shells consist mainly of amorphous (noncrystalline) silicon dioxide (SiO_2) and have a very porous structure.

[see also: Kainer, Franz (1951): Kieselgur, ihre Gewinnung, Veredlung und Anwendung. 2., umgearb. Aufl. Stuttgart: Enke (Sammlung chemischer und chemisch-technischer Vorträge, N.F. H. 32.)]

1,4-Dichlorobenzene (*para*-dichlorobenzene) is the current chemical nomenclature for (obsolete) dichlorobenzene, *p*-dichlorobenzene, or paradichlorobenzene. It is an aromatic chemical compound consisting of a benzene ring with two chlorine atoms as substituents. The chlorine atoms can be in three different positions: 1,2-dichlorobenzene, 1,3-dichlorobenzene, and 1,4-dichlorobenzene or the so-called constitutional isomers. The dichlorobenzene with chlorine atoms at position 1,4 is a crystalline solid that, like naphthalene, sublimates at room temperature and develops a strong odor that can retain harmful insects (especially moths) and scare them away. 1,4-Dichlorobenzene is a by-product of monochlorobenzene, which is an intermediate in the production of paints. This invention was marketed as "Dichlorobenzene Agfa" in 1913 by the Aktiengesellschaft für Anilin-Fabrikation in Berlin, and from 1914 onward it was marketed under the product name "Globol" by Fritz Schulz jun. from Leipzig.

Dichlorobenzene $\rightarrow$ 1,4-Dichlorobenzene (*para*-dichlorobenzene)

"Dichlorbenzol Agfa" $\rightarrow$ 1,4-Dichlorobenzene (*para*-dichlorobenzene)

1,2-Dichloroethane   (ethylene dichloride) is an oily, colorless, flammable, and volatile liquid with an odor similar to chloroform. It is used as a solvent for resins, waxes, oils, and asphalt. It is also used in the production of plastics (vinyl chloride). Dichloroethane causes organ damage and is suspected of being carcinogenic. It is ozone depleting and hazardous to water, which is why its use has been severely restricted.

DDT   is the abbreviation of the chemical compound dichlorodiphenyltrichloroethane and consists of the isomers *p,p'*-DDT (approximately 77 %) and *o,p'*-DDT (approximately 15 %). Its insecticidal effect was discovered in 1939 by the Swiss Paul Hermann Müller, who received the Nobel Prize in Medicine in 1948. It was used as a contact and feeding poison and was the most widely used insecticide in the world for decades because of its good efficacy against insects, low toxicity to mammals, and the simple manufacturing process. In the USA, the export of DDT was banned in 1981 and production ceased in 1982. In the Federal Republic of Germany, a ban on the production and distribution of DDT has been in force since July 1, 1977. DDT is still used to control malaria on the recommendation of the World Health Organization (WHO). DDT is highly hazardous to health. Acute poisoning manifests itself primarily in neurotoxic effects, such as tongue numbness, dizziness, twitching of the facial muscles, seizures, and paralysis. The WHO's International Agency for Research on Cancer (IARC) classified DDT as "probably carcinogenic to humans" in 2015. A study in 2014 showed that DDT may be involved in the development of Alzheimer's disease. DDT endangers the environment and is one of the chemical compounds that accumulate on the surface of plastic waste floating in the ocean.

[see also: Simon, Christian (1999): DDT. Kulturgeschichte einer chemischen Verbindung. Basel: Christoph-Merian-Verlag.]

DDD   (dichlorodiphenyldichloroethane) and **DDE** (dichlorodiphenyldichloroethene) are degradation products of the DDT.

Dichlorvos   (DDVP) is the name for an insecticide with contact and feeding poison effect from the group of phosphoric acid esters. It is a viscous, volatile, and colorless to yellowish brown liquid with an aromatic odor. Dichlorvos is flammable, harmful to the environment, and dangerous to health. In the EU, the use of biocidal products containing the active substance dichlorvos has been reduced; since November 1, 2012, it is no longer allowed. Dichlorvos disrupts the functioning of nerve cells. The substance is absorbed by inhalation or through the skin.

Double acidic potassium   is an obsolete chemical name for potassium dichromate ($K_2Cr_2O_7$). Potassium dichromate is a red water-soluble substance and, like all chromium compounds, it is toxic. It is used in the chemical industry as a strong oxidizing agent, but its importance has decreased.

Dowfume is the trade name for a mixture of the gases ethylene dichloride, ethylene dibromide, and carbon tetrachloride. It was produced by the US company Dow Chemical Company, which was dissolved in 2017, in different formulations (Dowfume C, G, 75, mc-2, mc-33, and EB-5). The liquid mixture was placed in open cups in storage rooms. Because of its toxicity, it may no longer be used. Oral ingestion, inhalation, and absorption through the skin are toxic. In addition, dowfume is carcinogenic, ethylene dichloride is flammable, and the fumes are explosive.

[see also: Goldberg, Lisa (1966): A History of Pest Control Measures in the Anthropology Collections, National Museum of Natural History, Smithsonian Institution, *JAIC*, 1996, (35):23–43.; Zycherman, Lynda A.; Schrock, J. Richard (1988): A Guide to Museum Pest Control, FAIC and Association of Systematics Collections, Washington DC, 1988.]

E

"Eryl" was a product of the company I. Ehrlich from Munich. There was a strong smell of → "Carbolineum".

[see also: Kleine, R. (1926): (first name unknown, author's note). Control experiments of *Calandra granaria* with Eryl. In: *Mitteilungen der Gesellschaft für Vorratsschutz e.V.*, No. 6, Volume 2, November 1926, 69–80.]

Ethanol Ethyl alcohol is a monohydric alcohol with the formula C_2H_5OH. Ethanol is a product of the alcoholic fermentation of fruits and fruit juices, as well as sugar solutions or starch from corn, cereals, or potatoes. Ethanol is obtained from the fermented material by distillation. This earned it the now historical name "wine spirit" or "spirit" from the Latin "spiritus vini" or spirit of wine. In the second half of the 20th century, the synthetic production of ethanol was introduced. The substance is an important solvent and extractant, as well as an additive for fuels. Pure ethanol is a colorless, highly flammable liquid with a burning taste and a characteristic, spicy (sweetish) odor. For technical purposes, a completely denatured and thus tax-privileged ethanol, which is also called "methylated spirits", is on the market. Methyl ethyl ketone (MEK) and denatonium benzoate, which tastes extremely bitter, are usually used as denaturants. Taken in larger quantities and over a longer period of time, ethanol acts as a liver toxin. It can also result in acute fatal poisoning.

Ether (diethyl ether) are organic compounds that contain functional groups (symbol R) bound via oxygen. This results in the general formula R_1-O-R_2. In colloquial language, "ether" is the name for the simplest ether, the diethyl ether with the formula $C_5H_2-O-C_2H_5$. This colorless, rapidly evaporating, and flammable liquid was formerly used as an anesthetic.

Ethylene dichloride → 1,2 Dichloroethane

Ethylene oxide   (oxirane) Ethylene or ethylene oxide is a colorless, highly flammable gas with a sweet odor. It kills bacteria, viruses, and fungi. It is toxic
to humans and carcinogenic even in low concentrations. Ethylene oxide is
insidious because it has a high odor threshold, and its narcotic and lethal dose
cannot be smelled.

"Eulan"   is an abbreviation and is composed of the Greek word *eu* "good" and
the Latin word *lana* "wool". Farbenfabriken Bayer AG launched the first
"Motteneulan" in 1920 and had the product name "Eulan" legally protected.
In the meantime, more than 30 preparations with very different compositions
have been launched on the market by Bayer AG. "Eulane" are textile finishing
agents that are used to protect against moths. Analytically, permethrin could
be detected as an active ingredient in "EULAN SPA". DDT, lindane, PCP, as
well as arsenic, mercury, and lead compounds were detected in other "Eulan"
preparations.

[see also: Unger, Achim (2012): "Eulanisierte" Textilien – Semi-foursquare eine
Gefahr für Mensch und Material?
In: *Beiträge zur Erhaltung von Kunst- und Kulturgut*, **2012, (2), 25–39.**]

F

"Flit"   is the product name for an insecticide invented by chemist Franklin C.
Nelson and approved in 1923. It was made based on mineral oil and contained
pyrethrum as an active ingredient. Under the trade name "Flit", it was put
on market in 1926 by the Deutsch-Amerikanischen Petroleum-Gesellschaft in
Hamburg.

[see also: Wilhelmi, Julius; Kunike, Hugo (1927): Versuche und Untersuchungen
über die Wirksamkeit des Petroleum-Raffinates "Flit"bei der Fliegen- und
Stechmückenbekämpfung. In: *Zeitschrift für Desinfektions- und Gesundheitswesen*, 1927, 19 (3), 98–99.]

After the discovery of the insecticidal effect of DDT, "Flit" was first equipped with
mixtures of pyrethrum and DDT, and at the end of the 1940s only DDT was
used as an active ingredient in a concentration of 5 %.

[see: Roth, K.; Vaupel, E. (2017): Von Insekten, Chrysanthemen und Menschen.
Chemie in unserer Zeit. 2017, Heft 3, 162–184.]

Flowers of sulfur   or sulfur bloom is a powdery form of sulfur. It is produced
by cooling sulfur vapors, e.g., in a distillation apparatus. In nature, flowers of
sulfur is found in volcanoes.

Formaldehyde   is the common name for the chemical compound methanal, the
simplest aldehyde of formic acid. It is a colorless, pungent-smelling gas that
dissolves well in water. A polymer of formaldehyde is paraformaldehyde. The
aqueous (about 40 %) solution of formaldehyde is called "formalin". With
this trade name, it came to the market from 1893 by Schering and as "Formol" by Hoechst. The decay of paraformaldehyde (depolymerization) can
also be caused by certain substances, such as barium peroxide. It was on this
process that the fumigant "Autan" was based. Formaldehyde is used, among

other things, as a fungicide and fumigant in intensive livestock farming. Its use for fumigation of museum collections is limited because it causes irreversible changes in leather. This substance causes acute toxic reactions, such as allergies, skin, and respiratory or eye irritation, and has been classified as a probable carcinogen in humans since April 1, 2015.

"Formalin" → Formaldehyde

"Formol" → Formaldehyde

Fowler's solution or Fowler's arsenic trioxide drops were first produced by the physician Thomas Fowler in 1786 and used against malaria. This medicine was also used internally for skin and nerve diseases, as well as for the preservation of animal skins.

[see also: Neumüller, Otto-Albrecht (Hg.) (1973): Römpps Chemie-Lexikon. 7. Auflage. Stuttgart: Franckh'sche Verlagshandlung, 1973, Band 2, D-G, 1185–1186.]

Fluid (from Latin *fluidus* = flowing) is a supercritical fluid. It is a condition that occurs above the critical temperature and pressure of some gases and liquids. In the supercritical state, there are no differences between the gas and liquid phases. An example is supercritical carbon dioxide.

G

Glacial acetic acid (ethanoic acid, pure) is the concentrated anhydrous acetic acid that solidifies at 16 °C. The colorless and corrosive liquid smells typical of vinegar.

"Globol" → 1,4-Dichlorobenzene

Glycerol is the common name for propane-1,2,3-triol. Glycerol is the simplest trivalent alcohol, i.e., it has three –OH groups in the molecule. It is a colorless, slightly viscous hygroscopic liquid with a sweet taste (from Greek *glykós* = sweet). Glycerol plays an important role in human metabolic processes. As a humectant, it has multiple uses in cosmetic products, pharmaceuticals, and tobacco products. Nitroglycerin, an explosive substance, requires glycerin for its production. Glycerin is safe for health.

H

Hydrocyanic acid gas is the obsolete name for the vapors of hydrogen cyanide.

Hydrogen cyanide is gaseous hydrocyanic acid (prussic acid).

Hydrogen peroxide is a pale blue liquid when it is in a higher concentration. When it is diluted, it becomes colorless, and is composed of hydrogen and oxygen atoms (H_2O_2). Hydrogen peroxide decomposes, releasing oxygen, which explains its bleaching and disinfecting properties. A 3% solution is used for skin disinfection. It can also be used to combat mold infestation during interior renovation. However, highly concentrated hydrogen peroxide is a very strong oxidizing agent that can cause violent reactions with substances, such as metal shavings and cleaning rags, posing acute or chronic health hazards.

Hydrogen sulfide is a foul-smelling, colorless, highly toxic gas with the formula H_2S. It is flammable and forms explosive mixtures with air. Hydrogen sulfide is produced, among other things, during the decomposition of sulfur-containing amino acids in the proteins of egg whites and yolks. Depending on the length of exposure and concentration, symptoms of poisoning include the destruction of the red blood pigment hemoglobin, damage to the nervous system, drowsiness, dizziness, convulsions, loss of consciousness, and possibly death. Long-term exposure to low doses can cause fatigue, loss of appetite, headaches, irritability, poor memory, and concentration.

I

"Illo-Spezial-T" is the trade name for a preparation containing the active ingredient tetrachloroethene, also known as tetrachloroethylene or perchloroethylene. It was developed and put on the market by Chemische Fabrik "Illo", Hans Haag, Verwaltung in Berlin W 50, Passauerstr. 3.

Insect powder → "Zacherlin"

Isovaleric acid (3-methylbutanoic acid) is an isomer of valeric acid, a short-chain organic acid that, like citric and acetic acids, belongs to the carboxylic acids. Isovaleric acid is contained in larger quantities in the roots of valerian (*Valeriana officinalis*), from which it is extracted. It is a colorless, oily, and water- and alcohol-soluble liquid with an unpleasant odor. Highly diluted valeric acid and its compounds (esters) are used as a remedy, especially as a sedative and mild sleep aid.

K

Kerosene is a mixture of liquid hydrocarbons with a boiling range of 150–280 °C, which is produced by fractional distillation of crude oil. Kerosene has relatively low volatility, is not easily ignited, but is still flammable. The properties of kerosene can vary depending on the source of the crude oil. Kerosene, derived from natural deposits, has been used as a fuel and a component in drugs since ancient times. From around 1870, it was mainly used as a fuel for kerosene lamps, eventually being replaced by electric lighting at the beginning of the 20th century. Additionally, kerosene was used in the preservation of wood from the 18th to the 20th century. Today, various purified grades with narrower boiling ranges are used as solvents, cleaning agents for metal surfaces, and as additives in some engine fuels. However, kerosene is hazardous to health.

L

Laudanum (tincture of opium) → Opium

Lavender oil is an essential oil obtained from the flowers of a true lavender (*Lavendula vera* L.). It is colorless or slightly yellowish, runny, and has a pleasant and strong lavender smell. Lavender oil has weak antimicrobial effects. Internally, it is used as a mild sedative for restlessness and difficulty in falling asleep. In the perfume and soap industries, lavender oil is important as a fragrance, and it is also used as a repellent against insects, especially clothes moths, as well as to scare away cats. However, the applications are likely to be more often the cheaper spik oil.

[see also: Rochusen, Frank (1920): Ätherische Öle und Riechstoffe. Sammlung Göschen, Berlin und Leipzig: Vereinigung wissenschaftlicher Verleger, 1920, 82.]

Lead arsenate → Lead hydrogen arsenate

Lead hydrogen arsenate , with the formula $PbHAsO_4$, is a white toxic powder insoluble in water. It was formerly used as a pesticide in fruit growing and arable farming, especially as a feeding poison. It has been banned in Germany since 1928 because it is carcinogenic, toxic to reproduction, and to the environment.

Lead sulfate (lead(II) sulfate) is the obsolete name for lead(II) sulfate ($PbSO_4$). It is a white solid with a relatively high density of 6.35 g/cm^3. In water, lead sulfate is almost insoluble. In the past, it was used as a low-quality paint. Currently, the production and use of lead sulfate are severely restricted. Like all lead compounds, this compound is toxic and hazardous to the environment.

Linseed oil varnish is made of linseed oil. The drying of linseed oil is accelerated by special additives (siccatives). At present, cobalt compounds are used for this purpose. Linseed oil varnish is an effective binder for oil paints, particularly for artist oil paints, and can also function as a primer, such as for wood. The ecological, health, and technical properties of linseed oil varnish are positive.

Lost is a name for dichlorodiethyl sulfide, which is composed of the first letters of the surnames of two German chemists, Wilhelm **Lommel** and Wilhelm **Steinkopf**. They were colleagues of Fritz Haber at the Kaiser-Wilhelm-Institut für physikalische Chemie und Elektrochemie in Berlin-Dahlem. This active ingredient, which is also called dissolved sulfur or mustard gas, was proposed by the two chemists for use as a chemical warfare agent. Lost is a group of 28 chlorinated, organic, sulfur, or nitrogenous compounds, the so-called lost group. In cooperation with the Chemische Fabrik Bayer in Leverkusen und Elberfeld, dissolved sulfur was used in the First World War.

[see also: Schnedlitz, Markus (2008): Chemische Kampfstoffe. Geschichte, Eigenschaften, Wirkung; Studienarbeit. zugl.: Wiener Neustadt, FH, Seminararbeit, 2008. 1. Aufl. München: Grin-Verlag, 2008, 30.]

M

Mustard plaster was a well-known home remedy until the 19th century. It is used against various ailments and is still used today in hydrotherapy. Pierer's encyclopedia mentioned a mixture of sourdough and coarsely crushed black mustard seeds in a ratio of 1:1, as well as another mixture of black mustard powder, rye flour, and vinegar or just mustard powder and lukewarm water. Porridge was spread on the canvas and placed on the skin or wrapped. Nowadays, ready-made mustard plasters are also available. Surrogates are mustard paper and blotting paper soaked in mustard spirit.

[see also: Pierer, Heinrich August (1857–1865): Universal-Lexikon der Vergangenheit und Gegenwart oder neuestes encyclopädisches Wörterbuch der Wissenschaften, Künste und Gewerbe. Altenburg, 1857–1865, Bd.15, 844.]

Mercuric chloride   → (Mercury(II) chloride)

Mercury(II) chloride   is a colorless, odorless solid of high toxicity, which is poorly soluble in water. The salt also bears the trivial name sublimate because it sublimates slightly when heated. It can be produced by heating mercury(II) sulfate with sodium chloride. Mercury(II) chloride has an antifungal effect and was formerly used for the treatment of seeds and for pressure impregnation of wood (so-called antifungal chloride or kyanization after the inventor John Kyan). Because it is also effective against various pathogens, very weak solutions were used to disinfect wounds, as well as a drug against syphilis. The dust of mercury(II) chloride irritates the respiratory tract, eyes, and skin. If swallowed, mercury(II) chloride causes abdominal pain and diarrhea. Repeated exposure to mercury(II) chloride can result in chronic poisoning with kidney damage. The lethal dose for humans is 1 mg/kg body weight. So, in toxicity, it is equal to arsenic trioxide.

Mercury sulfide   is also called mercury(II) sulfide or cinnabar. It is a chemical compound of mercury and sulfur with the molecular formula HgS. It is solid, and the crystalline red powder is odorless. Mercury sulfide (cinnabarite) is used as a red pigment (vermilion). The noncombustible substance is practically insoluble in water. Unlike other mercury compounds, HgS is nontoxic due to its heavy solubility. Acute health hazards include irritation to mucous membranes and skin, as well as chronic allergic skin diseases. There is evidence suggesting potential for cancer-causing effects.

[see also: Institut für Arbeitsschutz der Deutschen Gesetzlichen Unfallversicherung (IFA), GESTIS-Stoffdatenbank. www.dguv.de/ifa/gestis/gestis-stoffdatenbank/index.jsp, last visited 19.1.2021.]

Methanol   (methyl alcohol, wood spirit) was known in earlier chemical nomenclature under the name methyl alcohol. It is an organic chemical compound and the simplest representative of a group of alcohols with the formula CH_3OH. Because of the earlier production of beech wood by means of dry distillation (pyrolysis), methanol received the now historical name "wood spirit". It is a colorless liquid with a pleasant odor. Methyl alcohol is unfit for consumption and can cause severe poisoning. If you survive acute methanol poisoning, you may experience vision loss.

Methoxychlor   is a mixture of several structural isomeric chemical compounds from the group of chlorinated diphenylmethane derivatives with the molecular formula $C_{16}H_{15}Cl_3O_2$. It is a solid, crystalline substance that is colorless to yellowish and has a fruity smell. Methoxychlor is flammable but flame retardant and practically insoluble in water. The substance poses acute and chronic health hazards. Methoxychlor is hazardous to water.

[see also: Institut für Arbeitsschutz der Deutschen Gesetzlichen Unfallversicherung (IFA), GESTIS-Stoffdatenbank www.dguv.de/ifa/gestis/gestis-stoff-datenbank/index.jsp.

Methylated spirits → Ethanol

Methyl bromide (methylbromid) → Bromomethane

Mirbane oil → Nitrobenzene

Morphine → Opium

Mottenether ("moth ether") consists of 7.5 parts naphthalene, 2.5 parts camphor, 50 parts petrol, 40 parts turpentine oil, 1 part nitrobenzene, and moth paper. As a carrier material, the paper is soaked with a concentrated solution, or the mixture is melted over the steam bath in a covered crucible.

[see also: Frerichs Georg, Arends Georg, Zörnig Heinrich (Hrsg.) (1927): Naphthalinum. In: Hagers Handbuch der Pharmazeutischen Praxis. Springer, Berlin, Heidelberg, 1927, 200–201.]

Musk (deer) is a fragrance extracted from the glands of male musk deer. The strongly fragrant secretion is dried and used in perfume production and as a flavoring in food. The production is synthetic.

[see also: Schroeck, Lucas; Franz, Anselm; Hafner, Melchior (1682): Historia Moschi. Ad normam Academiae Naturae Curiosorum. Augustae Vindelicorum: Theophilius Göbelius.]

N

Naphtha or crude benzene is the name for a fraction of the distillation of crude oil or coal tar oil. It is one of the most important raw materials and is mainly used in the production of various types of petrol as well as a solvent. Naphtha is flammable, harmful to health, and hazardous to water.

Naphthalene is a bicyclic aromatic hydrocarbon consisting of two benzene rings. The colorless solid is slightly soluble in water and highly soluble in some organic solvents. It undergoes sublimation even at room temperature, emitting a distinctive odor, and is primarily obtained from coal tar oil. Discovered in 1819, naphthalene was formed into balls and used as an insecticide, especially against clothes moths (*Tineola* sp.). After both the environmental and health effects of naphthalene were recognised, it was classified as highly carcinogenic. As a result, naphthalene in mothballs was replaced by the less harmful 1,4-dichlorobenzene. Naphthalene played an important role in the production of some insecticides, such as chloronaphthalenes ("Xylamon LX Hell"). Today, naphthalene is used as a raw material for the production of some dyes and pharmaceuticals.

Naphthols are hydroxy derivatives of naphthalene, i.e., they are molecules of naphthalene in which one or more hydrogen atoms are substituted by OH groups. The term "naphthol" is used for monohydroxynaphthalene, with two isomers (1-naphthol and 2-naphthol). Both are important raw materials for paint production. In addition, they exhibit antiseptic properties and have been used in pesticides.

Natrium arsenicum   resp. **arsenicosum** is presumably a sodium salt from the group of arsenites, and natrium arsenicosum is a sodium salt from the group of arsenates.

(Friendly verbal communication from Achim Unger, January 21, 2021.)

Nicotine   is a substance present in the leaves of tobacco plants. In smaller quantities, it is also found in other plants, such as tomatoes and potatoes. Pure nicotine is a colorless, oily liquid that turns brown when exposed to air. In the past, it was used for plant protection against insects, including aphids. In plants, this substance is well tolerated and biodegradable. However, because of its high toxicity, the use of nicotine has been banned since the 1970s.

Nitrobenzene   is also known as "oil of mirbane" and is the simplest aromatic nitro compound, a benzene derivative with the molecular formula $C_6H_5NO_2$. The oily liquid is colorless to pale yellow with a pleasant bitter almond oil-like odor. It was used, among other things, in insect collections for preventive protection and disinfection. In addition, nitrobenzene was also used to perfume soaps. However, its use in cosmetic products has been prohibited. Nitrobenzene can provoke severe signs of poisoning, and severe poisoning can lead to death within a few hours. Alcohol enhances the effect of nitrobenzene. It is classified as carcinogenic and hazardous to the environment and reproduction.

[see also: Lueger, Otto (1904): Lexikon der gesamten Technik und ihrer Hilfswissenschaften. Bd. 7, Stuttgart, Leipzig, 1909, 425–426.]

Nitrocellulose lacquers   are cellulose nitrates dissolved in acetone, ethyl acetate, and other solvents. Nitrocellulose varnish has good mechanical strength and is therefore primarily used for finishing interior woods, especially furniture or musical instruments, as well as for indoor metals. Residues or waste must be disposed of in hazardous waste and must not enter the groundwater under any circumstances.

O

Opium   (poppy tears) is an intoxicant and anesthetic obtained from the milky juice of unripe seed pods of the opium poppy (*Papaver somniferum*). It used to be called poppy seed juice. The dried juice, the raw opium, is a dark mass. Its main active ingredients are morphine and codeine. Opium dissolved in ethanol gives the so-called opium tincture, which is known as laudanum. It was widely used in medicine until the early 19th century. Although opium and its derivatives are still used in medicine, their production and use are subject to strict controls. The abuse of opium leads to altered behavior and dependence; long-term abuse causes severe psychological and physical damage, and an overdose is fatal.

[see also: Ball, Philip (2014): The devil's doctor. Paracelsus and the world of Renaissance magic and science. London: Cornerstone Digital.]

P

"P.84." → Sodium silicate

"Paracide crystals" is a product name for *p*-dichlorobenzene.

***para*-Dichlorobenzene** → 1,4-Dichlorobenzene

***p*-Dichlorobenzene** → 1,4-Dichlorobenzene

Paraffin, also known as paraffin wax, is a mixture of linear saturated hydrocarbons with chain length $C_{14}C_{30}$. These hydrocarbons are obtained by distillation from crude oil or lignite tar oil after distillation of more volatile constituents (distillation fractions up to approximately 350 °C). Depending on the molecular mass, these are thin and viscous substances (paraffin oils), waxlike substances (paraffin petroleum jelly), or a solid substance called paraffin. Paraffin is odorless and tasteless, and soluble in gasoline, tetrachloromethane, chloroform, and other solvents. It is water repellent, flammable, fusible, electrically insulating, and resistant to many chemicals, even strong acids. The uses of paraffin are very diverse due to its properties. The main areas of application are sealing, care (cosmetics), and preservation. It is nontoxic and is classified as harmless to humans and nature.

Paraffin wax → Paraffin

Paraform is the trade name for paraformaldehyde (see also "Autan").

Paraformaldehyde → Formaldehyde

Patchouli oil is a good and intensely fragrant essential oil obtained by steam distillation of the green parts of the patchouli plant (*Pogostemon cablin*), which is native to India and Malaysia. Patchouli oil has several healing effects. It is also an effective repellent against insects. It is primarily used in the perfume industry.

[see also: Wiesner, Julius (1927): Die Rohstoffe des Pflanzenreiches. Bd. I, II. 4. Aufl. Leipzig, Wilhelm Engelmann 1927, 91.]

Pentylacetate is a compound of acetic acid and pentyl alcohol. Other names include acetic acid, amyl ester, amyl acetate, amyl acetic ester, pentanol acetate, and pentyl ethanoate. It is a colorless, aromatic-smelling liquid with a pear or banana smell. Acetic acid pentyl ester is used in the food industry as a pear flavor.

Perchloroethylene → Tetrachloroethylene

Persulfates are the colloquial abbreviation for peroxodisulfates. They are salts of peroxodisulfuric acid $H_2S_2O_8$. The best known peroxodisulfates are potassium peroxodisulfate ($K_2S_2O_8$) and ammonium peroxodisulfate [$(NH_4)_2S_2O_8$]. These are strong oxidizing agents, and care must be taken when applying them to paper.

"Perthan" is the product name for a chlorinated hydrocarbon-based insecticide with an active ingredient structurally related to dichlorodiphenyldichloroethane (DDD). It is an odorless crystalline solid designed to control pests on fruits and vegetables. In addition, "Perthan" was used in the textile industry against clothes moths and carpet beetles. Similar to DDT, it is believed to be carcinogenic and therefore not authorised in the European Union and Switzerland as a product containing this active ingredient as a pesticide. In addition, it is hazardous to water.

Petrol   (gasoline) is a liquid mixture that is primarily composed of linear hydrocarbons produced through fractional distillation of crude oil or lignite tar oil. This fraction is heated to a temperature range of 50–200 °C, and then refined into various types of gasoline (light petrol, heavy petrol, motor petrol, etc.).

Petroleum ether   (light ligroin) is a colorless mixture of various saturated hydrocarbons, and not a representative of the chemical group of substances "ether". It is a low-boiling light petrol, also called "wound petrol", and has its boiling range at about 25–80 °C (see also petrol). Petroleum ether is used as a solvent in laboratory work to remove grease stains on paper, textiles, etc. It is flammable, and its vapor forms an explosive mixture with air. Prolonged exposure to petroleum ether can result in irritant effects on the skin. It can cause drowsiness and dizziness. It is fatal if it enters the respiratory tract. Petroleum ether is hazardous to water.

Pepper   (black pepper) are seeds of the pepper bush (*Piper nigrum*) and are a well-known spice. Pepper bushes are native to southeastern Asia, and they are also grown in various areas of India. The pungent taste of pepper is caused by an organic base called piperine (content approximately 7 %). The scent of pepper is caused by an essential oil (content approximately 1–2.3 %).

(see: Mercks Warenlexikon für Handel, Industrie und Gewerbe (1920): Manuscriptum, Siebente völlig neu bearbeitete Auflage. Leipzig: Gloeckner 1920. Reprint Recklinghausen 1996. 328 u. 344.)

Phenol   (hydroxybenzene) is the frequently used common name for the most important benzene derivative, hydroxybenzene. It is a colorless, crystalline fabric with a typical pungent odor that turns red when exposed to air. Its aqueous solution is called carbolic acid, also known as carbolic. Weak phenolic solutions were previously used, among other things, for disinfection in hospitals as well as of cesspools and for the preservation of wood. Phenol and its derivatives (phenolates) are toxic and are suspected of damaging genetic material. It is an important chemical in industry, which is particularly used for the production of various plastics.

Phosphorus paste   or phosphorus dwarf, was used to control rodents.

[see also: Mayers Konversationslexikon (1896): "Phosphorlatwerge (Phosphorpaste), Mischung von Mehl und Phosphor, auch wohl mit etwas Fett, zur Vertilgung von Ratten und Mäusen".[1] 5. Auflage, Leipzig und Wien: Bibliographisches Institut, 1896, Bd. XIII, 875]

In the pharmaceutical literature of the 19th century, phosphorus dwarfs can be found as

(experimental) medicine for animals. Another recipe for phosphorus dwarfs is: "2 Quentchen Phosphor werden im Mörser in 6 Lot warmen Wassers geschmolzen, hierzu werden schnell 9 Lot Weizenmehl eingerührt. Nach dem Erkalten noch 8 Lot geschmolzene Butter und ¼ Lot pulverisierter Zucker gerührt".[2]

[see also: Hertwig, Heinrich, Carl (1847): Praktische Arzneimittellehre für Thierärzte. 3. Vermehrte Auflage Berlin: Veith & Comp., 1847, Anmerkung, 520.]

Picric acid   is the common name for 2,4,6-trinitrophenol (TNP). Their salts are called picrates. The yellow crystals of picric acid taste extremely bitter, hence the name (Greek *pikrós* = bitter). Picric acid was the first explosive used for grenades in the First World War. Because of the explosions, which were difficult to control, it was soon replaced by trinitrotoluene (TNT).

Pine oil   is an essential oil obtained from the dry distillation of the root wood of some pine species, especially *Pinus sylvestris* and *Pinus Lederbourii*. The main products of this dry distillation are charcoal and wood tar. Kien oil is sometimes confused with pine needle oil (extracted from pine needles) or turpentine oil (obtained by distilling turpentine from wood of various conifers).

[see also: Birnbaum, Carl; Merck, Klemens (Hg.) (1884): Klemens Merck's Warenlexikon für Handel, Industrie und Gewerbe. Beschreibung der im Handel vorkommenden Natur- und Kunsterzeugnisse unter besonderer Berücksichtigung der chemisch-technischen und anderer Fabrikate, der Drogen- und Färbewaren, der Kolonialwaren, der Landesprodukte, der Material- und Mineralwaren. 3., gänzlich umgearb. Aufl., 2., rev. Abdr. Leipzig: Gloeckner, 576 f.; Wiesner, Julius (1927): Die Rohstoffe des Pflanzenreiches, Bd. I, II. 4. Aufl. Leipzig, Wilhelm Engelmann 1927, 97.]

Potassium carbonate   is a potassium salt of carbonic acid with the molecular formula K_2CO_3. Potassium carbonate also bears the historical name "potash" because this hygroscopic salt was transported in closed pots. It is a white powder, highly soluble in water, and the solutions of which have a strong alkaline reaction. It is used, among other things, as a melting agent in the production of glass (potash glass) and soft soaps as well as water glass.

Potassium persulfate   (potassium peroxodisulfate) → (Persulfates)

Potash   → potassium carbonate

Propylene   (propene) was the name for propene in earlier chemical nomenclature and is a colorless combustible gas. It is an unsaturated, linear hydrocarbon with three carbon atoms in the molecule. Propylene is obtained by reacting light petrol, which is produced during kerosene refining. It is flammable and has limited solubility in water. With air, it forms explosive mixtures. Propylene is not very toxic but has a narcotic effect in high concentrations.

Prussic acid   (hydrogen cyanide), with the formula HCN, is a colorless to slightly yellowish, bitter almond-smelling, flammable, very volatile, and water-soluble liquid. Hydrogen cyanide is an unstable acid. It is displaced from its salts (cyanides, e.g., potassium cyanide KCN) by carbon dioxide from the air and in this form is also referred to as hydrogen cyanide. Its volatility allows it to be used to control pests and vermin. To simplify the handling of hydrocyanic acid, a carrier material, such as diatomaceous earth, was impregnated with hydrocyanic acid and marketed under the trade name "Zyklon B". "Zyklon B" was misused by the Nazis and used for the mass murder of Jews in the Auschwitz extermination camp. Hydrocyanic acid is produced technically, but it is also found in the form of the glycoside amygdalin in several natural products, such as bitter almonds and stone fruit seeds. Apricot kernels contain a particularly large amount of amygdalin. Hydrocyanic acid is extremely toxic, and 1 mg/kg body mass is lethal. Therefore, excessive consumption of almonds and/or apricot kernels can also be dangerous. Hydrogen cyanide is classified as highly hazardous to water. The name prussic acid is derived from the earlier production from the pigment Berlin blue, an iron(III) hexacyanidoferrate(II/III).

Pyrethrins   are ingredients of pyrethrum.

Pyrethrum   is an insecticide whose effect was already known to the Romans. This is also where the epithet "Persian insect powder" comes from. The insecticidal active ingredients of pyrethrum are pyrethrin I and II, cinerin I, and jasmolin I and II. Pyrethrum is mainly extracted from the flowers of the Dalmatian usury flower (*Tanacetum cinerariifolium*). It is a contact poison that has a neurotoxic effect on insects. The pyrethrins irritate the eyes and mucous membranes but are only slightly irritating to the skin. In Germany, Austria, and Switzerland, pyrethrum is not approved as an active ingredient in plant protection products, but in these countries there are approvals for pyrethrins, i.e., for pure components of pyrethrum. These are also used in means for fleas, head lice, cockroaches, etc.

Q

Quendel oil   with the trivial names breckland or wild thyme (*Oleum serphyllii*) is an essential oil extracted from the flowering herb of quendel (*Thymus serpyllum*), a plant similar to thyme. Quendel oil is colorless or golden yellow and smells pleasantly like lemon balm and thyme. The main constituents are the terpenes cymene and thymol with its isomer carvacrol.

[see also: Rochussen, Frank (1920): Ätherische Öle und Riechstoffe. Sammlung Göschen, Berlin und Leipzig: Vereinigung wissenschaftlicher Verleger 1920, 96.]

Quinine   is a heterocyclic, nitrogenous hydrocarbon that is assigned to the alkaloids. Quinine is found in the bark of several species of cinchona trees native to South America (Bolivia and Venezuela). The antipyretic effect of cinchona bark had long been known to the indigenous people there. This knowledge did not reach Europe until the 17th century. In France, quinine was isolated in pure form from cinchona bark in the 18th century and used in medicine as an effective remedy against malaria. It is a white, very poorly water-soluble crystalline powder with a bitter taste, which is also used as a bitter substance. Soft drinks containing quinine have been used to prevent malaria during colonial times. The bitter-tasting quinine is a popular additive for bitter drinks in the food industry. According to the GHS hazardous substance labeling, it is classified as highly hazardous to health in high doses.

[see also: Müller-Jahncke, Wolf-Dieter; Friedrich, Christoph; Meyer, Ulrich; Müller-Jahncke-Friedrich-Meyer (2005): Arzneimittelgeschichte. 2., überarb. und erw. Aufl. Stuttgart: Wiss. Verlag-Ges., 62 f.]

R

Rosemary oil   is a pleasantly fragrant essential oil, obtained by distillation from the leaves of the rosemary plant (*Rosmarinus officinalis*). It is mainly native to the Mediterranean countries. The components of rosemary oil include cineole, camphor, pinene, and numerous other substances. It is used in soap and perfume production and is also effective against numerous bacteria, yeasts, and molds. On the skin, it promotes blood circulation and is therefore also used in medicine.

[see also: Rochusen, Frank (1920): Ätherische Öle und Riechstoffe. Sammlung Göschen, Berlin und Leipzig: Vereinigung wissenschaftlicher Verleger, 1920, 94.]

S

Saltpeter   is an umbrella term for various salts of nitric acid HNO_3, such as ammonium nitrate, barite nitrate (barium nitrate), Chilean nitrate/sodium nitrate, potassium nitrate, and calcium nitrate. However, the term was mainly used for the sodium nitrate $NaNO_3$ (Chilean saltpeter), which used to play an important role as a nitrogen fertilizer. Because of advances in agriculture, the use of nitrogen fertilizers has decreased significantly.

Steinöl   ("rock oil") is the obsolete name for kerosene.

[see also: Zinke, Georg, Gottfried (1802): Kunst allerhand natürliche Körper zu sammeln, auf eine leichte Art für das Kabinett aufzubereiten und vor der Zerstörung feindlicher Insecten zu sichern. Jena 1802, 148.]

Sulfur dioxide   (sulfur(IV) oxide) is a gaseous substance with molecular formula SO_2. This gas is nonflammable, colorless, and has a pungent odor. Under pressure, the gas liquefies. In water, it is soluble under hydrolysis and forms sulfurous acid. The aqueous solution has a strong acidic reaction. Sulfur dioxide is hygroscopic and forms aerosols. It exhibits oxidizing and reducing effect. The substance poses acute or chronic health hazards.

[see also: Institut für Arbeitsschutz der Deutschen Gesetzlichen Unfallversicherung (IFA), GESTIS-Stoffdatenbank www.dguv.de/ifa/gestis/gestis-stoff-datenbank/index.jsp.

Sulfur ether   is an obsolete chemical name for diethyl ether, the former anesthetic. Because diethyl ether is produced by chemically reacting ethanol with sulfuric acid, it was also called "sulfur ether".

Sulfur flour   (sulfur powdered) → Flowers of sulfur

Sulfur mercury   → Mercury sulfide

Sulfuric acid,   with the formula H_2SO_4, is the most important and generally one of the strongest acids in technology. It is a colorless, oily, very viscous, and hygroscopic liquid. It is mainly used for the production of other mineral acids, such as hydrochloric or phosphoric acid, and for fertilizers. On the skin and mucous membranes, depending on the concentration, sulfuric acid has a strong irritant, as well as corrosive, up to irreversible destruction of tissues.

Sulfurous acid   is a weak volatile acid with the molecular formula H_2SO_3. It exists only in aqueous solution, which is formed when sulfur dioxide is dissolved in water. Sulfurous acid and its salts (sulfites) are reducing agents. The acid can accept an oxygen atom, whereby it is oxidised to sulfuric acid (H_2SO_4). Sulfurous acid is involved in the formation of acid rain. Because the combustion of fossil fuels (coal, kerosene, and its products) produces sulfur dioxide (SO_2), it is released into the air. It reacts with rainwater to form sulfurous acid, which then rains (diluted) to the earth.

Schweinfurt green   is a green, toxic pigment, a copper(II) arsenite acetate, which was first produced and marketed in Vienna in 1805 and in Schweinfurt since 1815. Schweinfurt green has many synonyms, e.g., Mitis Green, Parisian Green, Urania Green, New Green, etc. Even after the toxicity was known, Schweinfurt green was used as a paint color, which resulted in its use as a pesticide. In the USA, in 1868, J. P. Wilson patented the mixture of one part Paris Green with two parts mineral oil for use against the Colorado potato beetle. Schweinfurt green was also used in other insecticide formulations, for example mixed with wood ash. It was the first chemical insecticide to be applied on a large scale. In 1882, it was banned as a paint in Germany, and from 1887, there were bans on processing in aqueous binders and in pastel. After that, it was still used as an insecticide and as a ship's paint. In Germany, the use of arsenic-containing agents in viticulture was prohibited by a law of November 1942. Schweinfurt green is highly hazardous to health and water.

Sodium arsenate,   with the molecular formula Na_3AsO_4, is a crystalline, colorless, and odorless solid. It is nonflammable and highly soluble in water. This very dangerous contact poison has been used as an insecticide in wood and pest control products. It is harmful to the environment, especially to water and soil. Even in low concentrations, sodium arsenate has a lethal effect.

Sodium arsenite   is an inorganic chemical compound with the molecular formula $NaAsO_2$. It is the sodium salt of arsenic acid (metaform) and is considered one of the most toxic arsenic compounds, along with hydrogen arsenic. It is a colorless, crystalline solid that dissolves in water. Sodium arsenite is a contact poison that causes irritation and inflammation of the eyes, lungs, and mucous membranes. It is mutagenic, carcinogenic, and highly hazardous to water.

Sodium cyanide   (cyanodium) is the sodium salt of hydrocyanic acid. It is a solid substance that is either crystalline or white powder. The smell is faintly bitter almond-like. Sodium cyanide is easily soluble in water and develops highly toxic hydrogen cyanide with acid. The substance is acutely and chronically hazardous to health and hazardous to water.

(see also: Institut für Arbeitsschutz der Deutschen Gesetzlichen Unfallversicherung: GESTIS-Stoffdatenbank.)

Sodium fluoride   is a sodium salt of hydrofluoric acid with the formula NaF. It is a colorless and odorless solid that is soluble in water. Sodium fluoride is used as a wood preservative. It is toxic; therefore, inhalation of dusts should be avoided.

Sodium persulfate   (sodium peroxodisulfate) → Persulfates

Sodium silicate was also listed under the trade name "P.84". It is a sodium salt of orthosilicic acid, which comes in a variety of forms and forms a number of salts and esters. Sodium silicate is water soluble and, in addition to potassium silicate, is an important type of so-called water glass (soda water glass). It can be used in a variety of ways and is used in industry as a component of adhesives, additives in detergents and cleaning agents, cements, binders, etc. Before sufficient cooling options were developed, the water glass had an importance in households as a preservative for eggs. If large doses are ingested, irritation of the mucous membranes can occur, similar to that caused by caustic soda.

Spike oil is an essential oil that is extracted from the flowers of spike lavender (*Lavendula latifolia*). In France and Spain, this type of lavender is cultivated for the extraction of spike oil. Spike oil is mainly used as a substitute for real lavender oil. These two essential oils are well known, and their names have often been mixed.

[see also: Rochussen 1920: Rochussen, Frank (1920): Ätherische Öle und Riechstoffe. Sammlung Göschen, Berlin und Leipzig: Vereinigung wissenschaftlicher Verleger, 1920, 96.]

Spirit is the derivation of "spiritus vini" or spirit of wine ethanol.

Strychnine is a substance contained, among other things, in the emetic nut (*Strychnos nux vomica*) from which it is isolated. It is a colorless, crystalline solid with high toxicity. In the 19th century, strychnine was used as rat poison. It causes muscle rigidity even in very small doses. The dose of approximately 1 mg/kg body weight is lethal to humans.

"Sturmsches Mittel" , also known as Esturmit, was a powdered agent that was tested for use in plant protection. It contained an arsenic compound, obviously calcium(II) arsenate with the formula $Ca_3(AsO_4)_2$. Stellwaag, who published on the "Sturmsches Mittel", alternately used various misleading terms for the arsenic ingredient, such as "calcium arsenic", "calcium arsenate", and even "calcium arseniate".

[see also: Stellwaag, Fritz (1927): Der Gebrauch der Arsenmittel in deutschem Pflanzenschutz. Ein Rückblick und ein Ausblick unter Verwertung der ausländischen Erfahrungen. In: *Zeitschrift für angewandte Entomologie*, 1927, Bd. 12, Heft 1, 35–36.]

Sublimate → mercury(II) chloride

T

Tannins (French: tannins) are vegetable tannins that are widespread in plants and belong to the plant secondary substances. For example, the bark of oak trees and the so-called galls on oak leaves are particularly rich in tannins, but tannins are also found in red grapes and quinces. They are usually extracted from the plant material with warm or hot water; the extract is thickened, purified, and dried. Tannin is sold as a yellowish powder. Chemically, it is a mixture of water-soluble polycyclic hydroxy derivatives of phenol (e.g., resorcinol and quinoline). Tannins cause the formation of cross-links between macromolecules of proteins and thus ensure the conversion of hides into leather, which has a much higher durability. Therefore, the technical use of tannins lies mainly in leather production (tannery). Tannins react with corrosion products of iron to form bluish black iron tannates. This reaction is used to effectively protect rusted iron objects. Tannin solutions are also suitable as wood stains. In veterinary medicine, tannins are used as a remedy for diarrhea.

Tansy   (*Tanacetum vulgare*) is a species of flowering plant in the family Asteraceae. The strongly fragrant leaves as well as the flowers of tansy contain insect-repellent active ingredients and were formerly scattered to keep vermin away. Tansy was even placed in coffins in North America, and shrouds were soaked with tansy extract. Tansy was also planted to scare away the Colorado potato beetle. Dried tansy is used in beekeeping as a smoking agent. The essential tansy oil was also extracted from tansy in Germany, which was used in perfumery, liqueur production, and medicine.

[see also: Rochussen, Frank (1920): Ätherische Öle und Riechstoffe. Sammlung Göschen, Berlin und Leipzig: Vereinigung wissenschaftlicher Verleger 1920, 92.]

Tar   is the name given to products that are produced by thermal decomposition (pyrolysis) of various organic natural products. The word "tar" is also used to refer to the raw material, e.g., wood tar, coal tar, etc. They are dark brown to black solid substances. Together with lignite tar, these two tars are the most industrially important tars. Depending on their origin, the tars are mixtures of various hydrocarbons. Depending on their origin and resulting composition, they are further processed into petrol and other organic solvents, for example. These include paraffin, mineral oils, kerosene jelly, and tar dyes, as well as greasy tars from which greasy acids are extracted. Some tars are still important for industrial wood preservation, e.g., for railway sleepers or overhead line pylons. But the use of tar in public areas is prohibited because it poses a risk to both the environment and health. Upon contact with water, their components can enter the groundwater. For this reason, tar has been banned in the FRG since 1984 and in the former GDR since 1990. This applies to the use in public road and path construction, where it has been completely replaced by bitumen. Mixtures of bitumen and tar were also used, the so-called carbobitumen or pitch bitumen. This mixed form, like pure tar, is to be classified as harmful to health and must be disposed of separately. Long-term exposure of the tar to the skin can cause skin changes that can develop into cancer. However, tar preparations are used in medicine as externally applicable medicines for skin diseases, as they have an antipruritic, germicidal, and circulation-promoting effect.

Tar oil   is an oil obtained during the distillation of coal tar, which is used to preserve wood. The Avenarius company launched it in 1888 under the trade name "Carbolineum".

Tetrachlorobenzenes   are three isomeric chlorine derivatives of benzene, each with four chlorine atoms as substituents of hydrogen atoms of benzene. They are solid, fusible substances, insoluble in water and soluble in organic solvents.

They are especially important for the production of insecticides, herbicides, and other chemical products.

Tetrachloroethene, with the trivial name perchloroethylene, colloquially perchlorinate, is a colorless, chloroform-smelling, nonflammable, volatile liquid. It is a solvent used in the textile industry. Thanks to its high grease dissolving capacity, it is used as a degreasing agent in textile cleaning as well as in the film, optical, and metal industries. Because of its wide use in industry and commerce, it is one of the substances that contaminates groundwater. The vapors have a narcotic effect; they irritate the eyes and the respiratory tract. Long-term contact, even with low concentrations, leads to liver and kidney damage. Tetrachloroethene is suspected of being toxic to reproduction and carcinogenic.

Tetrachloroethylene → Tetrachloroethene

Tetrachloromethane, with the formula CCl_4, is a colorless, sweet-smelling, nonflammable liquid that is well miscible with ethanol, ether, or petrol. It is a good solvent for greases, waxes, and oils, and was formerly used for fire extinguishers and in the early to mid-20th century for dry cleaning of textiles. As a fire extinguishing agent, it became problematic because of its toxicity and the contact that occurs in the event of a fire. Tetrachloromethane also damages the ozone layer, which means that this active ingredient may be used to a very limited extent and is only permitted for research purposes. Its narcotic vapors cause damage to the liver and kidneys during long-term contact, even in low concentrations. Tetrachloromethane is suspected of being carcinogenic.

Theïn is also called theine or caffeine and belongs to one of the plant secondary substances, the alkaloids. The trivial name caffeine was given to the substance because it is contained in coffee. In the past, due to ignorance of the identical chemical composition, no distinction was made between the terms theine and caffeine. Chemically, it is 1,3,7-trimethylxanthine. It is a white, odorless, crystalline powder with a bitter taste. Caffeine stimulates the activity of nerves and is a component of stimulants, such as coffee, tea, Coca-Cola, mate, guaraná, energy drinks, and, to a lesser extent, cocoa.

[see also: Gossauer, Albert (2006): Struktur und Reaktivität der Biomoleküle. Eine Einführung in die organische Chemie. Zürich: Verlag Helvetica Chimica Acta, 2006, 184.]

Thyme camphor → Thymol

Thymol also bears the names Thymolum PhEur, Thymol INCI, thyme camphor, or thyme acid. It is an organic compound crystallizing in large, transparent cubes and octahedra, belonging to the phenols, with the molecular formula $C_{10}H_{14}O$, which is very difficult in water and in 96 % ethanol is very easily soluble. Thymol can also be easily dissolved in essential or greasy oils. This active ingredient is found in a number of essential oils, especially in thyme oil. It has fungicidal and antiseptic properties and hardly irritates the skin. Thymol is mainly used against colds, in toothpaste, as well as a veterinary drug.

Turpentine → Turpentine oil

Turpentine oil was an important, natural, organic solvent. It is obtained by distilling turpentine, a balm that flows out of conifers when the bark is carved. The residual substance after distillation is a resin called rosin. Turpentine oil is also called balsam turpentine oil, colloquially erroneously called turpentine, real turpentine oil, or also outdated turpentine ice. In Germany, turpentine oil was mainly extracted from pines. This production was discontinued due to lower demand, especially in the paint industry. As a substitute, white spirit (turpentine substitute) was introduced. The chemical composition of turpentine oil is not uniform; it depends on the type of tree. All turpentine oils can be characterised as mixtures of monocyclic terpenes. They are colorless to yellowish, volatile, strong-smelling, and flammable liquids with a boiling range of 155–175 °C, miscible with several organic solvents, but immiscible with water. In engineering, it is especially used for the manufacture of varnishes and for diluting oil paints. It is also used to remove grease stains and bleach substances that do not tolerate chlorine, such as ivory. Turpentine has an irritating effect and is harmful to health as well as hazardous to the environment.

"T-Gas" is a pesticide that does not contain hydrogen cyanide and consists of a mixture of nine parts ethylene oxide and one part carbon dioxide. It was mainly used for residential fumigation. "T-Gas" was developed and produced by Th. Goldschmidt AG, known today as Goldschmidt GmbH in Essen. The company was founded by Theodor Goldschmidt in Berlin in 1847 and later moved to Essen. Theodor Goldschmidt became an important manufacturer of special chemical products and worked closely with the Deutsche Gesellschaft für Schädlingsbekämpfung (Degesch) and the Deutsche Gold- und Silber-Scheideanstalt (Degussa). For the use of "T-Gas", Th. Goldschmidt A.G. Essen had specially founded the "T-Gas"-Gesellschaft m.b.H.

[see also: Ebbinghaus, Angelika (1999): Der Prozeß gegen Tesch & Stabenow. Von der Schädlingsbekämpfung zum Holocaust. In: *Zeitschrift für Sozialgeschichte des 20. und 21. Jahrhunderts*, 13, 1999, 2, 45.]

"Tillantin R" was a mercury-containing product that was used as a pesticide for seed dressing, among other things.

[see also: Taeger, Harald (1941): Die Klinik der entschädigungspflichtigen Berufskrankheiten. Berlin, Heidelberg: Springer Berlin Heidelberg.]

U

Urania-Grün ("Urania Green") → Schweinfurt green

V

Varnish is a non-pigmented paint that forms clear and transparent shiny films when dried. The German term "varnish" was derived from the French and Italian *vernice* = varnish. Varnishes consist of drying oil (linseed oil varnish), if necessary, mixed with dissolved resin or from one or more binders dissolved in solvents, which are mainly synthetic resins today. In painting, especially oil painting, varnish is applied as a transparent coating to protect paintings.

Vinyl acetate (acetic acid vinyl ester), with molecular formula $C_4H_6O_2$, is a colorless, light-sensitive, highly flammable liquid with a sweetish odor. It is a monomer prone to spontaneous polymerization. Vinyl acetate is mainly used for the production of polyvinyl acetate and polyvinyl alcohol. These polymers have various applications, such as solid resins, films, or solutions. For example, they are used as binders for paints, as raw materials for adhesives, in paper and textile industries, and in construction. Vinyl acetate is classified as carcinogenic.

W

Wine spirit is a historical name for ethanol. Today, undenatured pure ethanol with a concentration of around 95 % by volume is traded under this name. Ethyl alcohol is a monohydric alcohol with the formula C_2H_5OH. Ethanol is a product of the alcoholic fermentation of fruits and fruit juices, as well as sugar solutions or starch from corn, cereals, or potatoes. Ethanol is obtained from fermented material by distillation. This earned it the now historical name "wine spirit" or "spirit" from the Latin "spiritus vini" or spirit of wine. In the second half of the 20th century, the synthetic production of ethanol was introduced. The substance is an important solvent and extractant, as well as an additive for fuels. Pure ethanol is a colorless, highly flammable liquid with a burning taste and a characteristic, spicy (sweetish) odor. For technical purposes, a completely denatured and thus tax-privileged ethanol, which is also called "methylated spirits", is on the market. Methyl ethyl ketone (MEK) and denatonium benzoate, which tastes extremely bitter, are usually used as denaturants. Taken in larger quantities and over a longer period of time, ethanol acts as a liver toxin. It can also result in acute fatal poisoning.

Wormwood leaves are leaves of the wormwood plant (*Artemisia absinthium*), which is native to North Africa, the Caucasus, and India, among other places. Wormwood herb contains a high concentration of bitter substances and has been known as a medicinal plant since ancient times. Wormwood leaves have been said to have numerous healing effects, such as promoting appetite and digestion, as well as relieving headaches. Wormwood was also used to repel mice in libraries and against clothes moths. It is an ingredient in absinthe, an alcoholic beverage with extracts of wormwood, anise, lemon balm, and fennel, which became a fashionable drink especially in the 19th century.

"Wurm-Antorgan" , "Antorgan" or "Holzwurm-Antorgan", is an aqueous solution of 5–20 % ammonium fluoride and zinc fluoride (ammoniacal zinc fluoride solution), according to the literature. The preparation was used for wood preservation in the house, basement, and garden. It was preferably used for preventive protection and to control wood-destroying insects but also for indoor dry rot control.

(Friendly verbal communication from Achim Unger, September 30, 2017.)

X

"Xylamon"   is the product name of a series of wood preservatives. The first product with this name was marketed in 1923 by Vereinigte Alkaliwerke Westeregeln, Germany, with chloronaphthalenes as active ingredients. After 1945, Xylamon products also contained, among other things, the active ingredients pentachlorophenol (PCP) and lindane (γ-hexachlorocyclohexane).

The production of PCP was banned by the federal government in 1989; the active ingredient is considered to be highly carcinogenic in humans. Lindane has not been produced in Germany since 1984 and has also been considered carcinogenic in humans since 2015, according to the WHO.

"Xylamon-LX Hell"   was produced in the 1930s by DESOWAG Bayer-Holzschutz GmbH. It contained various chloronaphthalenes with insecticidal and fungicidal action. Because of the strong odor and efflorescence of chloronaphthalene crystals, some of the chloronaphthalenes were replaced by other highly chlorinated aromatic compounds during the 1950s. From the mid-1960s, the chloronaphthalenes in the "Xylamon" preparations were generally replaced by more modern active ingredients. In her doctoral thesis at the University of Greifswald, Anne Obst stated that from 1963 onward, "Xylamon Hell" contained 5.5 % pentachlorophenol (PCP) and 1.0 % γ-hexachlorocyclohexane (lindane).

(Friendly written message from Mr. Achim Unger dated June 21, 2018)

Z

"Zacherlin"   had the epithet Dalmatian or Persian insect powder and was named after its discoverer and producer Johann Zacherl, who produced this insect powder in Vienna and distributed it himself.

[see also: Sotriffer, Kristian (1996): Die Blüte der Chrysantheme. Die Zacherl—Stationen einer anderen Wiener Bürgerfamilie. Wien u.a. Böhlau, 13–31.; Offenthaler, Eva (2013): "Zacherlin wirkt staunenswert!"—Johann Zacherl und sein Pulver. Hg. v. Verlag der Österreichischen Akademie der Wissenschaften. Österreichisches Biographisches Lexikon, Biographie des Monats, 6. Available online at www.oeaw.ac.at/inz/forschungen/oesterreichisches-biographisches-lexikon/, last visited March 14, 2017.]

It contained extracts from the usury flower species Tanacetum coccineum (Persian insect flower) u. *Tanacetum cinerariifolium* (Dalmatian insect flower), which contain the insecticidal substances pyrethrin I and II, jasmolin I and II, and cinerin I and II.

[see also: Unger, Achim; Schniewind, Arno P.; Unger, Wibke (2001): Conservation of wood artifacts. A handbook. Berlin: Springer. Natural science in archaeology. Available online at www.loc.gov/catdir/enhancements/fy0815/2001020310-d. html, 246–248.]

Zaponlacke   are low-viscosity, transparent nitro paints (nitrocellulose lacquers) based on highly viscous cellulose nitrates dissolved in highly volatile solvents, such as amyl and ethyl acetate. They are transparent but may also contain pigments. Transparent Zaponlacke are usually used to protect metals, such as brass, bronze, or silver, from oxidation. They are not suitable for iron and steel, as they do not have a rust-protecting effect. They are also used as varnish, as well as for painting wood, glass, and leather. Acute or chronic damage to human health can result from the ingestion, inhalation, or skin absorption of these harmful substances. They irritate the eyes and skin, and the solvent fumes can cause drowsiness. They are highly flammable and hazardous to water.

Zinc chloride   (zinc(II) chloride), with the molecular formula $ZnCl_2$, is a white, granular powder formed by the reaction of zinc with chlorine or zinc sulfate with calcium chloride. It is very soluble in water, as well as in ethanol and some other solvents. It is used for impregnating wood, refining oil, making parchment paper, bleaching paper mixed with chlorinated lime, dyeing, etc. It has a burning taste and a strong corrosive effect.

Zinc oxide   (zinc(II) oxide), with the formula ZnO, is both a colorless crystalline substance and a loose white powder due to the refraction of light in its small crystals. Zinc oxide is used as a pigment under the name zinc white. Medical preparations for skin and wound treatment often contain zinc oxide because of its antiseptic effect. Although zinc oxide is insoluble in water, it is considered hazardous to water.

Zyklon B   → Hydrogen cyanide

Notes

1 Phosphorus dwarfs (phosphorus paste), mixture of flour and phosphorus, also with a little grease, for the destruction of rats and mice (translation by the author).

2 Two quants of phosphorus are melted in a mortar in six lots of warm water, and nine lots of wheat flour are quickly stirred in. After cooling, stir another eight lots of melted butter and 1/4 lot of powdered sugar (translation by the author).

Bibliography

Aall, Hans (1925): Arbeide og ordning i kulturhistoriske Museer. Kort Veiledning. Oslo: Utgitt med Statsbidrag.

AAMNH-DAA (1901): Correspondence. Generalverwaltung der Königlichen Museen zu Berlin. Fragebogen zur Schädlingsbekämpfung. Museum für Völkerkunde, Box 13, Folder 2. Loseblattsammlung, 2- seitig, ohne Paginierung.

Abelshauser, Werner (Hrsg.) (2003): Die BASF. Eine Unternehmensgeschichte. 2. Aufl. München: Beck. Available online at http://hsozkult.geschichte.hu-berlin.de/rezensionen/2003-3-009.

Aberle, Brigitte.; Koller, Manfred (1968): Konservierung von Holzskulpturen. Probleme und Methoden. Wien: Institut fiir Osterreichische Kunstforschung des Bundesdenkmalamtes, 43 Seiten.

Achsel, Bettina (2012): Das Manuale von Giovanni Secco Suardo von 1866/1894. Zugl.: Dresden, Hochschule für Bildende Kunst, Dissertation, 2011 u.d.T.: Achsel, Bettina: Kommentierte Übersetzung von Giovanni Secco Suardo, "Il manuale ragionato per la parte meccanica dell'arte del ristauratore dei dipinti" (1866) und "Il Restauratore dei dipinti" (1894). Göttingen: V & R Unipress.

Aderhold, Rudolf (1906): Die Kaiserliche Biologische Anstalt für Land- und Forstwirtschaft in Dahlem. *Mitteilungen aus der Kaiserlichen Biologischen Anstalt für Land- und Forstwirtschaft*, (1), 1–20.

Adorno, Theodor W. (1995): Ästhetische Theorie. 13. Aufl. Frankfurt am Main: Suhrkamp (Suhrkamp Taschenbuch Wissenschaft).

AIFM Wolfen, ohne Signatur (1912–1918): Loseblattsammlung. AGFA Jahresgeschäftsberichte. Actien-Gesellschaft für Anilin-Fabrikation. Blatt 105–106; 109–114, 12 Seiten.

AMPG (1917): V a ABT, Rep. 0005, Nr. 516. Haber-Sammlung von Johann Jaedicke. Loseblattsammlung. Anonymous. Auszug aus dem Senatsprotokoll in den Akten I 4.13. Gasforschung und Schädlingsbekämpfung. Protokollauszug vom 19.10.1917, Blatt 16, 1 Seite.

AMPG (ohne Jahresangabe): Haber-Sammlung, V a ABT, Rep 0005, Nr. 533. Auszug aus Akte ohne Angabe. Farbwerke Hoechst AG, vorm. Meister, Lucius & Brüning. Rezepte zur Herstellung von Gaskampfstoffen sowie zur Füllung von Nebelgranaten und Nebelapparaten.

Anderson, Benedict (1998): Die Erfindung der Nation. Zur Karriere eines folgenreichen Konzepts. Erweiterte Ausgabe. Berlin: Ullstein (Ullstein-Buch Propyläen-Taschenbuch, 26529).

Anderson, Sydney (02.04.1964): Mottenschutz für Felle. Archive of the American Museum of Natural History, SA:ji. Brief, 1 Seite, ohne Paginierung.

Andree, Christian (1969): Geschichte der Berliner Gesellschaft für Anthropologie, Ethnologie und Urgeschichte, 1869–1969. In: Festschrift zum 100- jährigen Bestehen der Berliner Gesellschaft für Anthropologie, Ethnologie und Urgeschichte 1860–1969, 9–139.

Andres Ad. (ohne Jahresangabe): Bekämpfung der Kleidermotte (*Tineola biselliella*) durch Blausäure. *Zeitschrift für angewandte Entomologie* Sonder-Abdruck; Bd. 4(3), 1–3.

Anonymous (1875): Der Kartoffelkäfer, Chrysomela (Doryphora) decemlineata. Berlin: Ernst Schotte & Voigt.

Anonymous (1888): Übersicht über die Amerikanischen Sammlungen des Königlichen Museums für Völkerkunde. Zusammengestellt für die 7. Tagung des internationalen Amerikanisten-Kongresses. Berlin: H.S. Hermann.

Anonymous (1889): Übersichtlicher Abriß der Sammlungen im Königlichen Museum für Völkerkunde: Den Mitgliedern des Deutschen Geographentages in ihrer 8. Berlin: Sitzung überreicht.

Anonymous (1906): Das Autan-Verfahren zur Formaldehyddesinfektion. *Pharmazeutische Zeitschrift*, LI(77), 769.

Anonymous (1913a): Dichlorbenzol "Agfa". *Vierteljahresschrift für praktische Pharmazie*, (10). Available online at www.digibib.tu-bs.de/?docid=00039116.

Anonymous (1913b): Personalnachrichten. *Zeitschrift für das Landwirtschaftliche Versuchswesen in Österreich*, (XVI), 39–40.

Anonymous (1916): The Technical Preservation of Antiquities. *The Museums Journal*, (15), 268–269.

Anonymous (1924): Direktoralbeamte der Staatlichen Museen am 15. Oktober 1924. Berichte aus den Preussischen Kunstsammlungen. Beiblatt zum Jahrbuch der Preussischen Kunstsammlungen. In: *Berliner Museen*, XLV. Jahrgang, Heft 4.

Anonymous (1927): Zweiter Lehrgang zur Bekämpfung der Gesundheitsschädlinge vom 14. bis 22. February 1927. Preußische Landesanstalt für Wasser-, Boden- und Lufthygiene, Berlin-Dahlem, Ehrenbergstr. 38–42. *Zeitschrift für Desinfektion und Gesundheitswesen*, 19(1), 49.

Ansprenger, Franz (1981): Auflösung der Kolonialreiche. 4., durchges. u. erw. Aufl. München: Dt. Taschenbuchverl. (Dtv, 4013).

Archiv des Museums am Rothenbaum–Kulturen und Künste der Welt (1914/1929/1932/1934): Findbuch, 101–1; Nr. 281. Loseblattsammlung. Fritz Schulz jun. Aktiengesellschaft Leipzig. Angebot von Globol. Brief vom 11.11.1914, 1- seitig, Brief vom 03.05.1929, 2- seitig, Brief vom 27.10.1932, Brief vom 29.01.1934, und vom 16.04.1934, je 1- seitig, ohne Paginierung.

Archive of the Regional Museum Yekaterinburg (28.04.1928): Serie 1928, Nr. 3026, J.-Nr. E. 356/28. Loseblattsammlung. Kümmel, Otto. Stellungnahme zu Areginal und Eulan. Brief vom 28.04.1928, 1-seitig.

Archive of the Regional Museum Yekaterinburg (01.11.1929): J.- Nr. E. 356/28, Brief Nr. 57. Loseblattsammlung. Kümmel, Otto. Gutachten von Carl Brittner über Eulan. Brief vom 01.11.1929, 2- seitig.

Arndt, Walther (1932a): Die Berufskrankheiten an naturwissenschaftlichen Museen. I. Vergiftungen. *Museumskunde*, Neue Folge, IV(2), 47–66.

Arndt, Walther (1932b): Die Berufskrankheiten an naturwissenschaftlichen Museen. II. Schädigungen durch Übertragung von Krankheitserregern. *Museumskunde*, Neue Folge, IV(3), 103–117.

Assmann, Aleida (2017): Das kulturelle Gedächtnis zwischen materiellem Speicher und digitaler Diffusion. In: Die Zukunft des Sammelns an wissenschaftlichen Bibliotheken. Wiesbaden: Harrassowitz Verlag, 1–18.

Assmann, Jan (2018): Das kulturelle Gedächtnis. Schrift, Erinnerung und politische Identität in frühen Hochkulturen. 8. Auflage in C.H. Paperback. München: C.H.Beck (C.H. Beck Paperback, 1307).

Austin, Michele, Firnhaber, Natalie, Goldberg, Lisa, Hansen, Greta und Magee, Catherine (2005): The Legacy of Anthropology Collections Care at the National Museum of Natural History. *Journal of the American Institute for Conservation/American Institute for Conservation of Historic and Artistic Works*, 44(3), 185–202.

Ayass, Wolfgang (1996): Arbeiterschutz. Quellensammlung zur Geschichte der deutschen Sozialpolitik, 3. Stuttgart, Jena, New York: Fischer.

BAL (1906–1916): 15/D. 1. Akte Statistik. Farbenfabriken vorm. Friedrich Bayer & Co. Elberfeld. Kiloverkäufe pharmazeutischer Produkte, Geldumsätze, Pharmazeutika, Riechstoffe, Pflanzenschutz, Photographika, Farben. Loseblattsammlung. 1 Tabelle, 2- seitig.

Ball, Philip (2014): The Devil's Doctor. Paracelsus and the World of Renaissance Magic and Science. London: Cornerstone Digital.

BArch (1927–1933): R 3602/2461. Berücksichtigung des Vorratsschutzes; Zahlreiche Artikel und Rezensionen. Online-Findbuch. Entwicklung und Tätigkeit allgemein; Historische Entwicklung.

BArch (2008): R 3602. Online-Findbuch. Biologische Reichsanstalt für Land- und Forstwirtschaft (Bestand). 1915–1945.

Bäumler, Ernst (1963): Ein Jahrhundert Chemie. Unter Mitarbeit von Gustav Ehrhart und Volkmar Muthesius. Düsseldorf: Econ Verlag.

Beßler, Gabriele (2012): Wunderkammern. Weltmodelle von der Renaissance bis zur Kunst der Gegenwart. 2., erw. Aufl. Berlin: Reimer.

Beck, Hanno (1985): Alexander von Humboldts Amerikanische Reise. Stuttgart: Thienemann Ed. Erdmann (Alte abenteuerliche Reiseberichte).

Berghahn, Volker Rolf (2014): Der Erste Weltkrieg. München: C.H. Beck.

Birnbaum, Carl; Merck, Klemens (Hrsg.) (1884): Klemens Merck's Warenlexikon für Handel, Industrie und Gewerbe. Beschreibung der im Handel vorkommenden Natur- und Kunsterzeugnisse unter besonderer Berücksichtigung der chemisch-technischen und anderer Fabrikate, der Droguen- und Farbewaren, der Kolonialwaren, der Landesprodukte, der Material- und Mineralwaren. 3., gänzlich umgearb. Aufl., 2., rev. Abdr. Leipzig: Gloeckner.

Blunck, Hans (1922): Über die Wirkung arsenhaltiger Gifte auf Ölfruchtschädlinge nach Beobachtungen an der Naumburger Zweigstelle der Biologischen Reichsanstalt. Dritte Mitgliederversammlung zu Eisenach vom 28. bis 30. September 1921. *Verhandlungen der Deutschen Gesellschaft für angewandte Entomologie*, 40–55.

Bochkarev, V.; Solovyev, V.; Wichmann, S. (2014): Universals Versus Historical Contingencies in Lexical Evolution. *Journal of the Royal Society, Interface*, 11(101), 20140841. c., 1–8.

Bolle, Johann (1882): Die Mittel zur Bekämpfung der Reblaus (*Phylloxera vastatrix*). Görz: Verlag der k.k. Seiden- und Weinbau-Versuchsstation.

Bolle, Johann (1892): Ausführliche Anleitung zur rationellen Aufzucht der Seidenraupe. Berlin: Gramsch.

Bolle, Johann (1898): Der Seidenbau in Japan. Budapest: Hartleben.

Bolle, Johann (1899): Der Seidenspinner des Maulbeerbaumes, seine Aufzucht seine, Krankheiten und die Mittel zu ihrer Bekaempfung. Vort. 1898. Wien: Selbstverl. (Vortraege des Vereins zur Verbreitung naturwissenschaftlicher Kenntnisse in Wien. XXXIX. Jahrgang. Heft 4).

Bolle, Johann (1919): Die Ermittlung der Wirksamkeit von insektentötenden Mitteln gegen die Nagekäfer des verarbeiteten Werkholzes. *Zeitschrift für angewandte Entomologie*, (Band 5), 105–117.

Bolle, Johann; Mewis, F.A. (1892): Ausführliche Anleitung zur rationellen Aufzucht der Seidenraupe. Berlin: A. Gramsch.

Bolz, Peter (1999): Entstehung und Geschichte der Berliner Nordamerika-Sammlung. *Indianer Nordamerikas*, 23–49.

Bolz, Peter (2001): Ethnologisches Museum: Neuer Name mit traditionellen Wurzeln. Die Umbenennung des Berliner Museums für Völkerkunde. *Baessler-Archiv. Beiträge zur Völkerkunde* ausgegeben am 25. February 2003, (Sonderdruck aus Band 49), 11–16.

Bolz, Peter (2005): Ethnologische Sammlungen in Berlin bis zur Eröffnung des "Königlichen Museums für Völkerkunde". Bastian-Symposium im Ethnologischen Museum Berlin. Unveröffentlichter Beitrag.

Bolz, Peter (2007): From Ethnographic Curiosities to the Royal Museum of Ethnology. Early Ethnological Collections in Berlin. In: Fischer, Manuela; Bolz, Peter; Kamel, Susan (eds.), Adolf Bastian and His Universal Archive of Humanity. The Origins of German Anthropology, Hildesheim: Georg Olms, 173–190.

Born, Hermann; Hausdörfer, Ute; Thieme, Franziska (2004/2005): Die Restaurierungswerkstätten. *Das Berliner Museum für Vor- und Frühgeschichte; Festschrift zum 175-jährigen Bestehen*, 36/37, 487–498.

Böttcher, (first name unknown, author's note) (1927): Entmottungsanlagen nach dem Zyklonverfahren. *Zeitschrift für Desinfektion und Gesundheitswesen*, 19(4), 143–146.

Boyle, (first name unknown, author's note) (1665): A Way of Preserving Birds Taken Out of the Egge, and Other Small Faetus's. *Philosophical Transactions 1665–1666*, (1), 199–201.

Braßler, Karl (1925): Areginal, ein neues Mittel gegen Sammlungs- und Bücherschädlinge. *Anzeiger für Schädlingskunde*, 1(6), 69–70. DOI: 10.1007/BF02628433.

Bracchi, Eva (2013): Der erste Chemiker in Sachen Kunst. *Jahrbuch Preußischer Kulturbesitz*, 49, 258–268.

Bracchi, Eva (2014): Friedrich Rathgen, Pionier der modernen archäologischen Restaurierung. *Berliner Beiträge zur Archäometrie, Kunsttechnologie und Konservierungswissenschaft*, 22, 5–13.

Brauer, August Bernhard (1907): Anleitung zum Sammeln, Konservieren und Verpacken von Tieren für das Zoologische Museum in Berlin. Dritte vermehrte Auflage. Berlin: A. Hopfer in Burg b.M.

Brocke vom, Bernhard (Hrsg.) (1996): Die Kaiser-Wilhelm-, Max-Planck-Gesellschaft und ihre Institute. Studien zu ihrer Geschichte. Kaiser-Wilhelm-Gesellschaft zur Förderung der Wissenschaften; Max-Planck-Gesellschaft zur Förderung der Wissenschaften. Berlin: De Gruyter. Available online at www.gbv.de/dms/faz-rez/F19970326NOTKE-100.pdf.

Brocke vom, Bernhard; Vierhaus, Rudolf (Hrsg.) (1990): Forschung im Spannungsfeld von Politik und Gesellschaft. Geschichte und Struktur der Kaiser-Wilhelm-/Max-Planck-Gesellschaft. Kaiser-Wilhelm-Gesellschaft zur Förderung der Wissenschaften; Max-Planck-Gesellschaft zur Förderung der Wissenschaften. Stuttgart: DVA. Available online at www.gbv.de/dms/faz-rez/900926_FAZ_0037_37_0001.pdf.

Brues, Charles Thomas (1909): The Insect Pests of Museums. *Proceedings of the American Association of Museums*, 33–54.

Buck-Heilig, Lydia (1989): Die Gewerbeaufsicht. Entstehung und Entwicklung. Wiesbaden: VS Verlag für Sozialwissenschaften (Studien zur Sozialwissenschaft, 87). Available online at http://dx.doi.org/10.1007/978-3-663-05750-5.

Bundesanstalt für Arbeitsschutz und Arbeitsmedizin (2010): Technische Regel für Gefahrstoffe 524 Schutzmaßnahmen bei Tätigkeiten in kontaminierten Bereichen, TRGS 524. In: Gemeinsames Ministerialblatt.

Bundesministerium der Justiz und für Verbraucherschutz (ausgegeben zu Bonn am 26.02.2002): Bundesgesetzblatt Teil I, G 5702., 869.

Burns, Ned J. (1941): Field Manual for Museums. Washington, DC: United States Government Printing Office.

Chiawara, Davison; O'Connell, Siona; Loubser, Maggie (2022): Potential Pesticide Contamination in Repatriated Artifacts in African Museums: The Need for the Adoption of Safety Protocols for Access and Use of Hazardous Artifacts. *Journal of the American Institute for Conservation*, 1–10. DOI: 10.1080/01971360.2022.2104576.

Chiwara, Davison (2022): Collections Conservation Practices and Possibilities of Contamination by Hazardous Pesticides: Towards a Non-Pesticide Approach of Conserving Organic Collections at the Natural History Museum of Zimbabwe. PhD in Heritage and Museum Studies. University of Pretoria, Faculty of Humanities, School of Arts, November 15.

Coleman, Laurence Vail (1927): Manual for Small Museums, with 32 Plates. New York, London: G.P. Putnam's Sons.

Cziesla, Erwin (2000): Spätpaläolithische Widerhakenspitzen aus Brandenburg. Eine Forschungsgeschichte. *Archäologisches Korrespondenzblatt*, 30, 173–186.

Dangeon, Marion (2013/2014): Conservation des collections naturalisées traitées aux biocides: étude de la collection Mammifères et Oiseaux du Muséum d'Histoire Naturelle de Neuchâtel. Bachelor of Arts. Haute École Arts, Appliqués-La Chaux-de-Fonds, Filière Conservation-Restauration, Neuchâtel.

Dann, Georg Edmund (Hrsg.) (1957): Die Medizinalordnung Friedrichs II. Eine pharmazie-historische Studie. Unter Mitarbeit von Wolfgang-Hagen Hein und Kurt Sappert. Eutin (Holstein): Internationale Gesellschaft für Geschichte der Pharmazie (Veröffentlichungen der Internationalen Gesellschaft für Geschichte der Pharmazie, Neue Folge, Band 12).

Department of Scientific and Industrial Research (Hrsg.) (1926): The Cleaning and Restoration of Museum Exhibits. Third Report Upon Investigations Conducted at the British Museum. London: Published Under the Authority of His Majesty's Stationery Office.

Deschka, Gerfried (1987): Die Desinfektion kleiner Insektensammlungen nach neueren Gesichtspunkten. *Steyrer Entomologenrunde*, 21, 57–61.

Dornheim, Andreas; Brügelmann, Walther (2006): Forschergeist und Unternehmermut. Der Kölner Chemiker und Industrielle Hermann Julius Grüneberg 1827–1894.

Dumont, Fritz (1914): Petroleum-Versorgung während des Krieges mit besonderer Berücksichtigung der örtlichen Verhältnisse von Danzig. Available online at http://resolver.staatsbibliothek-berlin.de/SBB0000635C00000000.

Ebbinghaus, Angelika (1998): Der Prozeß gegen Tesch & Stabenow. Von der Schädlingsbekämpfung zum Holocaust. *Zeitschrift für Sozialgeschichte des 20. und 21. Jahrhunderts*, 13(2), 16–71.

Ebers, Georg (Hrsg.) (1875): Papyros Ebers. Das Hermetische Buch über die Arzneimittel der alten Ägypter in hieratischer Schrift. Available online at http://digi.ub.uni-heidelberg.de/diglit/ebers1875bd2.

Eibner, Alexander (1928): Entwicklung und Werkstoffe der Tafelmalerei. München: B. Heller.

Eichengrün, Arthur (1905): Verfahren zur Entwicklung von gasförmigem Formaldehyd aus polymerisiertem Formaldehyd. Angemeldet durch Farbenfabriken, vormals Friedrich Bayer & Co., Elberfeld & Leverkusen. Veröffentlichungsnr: 181509; Klasse 12o; Gruppe 7. Prioriätsdaten: Zusatz zum Patent 177053 vom 13. Juli 1905.

Eichengrün, Arthur (1906): Über das neue Autan-Desinfektionsverfahren. *Pharmazeutische Zeitschrift*, LI(77), 852.

Escherich, Karl (1922): Die Stellung der angewandten Entomologie im Pflanzenschutz. Dritte Mitgliederversammlung zu Eisenach vom 28. bis 30. September 1921. *Verhandlungen der Deutschen Gesellschaft für angewandte Entomologie*, 17–25.

Evans, Richard J. (1991): Tod in Hamburg. Stadt, Gesellschaft und Politik in den Cholera-Jahren 1830–1910. 4.–5. Tsd. Reinbek bei Hamburg: Rowohlt.

Fabian, Johannes (2000): Out of Our Minds. Reason and Madness in the Exploration of Central Africa; The Ad. E. Jensen Lectures at the Frobenius Institute, University of Frankfurt. Berkeley: University of California Press. Available online at http://site.ebrary.com/lib/academiccompletetitles/home.action.

Falser, Michael (2008): Zwischen Identität und Authentizität. Zur politischen Geschichte der Denkmalpflege in Deutschland. Techn. Univ., Diss., Zugl.: Berlin, 2006. Dresden: Thelem. Available online at http://deposit.d-nb.de/cgibin/dokserv?id=3024092&prov=M&dok_var=1&dok_ext=htm.

Farbenfabriken, vormals Friedrich Bayer & Co., Elberfeld & Leverkusen (Hrsg.) (1918): Geschichte und Entwicklung der Farbenfabriken vorm. Friedrich Bayer & Co. Elberfeld in den ersten 50 Jahren. Pharmazeutisch-wissenschaftliche Abteilung. Das Autanverfahren. Unveröffentlichte Schrift. Unter Mitarbeit von Arthur Eichengrün. München.

Farber, Paul Lawrence (1977): The Development of Taxidermy and the History of Ornithology. *ISIS*, 68(244), 550–566.

Faulstich, Peter (2011): Aufklärung, Wissenschaft und lebensentfaltende Bildung. Geschichte und Gegenwart einer großen Hoffnung der Moderne. Bielefeld: Transcript-Verlag. (Theorie bilden, 25).

Fiedermutz-Laun, Annemarie (2007): The Scientific Legacy of Adolf Bastian (1826–1905). Compilation, Evaluation and Significance of Knowledge About the Life and Work of the Scholar. In: Fischer, Manuela; Bolz, Peter; Kamel, Susan (eds.), Adolf Bastian and His Universal Archive of Humanity. The Origins of German Anthropology, Hildesheim: Georg Olms, 55–74.

Fischer, Manuela; Bolz, Peter; Kamel, Susan (eds.) (2007): Adolf Bastian and His Universal Archive of Humanity. The Origins of German Anthropology. Ethnological Museum Berlin. Hildesheim: Georg Olms.

Florian, Mary-Lou (1997): Heritage Eaters: Insects and Fungi in Heritage Collections. London: James & James.

Fröhlich, Michael (1994): Imperialismus. Deutsche Kolonial- und Weltpolitik 1880–1914. Orig.-Ausg. München: Dt. Taschenbuch-Verlag. (Dtv, 4509).

Funk, Albert (2010): Kleine Geschichte des Föderalismus. Vom Fürstenbund zur Bundesrepublik. Paderborn: Schöningh.

Geist, Johann Friedrich; Kürvers, Klaus (1989): Das Berliner Mietshaus. München: Prestel.

Germann, Paul (1933): Bekämpfung der Museumsschädlinge. *Museumskunde*, Neue Folge, V, Sonderdruck (I), 9–11.

Giere, Peter; Bartsch, Peter; Quaisser, Christiane (2018): BERLIN: From Humboldt to HVac— The Zoological Collections of the Museum für Naturkunde Leibniz Institute for Evolution and Biodiversity Science in Berlin. In: Beck, Lothar A. (Hrsg.). Zoological Collections of Germany. Cham: Springer International Publishing (Natural History Collections), 89–122.

Gilberg, Mark (1987): Friedrich Rathgen: The Father of Modern Archeological Conservation. *Journal of the American Institute for Conservation*, 2(26), 105–120. Available online at http://cool.conservation-us.org/jaic/articles/jaic26-02-004.html.

Gilberg, Mark (1997): Alfred Lucas: Egypt's Sherlock Holmes. *Journal of the American Institute for Conservation*, 36(1), 31–48. DOI: 10.1179/019713697806113620.

Gmelin, Leopold (1827): Handbuch der theoretischen Chemie. Frankfurt: Franz Varrentrapp.

Goldberg, Lisa (1996): A History of Pest Control Measures in the Anthropology Collections, National Museum of Natural History, Smithsonian Institution. *Journal of the American Institute for Conservation/American Institute for Conservation of Historic and Artistic Works*, 35(1), 23–43.

Gossauer, Albert (2006): Struktur und Reaktivität der Biomoleküle. Eine Einführung in die organische Chemie. Zürich: Verl. Helvetica Chimica Acta.

Gottschalk, Sebastian; Hartmann, Heike; Hilden, Irene (2016): Deutscher Kolonialismus. Fragmente seiner Geschichte und Gegenwart. Berlin, Darmstadt: Stiftung Deutsches Historisches Museum; Theiss Verlag.

Grabowski, Jörn; Winter, Petra; Ebelt, Beate; Pilgermann, Carolin (Hrsg.) (2010): Kunst recherchieren. 50 Jahre Zentralarchiv der Staatlichen Museen zu Berlin. Staatliche Museen zu Berlin-Preußischer Kulturbesitz, Ethnologisches Museum. Berlin: Deutscher Kunstverlag.

Graven, Sonja (2019): Giftlos—erfolglos? Vier Jahre integrierte Schädlingsbekämpfung im Museum Mensch und Natur in München. *Restauro*, (2), 36–41.

Gründer, Horst; Hiery, Hermann (Hrsg.) (2017): Die Deutschen und ihre Kolonien. Berlin: Ein Überblick. be.bra verlag.

Gütebier, Thomas (2012a): Historische Schädlingsbekämpfungsmittel in Balgsammlungen—ein vielschichtiger Problemkomplex. Vortrag vom 20.03.2012. 50. Internationale Arbeitstagung vom 20–24. März 2012. München: Verband Deutscher Präparatoren e.V.

Gütebier, Thomas (2012b): Schädlingsbekämpfung in schwedischen Museen. Göteborg, 26.03.2012. Letter to Helene Tello.

Habermas, Jürgen (1993): Der philosophische Diskurs der Moderne. Zwölf Vorlesungen. 4. Aufl. Frankfurt am Main: Suhrkamp (Suhrkamp Taschenbuch Wissenschaft).

Halbwachs, Maurice (1991): Das kollektive Gedächtnis. Ungekürzte Ausg., 4–5. Tsd. Frankfurt am Main: Fischer-Taschenbuch-Verl. [Fischer-Taschenbücher, 7359. (Fischer Wissenschaft)].

Hansestadt Hamburg (1921): Entwurf des hamburgischen Staatshaushaltsplanes. Staats- und Universitätsbibliothek Hamburg. Loseblattsammlung. Artikel 87, 309.

Harmer, Sidney (1922): The Restoration and Preservation of Objects at the British Museum. *Journal of the Royal Society of Arts*, LXX(3618), 333–334.

Hase, Albrecht (ausgegeben 1916): Der Verbreiter des Fleckfiebers: Die Kleiderlaus. In: *Merkblatt der Deutschen Gesellschaft für angewandte Entomologie e.V.*, Nr. 1, Serie I. Berlin: Verlagsbuchhandlung P. Parey, 1–8.

Hase, Albrecht (ausgegeben 1917): Die Bettwanze und ihre Bekämpfung. In: *Merkblatt der Deutschen Gesellschaft für angewandte Entomologie e.V.*, Nr. 4, Serie I, Berlin: Verlagsbuchhandlung P. Parey, 1–8.

HAStK (1909–1912): Best. 614, A 73. Projektakte. Foy, Willi Beschreibung der Versuche zur Begasung von Sammlungsobjekten in einer Begasungsanlage à la Nielsson mit Johann Bolle im Rautenstrauch-Joest-Museum Köln.

HAStK (1910/1913): Best. 614, Nr. 438. Akte. Loseblattsammlung. Chemische Fabrik Griesheim-Elektron; Chemische Industrie-Gesellschaft. Ankauf von Imprägniermitteln. Brief von der Chemischen Fabrik Griesheim-Elektron vom 25.06.1910, Blatt 1, 1-seitig; 2 Briefe von der Chemischen Industrie-Gesellschaft Berlin vom 15.11.1913 und 08.12.1913; Blatt 2–3, je 1-seitig;

HAStK (1911): Best. Nr. 614, A 88. Projektakte. Loseblattsammlung. Bolle, Johann. Versuche in der Begasungsanlage des Rautenstrauch-Joest-Museums in Köln mit Sammlungsobjekten. Brief von Johann Bolle vom 05.07.1911, 3- seitig, Blatt 66–67; Postkarte vom 26.09.1911, Blatt 69, 1 seitig.

HAStK (1913/1914): Best. Nr. 614; A 432. Akte Schriftwechsel. Loseblattsammlung. Ac-tien-Gesellschaft für Anilin-Fabrikation. Recherche Schädlingsbekämpfungsmittel. 2 Briefe vom 20.11.1913 und 20.05.1914, je 1 Seite, ohne Paginierung.

HAStK (1914–1915): Best. Nr. 614, A 70. Akte Ankauf von Insektenvertilgungsmitteln. Loseblattsammlung. Fritz Schulz jun. Aktiengesellschaft Leipzig. Angebot von "Globol". Brief vom 13.11.1914, Blatt 5, 1- seitig; Brief vom 21.05.1915, Blatt 7, 2- seitig; 2 Bro-schüren, Blatt 2–3; 6, je 2- seitig.

HAStK (1922): Best. Nr. 614, A 70. Akte Ankauf von Insektenvertilgungsmitteln. Loseblatt-sammlung. Chemische Fabrik Flörsheim Dr. H. Noerdlinger. Anfrage zur Neubestellung von "Wurm-Antorgan". Brief vom 24.08.1922, 1- seitig.

Heerdt, Walter (1924): Zyklon B, ein verbessertes Blausäureverfahren. Vierte Mitglieder-versammlung zu Frankfurt a.M. vom 10. bis 13. Juli 1924. *Verhandlungen der Deutschen Gesellschaft für angewandte Entomologie*, 81–83.

Hellwald, Friedrich von (1887): Das Berliner Museum für Völkerkunde. *Vom Fels zum Meer*, (2), 101–112.

Hermannstädter, Anita (2002): Symbole kollektiven Denkens. Adolf Bastians Theorie der Dinge. In: Hermannstädter, Anita (Hrsg.). Deutsche am Amazonas – Forscher oder Abenteurer? Expeditionen in Brasilien 1800 bis 1914. Begleitbuch zur Ausstellung im Ethnologischen Museum, Berlin-Dahlem in Zusammenarbeit mit dem Brasilianis-chen Kulturinstitut in Deutschland. Ethnologisches Museum Berlin. 2., unveränd. Aufl. Münster: LIT. Veröffentlichungen des Ethnologischen Museums Berlin Fachreferat Amerikanische Ethnologie, Neue Folge, 2002(71), 44–55.

Hermannstädter, Anita (Hrsg.) (2002): Deutsche am Amazonas—Forscher oder Abenteurer? Expeditionen in Brasilien 1800 bis 1914; Begleitbuch zur Ausstellung im Ethnologischen Museum, Berlin-Dahlem in Zusammenarbeit mit dem Brasilianischen Kulturinstitut in Deutschland. Ethnologisches Museum Berlin; Ausstellung. 2., unveränd. Aufl. Münster: LIT. Veröffentlichungen des Ethnologischen Museums Berlin Fachreferat Amerikanische Ethnologie, Neue Folge 71; 9.

Hintzenstern, Ulrich von; Arens, Larissa (eds.) (2007): Notarzt-Leitfaden. 5. Aufl. München: Elsevier Urban & Fischer (Klinikleitfaden). Available online at http://deposit.d-nb.de/cgi-bin/dokserv?id=2995216&prov=M&dok_var=1&dok_ext=htm.

Hobsbawm, Eric J. (2008): Das imperiale Zeitalter 1875–1914. Flörsheim a. M.: Campus Verlag GmbH.

Hoeftmann, Friedrich Wilhelm (1868): Der Preußische Ordens-Herold. Zusammenstellung sämmtlicher Urkunden, Statuten und Verordnungen über die Preußischen Orden und Ehrenzeichen. Berlin: Königliche Buchhandlung von Mittler & Sohn.

Hoffmann, Almut (1997): Zur Geschichte des Fundes von Le Moustier. *Acta Praehistorica et Archaeologica*, 29, 7–16.

Homolka, Martina (2015a): Eulan—ein Biozid gegen Keratin-Schädlinge und seine Rel-evanz in musealen Sammlungen. Berlin: Stiftung Deutsches Historisches Museum (1. Produktgeschichte). Available online at www.dhm.de/publikation/eulan-ein-biozid-gegen-keratin-schaedlinge-und-seine-relevanz-in-musealen-sammlungen-produkt-geschichte/.

Homolka, Martina (2015b): Eulan—ein Biozid gegen Keratin-Schädlinge und seine Rel-evanz in musealen Sammlungen. Berlin: Stiftung Deutsches Historisches Museum (2. Lexikalischer Produktschlüssel). Available online at www.dhm.de/publikation/eulan-ein-biozid-gegen-keratin-schaedlinge-und-seine-relevanz-in-musealen-sammlungen-lexika-lischer-produktschluessel/.

Horkheimer, Max; Adorno, Theodor W. (2017): Dialektik der Aufklärung. Philosophische Fragmente. 23. Auflage, ungekürzte Ausgabe. Frankfurt am Main: Fischer Taschenbuch Verlag (Fischer-Taschenbücher Fischer Wissenschaft, 7404).

Hough, Walter (1889): The Preservation of Museum Specimens from Insects and the Effects of Dampness. For the Year Ending June 30, 1887; Report of the National Museum, Washington, DC, Zoological Pamphlets, 5.

HStAK (1907–1914): Best. Nr. 614, A 88. Projektakte. Bolle, Johann; Foy, Willi. Versuche zur Begasung mit Sammlungsobjekten in der Begasungsanlage des Rautenstrauch-Joest-Museums in Köln.

Hüntelmann, Axel Cäsar (2008): Hygiene im Namen des Staates. Das Reichsgesundheitsamt 1876–1933. Diss. Bremen. Univ., 2005. Göttingen: Wallstein. Available online at http://deposit.d-nb.de/cgi-bin/dokserv?id=3099685&prov=M&dok_var=1&dok_ext=htm.

Jäckel, R. (1927): Schädlingsbekämpfung mit Zyklon B (Blausäure). *Zeitschrift für Desinfektion und Gesundheitswesen*, 19(1), 37–41.

Jansen, Sarah (2003): "Schädlinge". Geschichte eines wissenschaftlichen und politischen Konstrukts 1840–1920. Techn. Univ., Diss. Braunschweig, 1997. Frankfurt/Main: Campus Verlag GmbH (Campus historische Studien, 25). Available online at www.gbv.de/dms/faz-rez/FD1200307281954595.pdf.

Junker, Horst (2004/2005): Zur Dokumentation archäologischer Sammlungen und Archivierung von Quellenmaterial am Museum für Vor- und Frühgeschichte. *Das Berliner Museum für Vor- und Frühgeschichte. Festschrift zum 175-jährigen Bestehen*, 36/37, 415–471.

Kadlubek, Günther; Hillebrand, Rudolf (1998): AGFA. Geschichte eines deutschen Weltunternehmens von 1867 bis 1997: Neuss: Verlag Rudolf Hillebrand, 1998.

Kainer, Franz (1951): Kieselgur, ihre Gewinnung, Veredlung und Anwendung. 2., umgearb. Aufl. Stuttgart: Enke-Verlag (Sammlung chemischer und chemisch-technischer Vorträge, N.F. H. 32).

Kaiser, Gerhard (2002): Wie die Kultur einbrach. Giftgas und Wissenschaftsethos im Ersten Weltkrieg. *Merkur*, 56(3), 210–220.

Kaltenbach, Angelika (ed.) (2011): Bezirk Steglitz-Zehlendorf, Ortsteil Dahlem. Bearb.-Stand: Januar 2011. Petersberg: Imhof (Denkmale in Berlin,/hrsg. Senatsverwaltung für Stadtentwicklung und Umweltschutz; Bezirk Steglitz-Zehlendorf).

Kalthoff, Jürgen; Werner, Martin (1998): Die Händler des Zyklon B. Tesch & Stabenow; eine Firmengeschichte zwischen Hamburg und Auschwitz. Hamburg: VSA-Verlag.

Kant, Immanuel (1794): Critik der reinen Vernunft. 4. Aufl. Riga: Hartknoch.

Karlsch, Rainer; Stokes, Raymond G. (2003): Faktor Öl. Die Mineralölwirtschaft in Deutschland 1859–1974. München: Beck. Available online at www.gbv.de/dms/faz-rez/FD1200304101799547.pdf.

Karlsch, Rainer; Wagner, Paul Werner (2010): Die AGFA-ORWO-Story. Geschichte der Filmfabrik Wolfen und ihrer Nachfolger. Berlin: VBB.

Keil, Ernst (1879): Die Gartenlaube. Das Wickersheimers'sche Conservierungsverfahren. Leipzig: Hg. v. Verlag von Ernst Keil. Available online at jttps://de.wikisource.org/w/index.php?title=Seite:Die_Gartenlaube_(1879)_844.jpg&oldid=-(Version vom 21.05.2018).

Kieß, Walter (1991): Urbanismus im Industriezeitalter. Von der klassizistischen Stadt zur Garden City. Berlin: Ernst.

Kingsley, Helen; Pinninger, David; Xavier-Rowe, Amber; Winsor, Peter (Hrsg.) (2001): Integrated Pest Management for Collections. Proceedings of 2001: A Pest Odyssey; A Joint Conference of English Heritage, The Science Museum and the National Preservation Office, 1–3 October. London: James & James.

Kirchner, Martin (1907): Die Gesetzlichen Grundlagen der Seuchenbekämpfung im Deutschen Reiche unter besonderer Berücksichtigung Preußens. XIV. Internationaler Kongreß für Hygiene und Demographie. Festschrift dargeboten von dem Preußischen Minister der geistlichen, Unterrichts- und Medizinalangelegenheiten. Jena.

k.k. Zentralkommission für Kunst- und historische Denkmale in Wien (Hrsg.) (1905): Enquete betreffend die Konservierung von Kunstgegenständen. Auszug aus dem stenographischen Protokoll. 10., 11. und 12. Oktober 1904. Wien: Rudolf Brzezowsky.

Klaus, Burkhard (2021): Geschichte des Vereins für Wasser-, Boden- und Lufthygiene e.V., Berlin. WaBoLu Homepage. Available online at https://wabolu.de/geschichte/.

Klaus, Marianne; Plitnikas, J.; Norton, Ruth; Almazan, T.; Coleman, S. (2005a): Preliminary Results from a Survey for Residual Arsenic on the North American Ethnographic. Poster Submission. 12–16 September. In: ICOM-CC Preprints Triennial Meeting, The Hague, Vol. I, 127.

Klaus, Marianne; Plitnikas, J.; Norton, Ruth; Ruth; Almazan, T.; Coleman, S. (2005b): Preliminary Results from a Survey for residual Arsenic on the North American Ethnographic Collections at the Field Museum. *Western Association for Art Conservation Newsletter*, 27(1), 24–26.

Klemm, Friedrich (1989): Geschichte der Technik. Der Mensch und seine Erfindungen im Bereich des Abendlandes. Orig.-Ausg. Reinbek bei Hamburg: Rowohlt (rororo, 7714).

Koch-Grünberg, Theodor (1903–1905): Tagebuch Rio Negro-Expedition 1903–1905. Völkerkundliche Sammlung der Philipps-Universität Marburg, VK Mr KG-B-I.2. Heft 1. Loseblattsammlung. Nachlass Theodor Koch-Grünberg, Unveröffentlichte Quelle, 4- seitig.

Kocka, Jürgen (Hg.) (1995): Bürgertum im 19. Jahrhundert. Deutschland im europäischen Vergleich; eine Auswahl. Göttingen: Vandenhoeck & Ruprecht (Kleine Vandenhoeck-Reihe).

Komander, Gerhild, H., M. (2004): Die Geschichte Berlins. Arndt, Walther. Hg. v. Verein für die Geschichte Berlins e.V., gegr. 1865. Available online at www.diegeschichteberlins.de/geschichteberlins/persoenlichkeiten/persoenlichkeiteag/434-arndt.html.

Konrád, György; Paetzke, Hans-Henning (2013): Europa und die Nationalstaaten. Essay. Dt. Ausg., 1. Aufl. Berlin: Suhrkamp.

Kornauth, Karl (1904): Über die Bekämpfung tierischer landwirtschaftlicher Schädlinge mit Hilfe von Mikroorganismen. Mitteilung der K.k. landwirtschaftlich-bakteriologischen und Pflanzenschutzstation in Wien. Nach einem Vortrag gehalten am 12. Februar 1904 in den Kursen für praktische Landwirte unter Benutzung von Versuchsergebnissen der k.k. landwirtschaftlichen bakteriologischen und Pflanzenschutzstation in Wien. *Zeitschrift für das Landwirtschaftliche Versuchswesen in Österreich*, VII. Jahrgang, 365–387.

Kraus, Michael (2004a): Bildungsbürger im Urwald. Die deutsche ethnologische Amazonienforschung (1884–1929). Marburg/Lahn: Curupira (Reihe Curupira, Bd. 19).

Kraus, Michael (ed.) (2004b): Koch-Grünberg, Theodor. Die Xingu-Expedition (1898–1900). Ein Forschungstagebuch. Köln: Böhlau Verlag.

Krause, Eduard (1882): Hr. E. Krause berichtet über ein neues Verfahren zur Conservierung der Eisen-Alterthümer. Sitzung am 11. November 1882. *Verhandlungen der Berliner anthropologischen Gesellschaft*, 533–537.

Krause, Eduard (1883): Mittheilungen über trapezförmige Feuersteinscherben. *Zeitschrift für Ethnologie und der Verhandlungen der Berliner Gesellschaft für Anthropologie, Ethnologie und Urgeschichte*, 15, 361.

Krause, Eduard (1897): Wunderliche Heilige. *Jahrbuch der Hamburgischen Wissenschaftlichen Anstalten*, (1 und 2), Heft 1, 8; Heft 2, 37–40.

Krause, Eduard (1899): Die Verwendung von Celluloid-Lack zur Conservirung von Althertümern, sowie von Holz, Stoffresten und Papier, namentlich alten Zeichnungen,

Drucken, Acten in Archiven usw. Verhandlungen der Berliner Gesellschaft für Anthropologie, Ethnologie und Urgeschichte. *Zeitschrift für Ethnologie und der Verhandlungen der Berliner Gesellschaft für Anthropologie, Ethnologie und Urgeschichte*, 31, 576–579.

Krause, Eduard (1900): Die ältesten Pauken. *Globus*, Bd. 78, 193–196.

Krause, Eduard (1901): Zur Frage der Rotfärbung vorgeschichtlicher Skelettknochen. *Globus*, (Sonder-Abdruck aus Bd 83, 23), 361–367.

Krause, Eduard (1903): Über die Herstellung vorgeschichtlicher Tongefässe. *Zeitschrift für Ethnologie; Organ der Berliner Gesellschaft für Anthropologie, Ethnologie und Urgeschichte*, 35(Heft II und III), 317–323.

Krause, Eduard (1904): Vorgeschichtliche Fischereigeräte und neuere Vergleichsstücke. Eine vergleichende Studie als Beitrag zur Geschichte des Fischereiwesens. Berlin: Borntraeger.

Kuckhan, T. S. (1770): Four Letters from T.S. Kuckhan, to the President and Members of the Royal Society, on the Preservation of Dead Birds. *Philosophical Transactions*, (60), 302–320.

LAKD M-V/LD (13.07.1912–25.10.1933): Objektakte Bad Doberan, Klosterkirche Münster. Verwaltung des Grossherzoglichen Museums und der Grossherzoglichen Kunstsammlungen Schwerin. Anfrage an das Kaiser-Friedrich Museum in Berlin. Erhaltung des Grabdenkmals des Herzoglichen Geheimen Rats Graf Samuel von Behr in der Kirche zu Doberan.

Landgraf, Theodor (1925): Grundsätze zur Schädlingsbekämpfung im Gartenbau. Von Gewerbe-Oberlehrer Theodor Landgraf, Wandsbek. *Führer durch die Gartenbau-Ausstellung*, 33–37.

LASA (28.07.1930): I 532, Nr. 600. I.G. Farbenindustrie Aktiengesellschaft, Abteilung Z III, Frankfurt am Main Stellungnahme zu Globol—Schädlingsnaphtalin—Areginal—Areginal U. Landesarchiv Sachsen-Anhalt, Brief, 1- seitig, ohne Paginierung.

Ledebur, Leopold, Freiherr von (1869): Aus der Ethnologischen Sammlung des Königlichen Museums zu Berlin. *Zeitschrift für Ethnologie und ihre Hülfswissenschaften als Lehre vom Menschen in seinen Beziehungen zur Natur und zur Geschichte*, Erster Jahrgang (III), 193–204.

Leechman, Douglas (1931): Technical Methods in the Preservation of Anthropological Museum Specimens. Hg. v. Canada Department of Mines. National Museum of Canada. Ottawa (Annual Report for 1929, Bulletin 67).

Lehmann, Detlev (1964): Die EULAN-Behandlung von Textilien und zoologischen Präparaten. Staatliche Museen Berlin. Islamische Abteilung. *Ergänzungsbände des Berliner Jahrbuchs für Vor- und Frühgeschichte*, (Band I), 67–72.

Lehmann, Dieter; Lehmann, Andrea (1985): Zwei wundärztliche Rezeptbücher des 15. [fünfzehnten] Jahrhunderts vom Oberrhein. Zugl.: Würzburg, Univ., Diss. 1983. Pattensen/Han.: Wellm. Würzburger medizinhistorische Forschungen, 34.

Lehmann, Jirina (2005): Geschichte der Konservierung und Restaurierung in Russland und in der Sowjetunion. im Buch von Professor M.W. Farmakowskij. *VDR Beiträge zur Erhaltung von Kunst- und Kulturgut*, (2), 47–62.

Lissauer, Abraham (1906): Sitzung vom 21. Juli 1906. *Zeitschrift für Ethnologie und der Verhandlungen der Berliner Gesellschaft für Anthropologie, Ethnologie und Urgeschichte*, 38(IV und V), 761–762.

Locke, John; Yolton, John W. (1794): The works of John Locke. [in nine volumes]. Repr. of the 1794 ed. London: Routledge/Thoemmes Press.

Löhr, Isabella; Wenzlhuemer, Roland (2013): The Nation State and Beyond. Governing Globalization Processes in the Nineteenth and Early Twentieth Centuries. Berlin, Heidelberg:

Springer (Transcultural Research-Heidelberg Studies on Asia and Europe in a Global Context). Available online at http://dx.doi.org/10.1007/978-3-642-32934-0.

Lowe, Derek B. (2017): Das Chemiebuch. Vom Schießpulver bis zum Graphen, 250 Meilensteine in der Geschichte der Chemie. Kerkdriel: Librero IBP.

Lucas, Alfred (1932): Antiques. Their Restauration and Preservation. Zweite überarbeitete Auflage. London: Edward Arnold & Co.

Lueger, Otto (Hrsg.) (1904): Lexikon der gesamten Technik und ihrer Hilfswissenschaften. Im Verein mit Fachgenossen herausgegeben von Otto Lueger. Zweite, vollständig neu bearbeitete Auflage. 8 Bände. Stuttgart, Leipzig: Deutsche Verlagsanstalt.

Lueger, Otto (Hrsg.) (1909): Lexikon der gesamten Technik und ihrer Hilfswissenschaften. Bd. 7. Stuttgart, Leipzig: Deutsche Verlags-Anst.

Mäder, Denis (2014): Wider die Fortschrittskritik. Mit einem Appendix zum Fortschritt als Human Enhancement. *Momentum Quarterly. Zeitschrift für Sozialen Fortschritt*, (3), 190–205.

Maertins, Katharina (2005): Rathgen-Forschungslabor. Unveröffentlichte Quelle. Berlin: Rathgen-Forschungslabor der Staatlichen Museen Berlin.

Malešević, Siniša (2013): Nation-States and Nationalisms. Organization, Ideology and Solidarity. Cambridge: Polity Press (Political Sociology Series).

Marion, Dangeon (2014): Conservation des collections naturalisées traitées aux biocides: étude de la collection Mammifères et Oiseaux du Muséum d'Historie Naturelle de Neuchatel. Bachelor of Arts HES-SO en Conservation. Haute École Arts, Appliqués-La Chaux-de-Fonds, Filière Conservation-Restauration, Neuchatel.

Marte, Fernando; Péquignot, Amandine; von Endt, David W. (2006): Arsenic in Taxidermy Collections: History, Detection, and Management. *Collection Forum*, 21(1–2), 143–150.

Martin, Petra (2005): Adolph Bernhard Meyer. Hg. v. Institut für Sächsische Geschichte und Volkskunde e. V. (Sächsische Biographie). Ohne Seitenangabe. Online verfügbar unter http://www.isgv.de/saebi.

März, Peter (2014): Nach der Urkatastrophe. Deutschland, Europa und der Erste Weltkrieg. Köln, Berlin: Böhlau Verlag; De Gruyter. Available online at www.degruyter.com/search?f_0=isbnissn&q_0=9783412216658&searchTitles=true.

Mason, Otis Tufton (1889): Report Upon the Work in the Department of Ethnology in the U.S. National Museum. For the Year Ending June 30, 1886. Part II. [S.l.]: Government Printing Office.

Max-Planck-Gesellschaft zur Förderung der Wissenschaften (2011): Wissenschaft im "Deutschen Oxford". Stadtrundgang durch das Wissenschaftsquartier Berlin-Dahlem. Hg. v. Max-Planck-Gesellschaft zur Förderung der Wissenschaften. Available online at www.mpiwg-berlin.mpg.de/PDF/Flyer_MPG_Spaziergaenge 2011.pdf.

Meyen, Franz Julius Ferdinand (1834): Reise um die Erde. Ausgeführt auf dem königlich preussischen Seehandlungs-Schiffe Prinzess Louise, commandirt von Capitain W. Wendt, in den Jahren 1830, 1831 und 1832; historischer Bericht. Berlin: Sander.

Meyer, Adolph Bernhard (1903): 3. Bericht über einige Neue Einrichtungen des Königlichen Zoologischen und Anthropologisch-Ethnographischen Museums in Dresden. XI. Desinfektionsapparat. Mit Tafel XIX. *Abhandlungen und Berichte des Königlichen Zoologischen und Anthropologisch-Ethnographischen Museums zu Dresden*, Bd. X, 1902/03 (5), 22.

Meyer, Andrea (2014): The Journal Museumskunde—"Another Link between the Museums of the World". *The Museum Is Open: Towards a Transnational History of Museums 1750–1940*. DOI: 10.1515/9783110298826.179.

Müller, Sascha (2010): Die historisch-kritische Methode in den Geistes- und Kulturwissenschaften. Würzburg: Echter.

Müller-Jahncke, Wolf-Dieter; Friedrich, Christoph; Meyer, Ulrich; Müller-Jahncke-Friedrich-Meyer (2005): Arzneimittelgeschichte. 2., überarb. und erw. Aufl. Stuttgart: Wiss. Verlag-Ges.

Münch, Ragnhild (1995): Gesundheitswesen im 18. und 19. Jahrhundert. Das Berliner Beispiel. Zugl.: Berlin, Freie Univ., Diss., 1992 u.d.T.: Münch, Ragnhild: Öffentliches Gesundheitswesen und soziale Fürsorge in Berlin zwischen staatlicher Repression und Reformkonzepten (18. und 19. Jahrhundert). Berlin: Akademie Verlag (Publikationen der Historischen Kommission zu Berlin). Available online at www.gbv.de/dms/faz-rez/F19950902ROST1-100.pdf.

Murray, Andrew (1877): The Museum Mite. *The American Naturalist*, Band 8(11), 479-482.

Nemecek, Natasa (2013): Friedrich Rathgen and His Impact on Slovenian Conservation in the Beginning of the Twentieth Century. In: *CeROArt* [En ligne]. Available online at http://ceroart.revues.org/3686.

Neugebauer, Wolfgang (2007): acta borussia nf. Preußen als Kulturstaat. Unter Mitarbeit von Bärbel Holtz, Rainer Paetau, Christina Rathgeber, Hartwin Spenkuch, Reinhold Zilch, Gaby Huch. Berlin: Berlin-Brandenburgische Wissenschaften (Acta Borussia).

Neugebauer, Wolfgang; Holtz, Bärbel (Hrsg.) (2010): Kulturstaat und Bürgergesellschaft. Preußen, Deutschland und Europa im 19. und frühen 20. Jahrhundert. Berlin-Brandenburgische Akademie der Wissenschaften. Berlin: Akademischer-Verlag.

Neumayer, Georg von; Ascherson, Paul (1875): Anleitung zu wissenschaftlichen Beobachtungen auf Reisen. Mit besonderer Rücksicht auf die Bedürfnisse der kaiserlichen Marine. Berlin: Verlag von Robert Oppenheim. Online verfügbar unter http://data.onb.ac.at/ABO/%2BZ102227706.

Newstead, Robert (1893): The Use of Boric Acid as a Preservative for Birds' Skins. In: *Museums Association, Report of Proceedings*, 104-106.

Nietzsche, Friedrich (2020): Friedrich Nietzsche: Der Antichrist. 1. Auflage. Hg. v. Gerald-Hermann Monnheim. Berlin: epubli.

Nilsson, Axel Rudolf (1907): Desinfektion fester Gegenstände. *Chemiker-Zeitung*, Band 31, 299.

Nipperdey, Thomas (1995): Deutsche Geschichte 1866-1918. Zweiter Band: Machtstaat vor der Demokratie. 3. Aufl. München: Beck.

NMA (1905 und 1907): Ämbetsarkiv 1 A, No. 61-18. Nordisk museet Stockholm. Bau einer Begasungsanlage. Nämndens Protokoll (Protokolle des Ausschusses), Nr. 8, § 62, Protokollnotiz vom 05.4.1905; Nr. 10, Protokollnotiz vom 12.04.1907.

Oddy, Andrew (1973): An Unsuspected Danger in Display. *Museum Journal*, 73, 27-28.

Odegaard, Nancy; Sadongei, Alyce (2005): Old Poisons, New Problems. A Museum Resource for Managing Contaminated Cultural Materials. Walnut Creek, CA: AltaMira Press.

Osterhammel, Jürgen (2016): Die Verwandlung der Welt. Eine Geschichte des 19. Jahrhunderts. 2. Auflage der Sonderausgabe. München: C.H. Beck (Historische Bibliothek der Gerda-Henkel-Stiftung).

Oswald von, Margareta (2022): Working Through Colonial Collections. An Ethnography of the Ethnological Museum in Berlin. Leuven: Leuven University Press.

Otto, Helmut (1979): Das chemische Laboratorium der Königlichen Museen in Berlin. *Berliner Beiträge zur Archäometrie*, (4), 1-298.

Pelizaeus, Ludolf (2008): Der Kolonialismus. Geschichte der europäischen Expansion. Wiesbaden: Marix-Verlag (Marixwissen).

Peltz, Uwe (2017): Das Chemische Laboratorium bis zur Gründung als "Zwillingsinstitute" im geteilten Berlin. *Berliner Beiträge zur Archäometrie, Kunsttechnologie und Konservierungswissenschaft*, (25), 55-94.

Pfister, Aude-Laurence (2008): L'Influence des Biocides sur la Conservation des Naturalis. Diplomarbeit. Haute École Arts, Appliqués-La Chaux-de-Fonds, Filière Conservation-Restauration, La Chaux-De-Fonds. Filière Conservation-Restauration.

Philosophical Transactions (1665): London: The Royal Society. Available online at https://royalsocietypublishing.org/journal/rstl.

Pinniger, David; Landsberger, Bill; Meyer, Adrian; Querner, Pascal (2016): Handbuch Integriertes Schädlingsmanagement in Museen, Archiven und historischen Gebäuden. Unter Mitarbeit von Annette Townsend. Berlin: Gebr. Mann Verlag; Rathgen-Forschungslabor Staatliche Museen zu Berlin.

Pinniger, David; Meyer, Adrian; Townsend, Annette (2015): Integrated Pest Management in Cultural Heritage. London: Archetype.

Pinniger, David; Townsend, Annette (2001): Pest Management in Museums, Archives and Historic Houses. London: Archetype.

Pinniger, David; Winsor, Peter (1998): Integrated Pest Management. Practical, Safe and Cost-Effective Advice on the Prevention and Control of Pests in Museums. London: Museums & Galleries Commission.

Plenderleith, Harold, James (1934): The Preservation of Antiquities. Oxford: University Press (I).

Purewal, Victoria Jane (2001): Analysis of the Pesticide Residues Present on Herbarium Sheets within the National Museums and Galleries of Wales. Proceedings of 2001: A Pest Odyssey. In: Kingsley, Helen; Pinniger, David; Xavier-Rowe, Amber; Winsor, Peter (eds.), Integrated Pest Management for Collections, London: James & James, 144.

Purewal, Victoria Jane (2012): Novel Detection and Removal of Hazardous Biocide Residues Historically Applied to Herbaria. Dissertation, University of Lincoln, Lincoln.

Rathgen, Friedrich (1896): Vortrag des Herrn Dr. Rathgen: Reiseerinnerungen. Mit 6 Abbildungen. Versammlung am 20.2.1896. *Polytechnisches Centralblatt. Organ der Polytechnischen Gesellschaft zu Berlin*, 57. Jahrgang der Gesamtfolge (11), 125–127.

Rathgen, Friedrich (1898): Die Konservirung von Alterthumsfunden. 1. Auflage. Berlin: W. Spemann (Handbücher der Staatlichen Museen zu Berlin).

Rathgen, Friedrich (1903): Konservierung von Altertumsfunden aus Eisen und Bronze. *Chemiker-Zeitung*, (56), 703–704.

Rathgen, Friedrich (1905): The Preservation of Antiquities. A Handbook for Curators. Translated by the Permission of the Authorities of the Royal Museums von George A. Auden und Harold A. Auden. Cambridge: University Press.

Rathgen, Friedrich (1908a): Luftdichte Museumsschränke. *Museumskunde, Zeitschrift für Verwaltung und Technik öffentlicher und privater Sammlungen*, Heft V, 97–102.

Rathgen, Friedrich (1908b): Mitteilungen aus dem Laboratorium der Königlichen Museen zu Berlin. IV. Die Verwendung von Tetrachlorkohlenstoff in der Konservierungspraxis. *Museumskunde, Zeitschrift für Verwaltung und Technik öffentlicher und privater Sammlungen*, Band IV, 90–91.

Rathgen, Friedrich (1910): Über Mittel gegen Holzwurmfraß. *Museumskunde, Zeitschrift für Verwaltung und Technik öffentlicher und privater Sammlungen*, Band VI, 23–27.

Rathgen, Friedrich (1911): Mitteilungen aus dem Chemischen Laboratorium der Königlichen Museen zu Berlin. VIII. Über die Verwendung von Kohlenstofftetrachlorid zur Abtötung von tierischen Schädlingen. *Museumskunde, Zeitschrift für Verwaltung und Technik öffentlicher und privater Sammlungen*, Band VII, 219–220.

Rathgen, Friedrich (1924): Die Konservierung von Altertumsfunden. Mit Berücksichtigung ethnographischer und kunstgewerblicher Sammlungsgegenstände. 2. Auflage. Berlin und Leipzig: Walter De Gruyter & Co. (Handbücher der Staatlichen Museen zu Berlin, II. und III. Teil).

Rathgen, Friedrich (1926): Friedrich Rathgen. Die Konservierung von Altertumsfunden/ Stein und steinartige Stoffe. 3. umgearb. Aufl. Berlin: De Gruyter (Handbücher der Staatlichen Museen zu Berlin, Teil 1).

Rauchensteiner, Manfried; Broukal, Josef (2015): Der Erste Weltkrieg und das Ende der Habsburgermonarchie 1914–1918. In aller Kürze. Wien, Köln, Weimar: Böhlau Verlag.

Réaumur de, René-Antoine Ferchault (1748): Divers Means for Preserving from Corruption Dead Birds, Intended to Be Sent to Remote Countries, So That They May Arrive There in a Good Condition. Some of the Same Means May Be Employed for Preserving Quadrupeds, Reptiles, Fishes, and Insects. *Philosophical Transactions*, (45), 304–320.

Reinbothe, Roswitha (2011): Geschichte des Deutschen als Wissenschaftssprache im 19. Jahrhundert. Vortrag bei einem Symposion an der Universität Bamberg am 15./16. Oktober 2009. Hg. v. Wieland Eins, Helmut Glück und Sabine Pretscher. Harrassowitz Verlag. Wiesbaden (Wissen schaffen–Wissen kommunizieren. Wissenschaftssprachen in Geschichte und Gegenwart). Available online at www.observatoireplurilinguisme.eu/ images/Education/Enseignement_superieur/reinbothe-geschichte_des_deutschen_als_ wissenschaftssprache.pdf.

Reinhard, Wolfgang (2008): Kleine Geschichte des Kolonialismus. 2., vollst. überarb. und erw. Aufl. Stuttgart: Kröner (Kröners Taschenausgabe, 475). Available online at http:// deposit.d-nb.de/cgi-bin/dokserv?id=3040366&prov=M&dok_var=1&dok_ext=htm.

Richter, Christian Gottlieb (1829): Anweisung Vögel auszustopfen, nebst Angabe aller dazu erforderlicher Hülfsmittel. Mit einem Vorwort von Brehm, Pastor in Renthendorf. Mit 2 Kupfertafeln. Jena: August Schmid.

Richter, Oswald (1907): Über die idealen und praktischen Aufgaben ethnographischer Museen (Fortsetzung). *Museumskunde, Zeitschrift für Verwaltung und Technik öffentlicher und privater Sammlungen*, Band III, 14–25, 99–120.

Richter, Oswald (1908): Über die idealen und praktischen Aufgaben ethnographischer Museen. *Museumskunde, Zeitschrift für Verwaltung und Technik öffentlicher und privater Sammlungen*, Band IV, 92–235.

Rochussen, Frank (1920): Ätherische Öle und Riechstoffe. 2. umgearb. Auflage. Berlin, Leipzig: Vereinigung wissenschaftlicher Verleger (Sammlung Göschen).

Rookmaaker, L.C.; Morris, P.A.; Glenn, I.E.; Mundy, P.J. (2006): The Ornithological Cabinet of Jean-Baptiste Bécoeur and the Secret of the Arsenical Soap. *Archives of Natural History*, 33(1), 146–158.

Rowley, Frederick Richard (1916): Demonstration of Objects Preserved in Arsenious Acid Glycerine Jelly. Read at the Ipswich Conference. *Museums Journal*, Band 16(4), 77–79.

Runeby, Nils (1997): Deutschland als technisches Vorbild. Möten och vänskapsband; [Deutsches Historisches Museum, 24.10–06.01. 1997, Nationalmuseum, 26.02–24.05. 1998, Norsk Folkemuseum . . .]. *Skandinavien och Tyskland* 1800–1914, 389–396.

Rytz, Walter (Hrsg.) (1922): Die Herbarien des Botanischen Instituts der Universität Bern (Schweiz). In: Mitteilungen der Naturforschenden Gesellschaft in Bern (V). Bern: K. J. Wyss Erben.

Sächsisches Staatsarchiv–Hauptsaatsarchiv Dresden (1902; 1903; 1904): 13842. Staatliches Museum für Tierkunde, Nr. 21. Gebundene Ausgabe. Königlich Zoologisches und Anthropologisch-Ethnographisches Museum Dresden. Jahresberichte der Verwaltung.

Savoy, Bénédicte (2015): Tatkräftiges Mitmischen. Alexander von Humboldt und die Museen in Paris und Berlin. In: "Mein Zweites Vaterland": Alexander von Humboldt und Frankreich, Bd. 40. Berlin [u.a.]: De Gruyter, 233–259.

Schiessl, Ulrich (1984): Historischer Überblick über die Werkstoffe der schädlingsbekämpfenden und festigkeitserhöhenden Holzkonservierung. *Maltechnik/Restauro*, (2), 9–40.

Schlenke, Manfred (Hrsg.) (1981): Preußen. Beiträge zu einer politischen Kultur. 5 Bände. Reinbek bei Hamburg: Rowohlt (. . . Wechselausstellung/Österreichische Galerie Belvedere, eine Ausstellung der Berliner Festspiele GmbH, 15 August–15 November, Gropius-Bau, ehemaliges Kunstgewerbemuseum, Berlin; Katalog in fünf Bänden/Gesamthrsg.: Berliner Festspiele GmbH).

Schmaltz, Florian (2005): Kampfstoff-Forschung im Nationalsozialismus. Zur Kooperation von Kaiser-Wilhelm-Instituten, Militär und Industrie. Vollst. zugl.: Bremen, Universität, Dissertation, 2004. Göttingen: Wallstein. Geschichte der Kaiser-Wilhelm-Gesellschaft im Nationalsozialismus, 11. Available online at www.h-net.org/review/hrev-a0f1o1-aa.

Schnedlitz, Markus (2008): Chemische Kampfstoffe. Geschichte, Eigenschaften, Wirkung; Studienarbeit. Zugl.: Wiener Neustadt, FH, Seminararbeit. 1. Aufl. München: Grin-Verlag.

Schneiders, Werner (2014): Das Zeitalter der Aufklärung. 5. Aufl. München: C.H.Beck (C.H.Beck Wissen). Available online at http://elibrary.chbeck.de/10.17104/9783406671265/das-zeitalter-der-aufklaerung.

Schnetzler, Jean Balthasar (1916): Sulphide as an Insecticide. *The Popular Science Monthly*, 9, 767.

Schreiber, Fritz (1923): Die Industrie der Steinkohlenveredelung. Zusammenfassende Darstellung der Aufbereitung, Brikettierung und Destillation der Steinkohle und des Teers. Wiesbaden, s.l.: Vieweg+Teubner Verlag. Available online at http://dx.doi.org/10.1007/978-3-663-05097-1.

Schroeck, Lucas; Franz, Anselm; Hafner, Melchior (1682): Historia Moschi. Ad normam Academiae Naturae Curiosorum. Augustae Vindelicorum: Theophilius Göbelius.

Schuchhardt, Carl (1912a): Die neue Zusammensetzung des Schädels vom Homo Mousteriensis Hauseri. *Praehistorische Zeitschrift*, Band IV, 443–446.

Schuchhardt, Carl (1912b): Die neue Zusammensetzung des Schädels vom Homo Mousteriensis Hauseri. *Amtliche Berichte aus den Königlichen Kunstsammlungen*, 34(1), Spalten 4–10.

Schulte von Drach, Markus C. (2015): Erster Giftgaseinsatz im Ersten Weltkrieg. Die schreckliche Erfindung des Patrioten Fritz Haber. In: *Süddeutsche Zeitung*, April 22, https://www.sueddeutsche.de/politik/erster-giftgaseinsatz-im-ersten-weltkrieg-die-schreckliche-erfindung-des-patrioten-fritz-haber-1.2385082.

Schulz, Andreas (1995): Weltbürger und Geldaristokraten. Hanseatisches Bürgertum im 19. Jahrhundert. In: *Schriften des Historischen Kollegs*, 1995, (Vorträge 40).

Scott, Alexander (1922): The Restoration and Preservation of Objects at the British Museum. *Journal of the Royal Society of Arts*, LXX(No. 3618), 327–339.

Seler, Eduard (1917): Sitzung vom 17. November 1917. *Zeitschrift für Ethnologie; Organ der Berliner Gesellschaft für Anthropologie, Ethnologie und Urgeschichte*, (49), 212–213.

Sirois, Jane; Poulin, Jennifer; Stone, Tom (2010): Detecting Pesticide Residues on Museum Objects in Canadian Collections. A Summary of Surveys Spanning a Twenty-Year Period. *Collection Forum*, 24(1–2), 28–45.

SMB-PK EM (01.01.1901 bis 30.04.1903): I/MV 0057, Bd. 5, Pars I c. Umzgugsakte. Königliches Museum für Völkerkunde zu Berlin Acta betreffend den Umzug und die Aufstellung der Sammlungen des Museums.

SMB-PK EM (08.08.1904a): I/MV 0058, E. Nr. 1119/04. k.k. Zentralkommission für Kunst- und historische Denkmale in Wien Ankündigung einer Enquete zur Bewahrung von organischen Materialien. Schreiben, gedruckt, 2 Seiten, ohne Paginierung.

SMB-PK EM (September 1904b): I/MV 0058, E. Nr. 111387/1904. k.k. Zentralkommission für Kunst- und historische Denkmale in Wien. Einladung und Programm zu einer Enquete über die Bewahrung von organischen Materialien. Schreiben, gedruckt, 1 Seite, ohne Paginierung.

SMB-PK EM (1910–1911): I/MV 0075; E. Nr. 1360/10. Acta betreffend die Restauration von Alterthümern. Loseblattsammlung Königliches Museum für Völkerkunde zu Berlin; kaiserlich königliche landwirtschaftlich-chemische Versuchsstation. Gemeinsame Versuche zur Schädlingsbekämpfung. Brief von k.k. Inspektor (Name unleserlich) vom 02.07.1910, 1- seitig, ohne Paginierung; Brief von Johann Bolle vom 19.01.1911, Blatt 2, 2 Seiten; 2 Briefe von Felix von Luschan vom 07.01.1911, Blatt 1, 1 Seite und vom 24.01.1911, Blatt 3–4, 2 Seiten; 1 Brief von August Eichhorn vom 21.02.1911, Blatt 5, 1 Seite.

SMB-PK EM (1911–1914): I/MV 0075; E. Nr. 1771/11. Acta betreffend die Restauration von Alterthümern. Loseblattsammlung. Königliche Museen zu Berlin. Aufbau einer Begasungsanlage. Anfrage von Eduard Seler vom 04.11.1911, 1- spaltig; Bericht von Eduard Krause vom 30.01.1912, 2- spaltig; Vorlage von Albert Grünwedel an die Generaldirektion vom 24.02.1912, 1- spaltig; Brief von Johann Bolle vom 30.01.1912, 4- seitig; 2 Briefe und 3 Fotos von Willi Foy vom 09.02.1912 und vom 21.02.1912, je 1- seitig; Brief von Alfred Hackman vom 08.02.2012, 3- seitig; Vorlage an die Generalverwaltung von Max Junker vom 24.02.1912, 2- spaltig; Max Junker zur Prüfung an Friedrich Rathgen vom 13.11.1913, 2- seitig; Bericht über Verschiebung des Baubeginns von Max Junker vom 15.05.1914, 2- spaltig.

SMB-PK EM (1912/1917/1918): I/MV 0075, E. Nr. 467/12. Acta betreffend die Restauration von Alterthümern. Loseblattsammlung. Staatliches Museum für Völkerkunde zu Berlin. Planungen zur Errichtung einer Begasungsanlage in Berlin-Dahlem. Notizen und Stellungnahmen von Albert Grünwedel vom 20.03.1912 und vom 05.01.1918, 1- spaltig, ohne Paginierung; Aktennotiz von Max Junker vom 23.11.1917, 1- spaltig, ohne Paginierung; Aktennotiz von Eduard Krause vom 19.03.1912, 1- spaltig, ohne Paginierung; Aktennotiz von Strebe, (Vorname unbekannt, Anmerk. d. Verf.) vom 04.03.1918, 1- spaltig, ohne Paginierung; Brief von Johann Bolle vom 04.03.1918, 2- seitig, ohne Paginierung; Firmenprospekt der Apparate-Bauanstalt Christ & Co. (Inh. Gustav Necker, Ingenieur), 4- seitig, ohne Paginierung.

SMB-PK EM (1913): I/MV 0075; Vol. 1, Pars IIc, E. Nr. 1225/13. Acta betreffend die Restauration von Alterthümern. Loseblattsammlung. Königliches Museum für Völkerkunde zu Berlin; Farbenfabriken, vormals Friedrich Bayer & Co., Elberfeld & Leverkusen Korrespondenzen. 2 Briefe vom 19.07.1913 und 26.07.1913, je 2- seitig, ohne Paginierung; 1 Merkblatt über Autan, 1 seitig, ohne Paginierung; 1 Prospekt über Autan, 4- seitig, ohne Paginierung.

SMB-PK EM (1914a): I/MV 0075; E. Nr. 25/14. Acta betreffend die Restauration von Alterthümern. Loseblattsammlung. Königliches Museum für Völkerkunde zu Berlin. Bau und Aufstellung eines Begasungskastens. Stellungnahmen von August Eichhorn vom 07.01.1914, Albert Grünwedel vom 20.01.1914, Eduard Krause vom 20.01.1914 und Carl Schuchhardt vom 24.01.1914; Kostenangebot der Firma Georg Bruns vom 22.01.1914 und der Firma Karl Kotte vom 23.02.1914, 5 Seiten, ohne Paginierung.

SMB-PK EM (1914g): Laufende Nummer 324. Bürojournal des Königlichen Museums für Völkerkunde zu Berlin. Eichhorn, August. Anzeige gegen den Konservator Eduard Krause sowie Löschung des Eintrags am 05.09.1914. Tabelle mit 2 Einträgen, 1 Spalte.

SMB-PK EM (1914h): I/MV 932; E. Nr. 324/14. Personalakte. Königliche Museen zu Berlin. Eduard Krause.

SMB-PK EM (1919): I/MV 0075, E. Nr. 800/19. Acta betreffend die Restauration von Alterthümern. Loseblattsammlung. Heerdt, Walter. Werbeschreiben mit Anhängen zur Mottenbekämpfung von der Deutschen Gesellschaft für Schädlingsbekämpfung m.b.H. Brief vom 05.08.1919, 1- seitig, ohne Paginierung; Allgemeine Bedingungen, 5- seitig, ohne Paginierung; Antragsformular auf Durchgasung, 1- seitig, ohne Paginierung; Auftragsformular zur Durchgasung, 1- seitig; ohne Paginierung.

SMB-PK EM (1924): I/MV 0015, Pars I 3, Vol. 1, E. Nr. 388/24. Projektakte. Loseblatt-sammlung. Falke, Otto. Aufstellung des Umzugsetats des Staatlichen Museums für Völkerkunde. Kostentabelle von 1924, 1- seitig, ohne Paginierung.

SMB-PK EM (1925): I/MV 0075, Acta betreffend die Restauration von Alterthümern. Bd. 1, Pars II c, E. Nr. 682/13. Actien-Gesellschaft für Anilin-Fabrikation. Einführung des Produktes Dichlorbenzol "Agfa". Loseblattsammlung. Briefe vom 15.04.1913; 25.04.1913; 10.10.1913; 27.10.1913; 29.10.1913; 30.10.1913, 7 Seiten; Prospekt 9 Seiten; beides ohne Paginierung.

SMB-PK EM (21.04.1928a): I/MV 0075, E. Nr. 356/28. Acta betreffend die Restauration von Alterthümern. Loseblattsammlung. Brittner, Carl Stellungnahme zu Areginal und Eulan. Bericht vom 21.04.1928, 1- seitig, ohne Paginierung.

SMB-PK EM (19.03.1928b): I/MV 0075, E. Nr. 356/28. Acta betreffend die Restauration von Alterthümern. Loseblattsammlung. Urallandesmuseum Jekaterinburg Anfrage zu Areginal und Eulan. Brief vom 19.03.1928, 1- seitig, ohne Paginierung.

SMB-ZA (1914, 1922–1923; 1928): I/BV 239. Bauakte Staatliche Museen zu Berlin-Preußischer Kulturbesitz, Ethnologisches Museum Einrichtung von Konservierung-swerkstätten im Untergeschoß des Hauptgebäudes in Berlin-Dahlem.

SMB-ZA (1915, 1919; 1922–1924): I/BV 286. Bauakte. Staatliche Museen zu Berlin-Preußischer Kulturbesitz, Ethnologisches Museum Neubau und Einrichtung des Asiatischen Museums, Dahlem; Inneneinrichtung.

SMB-ZA (24.07.1924–12.03.1928): I/BV 723. Bauakte. Staatliche Museen zu Berlin-Preußischer Kulturbesitz, Ethnologisches Museum Umbau, Einrichtung, Instandhal-tung–Gebäudekomplex Dahlem; Einrichten von Arbeitsräumen für den Konservator und Einbau einer Desinfektionsanlage.

SMB-ZA (1933): I/I M 26. Loseblattsammlung. Consolidierte Alkaliwerke Abteilung Hannover; Hübner, Paul Xylamon. Werbeschreiben für Xylamon. 1 Brief von den Consolidierten Alkaliwerken vom 31.05.1933, Blatt 8, 1- seitig; Broschüre zu Xylamon, Blatt 9, 2- seitig; Richtlinien zur Holzwurmbekämpfung mit Xylamon, Prospekt Nr. 41, Blatt 11–12, 4- seitig; Brief von Paul Hübner vom 27.03.1933, Blatt 10, 1- seitig.

SMB-ZA (Rechnungsjahr 1935): I/GV 1399. Beleg zur Verwaltungsrechnung Nr. 461–550. Kap. 154. Tit. 26. Bayer I.G. Farbenindustrie Aktiengesellschaft. Rechnung für Areginal.

Smith, John (1884): Some Observations on Museum Pests. *Proceedings of the Entomological Society, 1884–1889*, 1, 113–116.

Soëtard, Michel (2012): Jean-Jacques Rousseau. 1. Aufl. München: C.H.Beck (C.H.Beck Wissen).

Sotriffer, Kristian (1996): Die Blüte der Chrysantheme. Die Zacherl—Stationen einer anderen Wiener Bürgerfamilie. Wien u.a.: Böhlau Verlag.

Spenkuch, Hartwin (2015): "An die Spitze einer neuen Weltgestaltung gestellt". Zu Grundlinien der Entwicklung des Kulturstaats in Preußen (1807–1870). *Preußen als Kulturstaat im 19*. Jahrhundert, Bd. 20, 157–183.

Stansbury, Chas F.; Barkly, Henry; Campbell, W.H.; J.M.G.; H.S.; Farthing, John J. (1853): *Journal of the Society for Arts*, November 26, 1852–November 11, 1853, vol. 1.

Steinen von den, Karl (1894): Unter den Naturvölkern Zentral-Brasilien. Reiseschilderung und Ergebnisse der zweiten Schingú-Expedition, 1887–1888. Berlin: Geographische Verlagsbuchhandlung von Dietrich Reimer (Mit 30 Tafeln (1 Heliogravüre, 11 Lichtdruck-bilder, 5 Autotypien und 7 lithogr. Tafeln) sowie 160 Text-Abb.).

Steinen von den, Karl (1905): Gedächtnisfeier für Adolf Bastian. Am 11. März 1905. *Zeitschrift der Gesellschaft für Erdkunde zu Berlin*, Sonderabdruck, (3), 168.

Steinmetz, George (2017): Empire in Three Keys. Forging the Imperial Imaginary at the 1896 Berlin Trade Exhibition. *Thesis Eleven*, 139(1), 46–68. DOI: 10.1177/0725513617701958.

Stellwaag, Fritz (1927): Der Gebrauch der Arsenmittel in deutschem Pflanzenschutz. Ein Rückblick und ein Ausblick unter Verwertung der ausländischen Erfahrungen. *Zeitschrift für angewandte Entomologie*, Bd. 12(1), 35–36.

Stoltzenberg, Dietrich (1994): Fritz Haber. Chemiker, Nobelpreisträger, Deutscher, Jude. Eine Biographie. Weinheim: VCH.

Sutter, Hans-Peter (2003): Holzschädlinge an Kulturgütern erkennen und bekämpfen. Handbuch für Denkmalpfleger, Restauratoren, Konservatoren, Architekten und Holzfachleute. 4. überarb. und erw. Aufl. Bern: Paul Haupt.

Taeger, Harald (1941): Die Klinik der entschädigungspflichtigen Berufskrankheiten. Berlin, Heidelberg: Springer Berlin Heidelberg.

Tatershall, Creassy Edward Cecil (1924): To Preserve Woolen Textiles from Moth. *Museum Journal*, (23), 199–200.

Teibler, Claudia (2019): Absolut Biologisch. *Restauro*, (2), 42–43.

Tello, Helene (2006): Investigations on Super Fluid Extraction (SFE) with Carbon Dioxide on Ethnological Materials and Objects Contaminated with Pesticides. Diplomarbeit. Fachhochschule für Technik und Wirtschaft, Berlin. Fachbereich 5, Gestaltung, Studiengang Restaurierung/Grabungstechnik.

Tello, Helene (2016): Handle with Care. Der Einsatz historischer Biozide an Kunst- und Kulturgut und die Folgen für Materialien und Objekte. Tagung im Rahmen der Werkstattgespräche des Bayerischen Landesamtes für Denkmalpflege, 16. und 17. Oktober 2014. *Kontaminiert Dekontaminiert*, 18–24.

Tello, Helene; Unger, Achim (2010): Liquid and Supercritical Carbon Dioxide as a Cleaning and Decontamination Agent for Ethnographic Materials and Objects. *Smithsonian Contributions to Museum Conservation*, (1), 35–50.

Thilenius, Georg (1905): 2. Museum für Völkerkunde. Bericht für das Jahr 1905. *Jahrbuch der Hamburgischen Wissenschaftlichen Anstalten*, 23, 231–240. Available online at www.biodiversitylibrary.org/item/92087.

Thilenius, Georg (1916): Das Hamburgische Museum für Völkerkunde. *Museumskunde, Zeitschrift für Verwaltung und Technik öffentlicher und privater Sammlungen*, 16(i.e. 12) (Beiheft zu Band XIV), I–VIII; 1–154; Tafeln 1–8.

Thon, Theodor (1827): Handbuch für Naturaliensammler. oder gründliche Anweisung die Naturkörper aller drei Reiche zu sammeln, im Naturalienkabinet aufzustellen und aufzubewahren, namentlich Thiere aller Arten, Säugethiere, Vögel, Reptilien, Fische, Conchylien, Crustaceen, Insekten, Zoophyten und Eingeweidewürmer auszustopfen, zuzubereiten, zu versenden, so wie Pflanzen zu trocknen, Herbarien, Fruchtkabinette, Holzbibliotheken und Mineraliensammlungen anzulegen, einzurichten und in vollkommner Schönheit zu erhalten. Frei nach dem Französischen bearbeitet und vervollständigt. Ilmenau: Verlag von Bernhard Friedrich Voigt.

Tímár-Balázsy, Ágnes; Eastop, Dinah; Járó, Márta (2011): Chemical Principles of Textile Conservation. Abingdon: Routledge (Butterworth-Heinemann Series in Conservation and Museology).

Toothaker, Charles Robinson (1908): Fumigation. *Proceedings of the American Association of Museums*, 119–123.

Troschel, Ernst (1916): Handbuch der Holzkonservierung. Berlin: Springer.

Unger, Achim (2012): "Eulanisierte" Textilien—eine Gefahr für Mensch und Material? *Beiträge zur Erhaltung von Kunst- und Kulturgut*, (2), 25–39.

Unger, Achim (2018): Dekontaminations-Verfahren für biozidbelastetes Kulturgut und ihre Bewertung. Vortrag. Tagung Wood Art Conservation. Bern: Fachhochschule Bern.

Unger, Achim; Jakob, Georg; Debbert, Lothar (1988): Vor 100 Jahren: Gründung des ersten Museumlabors der Welt. *Neue Museumskunde*, (2), 132–135.

Unger, Achim; Schniewind, Arno, P.; Unger, Wibke (2001): Conservation of Wood Artifacts. A Handbook. Berlin: Springer (Natural Science in Archaeology). Available online at www.loc.gov/catdir/enhancements/fy0815/2001020310-d.html.

Unger, Achim; Tello, Helene; Lindex, Sörrn; Trommer, Bernhard; Behrendt, Stefanie (2006): "Grüne Chemie" hält Einzug in die Restaurierung. Versuche zur Reinigung, Entfettung und Dekontamination von Kunst- und Kulturgut mit flüssigem Kohlendioxid. *Restauro: Zeitschrift für Konservierung und Restaurierung*, 112, 384–394.

Unger, Achim; Weidner, Anke Grit; Tello, Helene; Mankiewicz, Johannes (2011): Neues zur Dekontamination von beweglichem Kunst- und Kulturgut mit flüssigem Kohlendioxid. *VDR-Beiträge zur Erhaltung von Kunst- und Kulturgut*, (2), 85–96.

Vanhöffen, Ernst (1918): Zur Erinnerung an August Brauer. *Mitteilungen aus dem Zoologischen Museum in Berlin*, 9. Band, (1. Heft), 1–12.

Vasold, Manfred (2002): Robert Koch, der Entdecker von Krankheitserregern. Heidelberg: Spektrum der Wissenschaft.

VDLUFA Verband Deutscher Landwirtschaftlicher Untersuchungs- und Forschungsanstalten (2013): 125 Jahre Verband Deutscher Landwirtschaftlicher Untersuchungs- und Forschungsanstalten e.V. Eine Dokumentation. Darmstadt: VDLUFA-Verlag (VDLUFA-Schriftenreihe, 69).

Verordnung (EU) Nr. 528/2012 des Europäischen Parlaments und Rates der Europäischen Union 27.06.2012. (2012): Biozidproduktarten und ihre Beschreibung vom 22.05.2012 gemäss Artikel 2 Absatz 1; Anhang V der Biozid-Verordnung (EU) Nr. 528/2012, (EU) Nr. 528/2012. In: Amtsblatt der Europäischen Union, L 167/1 vom 27. Juni.

Viktor, Adolf (1909): Tausend Topfscherben. In: Berliner Tagblatt, September 21. Quoted in Menghin (2005): Das Berliner Museum für Vor- und Frühgeschichte. Festschrift zum 175-jährigen Bestehen. Staatliche Museen zu Berlin-Preußischer Kulturbesitz, Menghin, Wilfried (Hrsg.). Acta Praehistorica et Archaeologica 36/37, 2005, 130.

Voß, Albert (1888): Merkbuch, Alterthümer auszugraben und aufzubewahren. Eine Anleitung für das Verfahren bei Ausgrabungen, sowie zum Konservieren vor- und frühgeschichtlicher Alterthümer. Unter Mitarbeit von Gustav von Gossler (Hrsg.). Berlin: Ernst Siegfried Mittler und Sohn.

Wagner, Fritz; Rieckenberg, Hans Jürgen; Glaubrecht, Martin; Jaeger, Hans; Hentig, Hans Wolfram von; Körner, Hans (1972): Neue Deutsche Biographie. Hess-Hüttig. 1–9. Berlin: Duncker & Humblot (9).

Warburg, Aby Moritz (2018): Gesammelte Schriften. [1. Auflage]. Berlin, Boston: De Gruyter.

Weber, Jörg; Unger, Achim (2018): Experimente zur Entfernung alter Holzschutz- und Holzfestigungsmittel mit Methyl-tert-butylether (MTBE) aus ungefassten und gefassten Holzproben. *VDR Beiträge zur Erhaltung von Kunst- und Kulturgut*, (2), 60–73.

Wegner, Dietrich; Hoffmann, Almut (2002): Der Schädel vom Combe Capelle im Museum für Vor- und Frühgeschichte wiederaufgefunden. *Archäologisches Korrespondenzblatt*, 7(3), 218–221.

Weidner, Herbert; Sellenschlo, Udo (2003): Vorratsschädlinge und Hausungeziefer. Bestimmungstabellen für Mitteleuropa. 6. Aufl. Heidelberg: Spektrum Akademie Verlag.

Weisband, Edward; Thomas, Courtney Irene Powell (2015): Political Culture and the Making of Modern Nation-States. London, New York: Routledge Taylor & Francis Group.

Werner, Frank (1997): Berlin: Neue alte Hauptstadt. *Der Bürger im Staat*, Heft 2, 74–79.

Westphal-Hellbusch, Sigrid (1969): 100 Jahre Ethnologie unter besonderer Berücksichtigung ihrer Entwicklung an der Universität. Fachhistorische Beiträge. In: Festschrift zum 100- jährigen Bestehen der Berliner Gesellschaft für Anthropologie, Ethnologie und Urgeschichte, 1869–1969, Band 1, 157–183.

Westphal-Hellbusch, Sigrid (1973): Zur Geschichte des Museums. Hundert Jahre Museum für Völkerkunde. *Baessler-Archiv*, XXI, 1–99.

Wiesner, Julius (1927): Die Rohstoffe des Pflanzenreiches. 4. Auflage. Leipzig: Wilhelm Engelmann (I, II).

Wilhelmi, Julius; Kunike, Hugo (1927): Versuche und Untersuchungen über die Wirksamkeit des Petroleum-Raffinates "Flit" bei der Fliegen- und Stechmückenbekämpfung. *Zeitschrift für Desinfektion und Gesundheitswesen*, 19(3), 98–99.

Wink, Michael; van Wyk, Ben-Erik; Wink, Coralie (2008): Handbuch der giftigen und psychoaktiven Pflanzen; mit 13 Tabellen. Stuttgart: Wiss. Verlag-Ges.

Winter, Petra; Grabowski, Jörn (Hrsg.) (2014): Zum Kriegsdienst einberufen. Die Königlichen Museen zu Berlin und der Erste Weltkrieg. Staatliche Museen zu Berlin-Preußischer Kulturbesitz, Ethnologisches Museum. Köln: Böhlau Verlag (Schriften zur Geschichte der Berliner Museen, 3).

Wood, Henry Trueman (1913): A History of the Royal Society of Arts with a Preface by Lord Sanderson. London: Murray.

Wray, L. (1908): The Preservation of Mammal Skins. *Museums Journal*, (8), 201–209.

Zacher, Friedrich (1916): Neue und wenig bekannte Schädlinge aus unseren Kolonien. II. Ein neuer Blattfloh als Gallenbildner an Kickxia. III. Einige Schädlinge des Tabaks an Kamerun. *Zeitschrift für angewandte Entomologie*, (III), 418–425.

Zacher, Friedrich (1921): Eingeschleppte Vorratsschädlinge. Sitzungsberichte. Sitzung vom. *Deutsche Entomologische Zeitschrift*, (4), 288–295.

Zacher, Friedrich (1922a): Eingeschleppte Vorratsschädlinge. *Verhandlungen der Deutschen Gesellschaft für angewandte Entomologie*, 55–58.

Zacher, Friedrich (1922b): Südamerikanische Kakaoschädlinge. *Tropenpflanzer*, 25, 119–121.

Zacher, Friedrich (1924a): Der Brotkäfer, ein schlimmer Haushaltsschädling. *Hof und Garten*, 46, 87–88.

Zacher, Friedrich (1924b): Methoden der Vorratsschädlingsbekämpfung. Vierte Mitgliederversammlung zu Frankfurt a.M. vom 10. bis 13. Juli 1924. *Verhandlungen der Deutschen Gesellschaft für angewandte Entomologie*, 45–50.

Zacher, Friedrich (1924c): Pelz- und Kleidermotten. *Der Rauchwarenmarkt*, 12(83), 2–3.

Zacher, Friedrich (1927): Die Vorrats-, Speicher- und Materialschädlinge und ihre Bekämpfung. Berlin: P. Parey.

Zadow, Mario (2003): Karl Friedrich Schinkel. Leben und Werk. 3., verb. Aufl. Stuttgart: Ed. Menges.

Zalewski, Przemyslaw Paul (2007): Altstadtsanierungen in Deutschland und in Europa bis zum Zweiten Weltkrieg. Eine Erinnerung an Motive und Method en. *Journal of Comparative Cultural Studies in Architecture*, (1), 28–36.

Zalewski, Przemyslaw Paul (2014): Einführung in die Voraussetzungen, Ziele und Strategien des Projektes. Arbeitspolitik und Arbeitsmarkt als Ausgangsvoraussetzungen. In: Biozidbelastete Kulturgüter. Grundsätzliche Hinweise und Texte zur Einführung in die

Problematik; Bericht über das EU-/ESF-Projekt "Kleine und Mittlere Unternehmen und Wissenschaft im Dialog. Dekontamination von Kulturgütern, 7–17.

Zalewski, Przemyslaw Paul; Tello, Helene; Meyer-Haake, Arne (2014): Biozidbelastete Kulturgüter. Grundsätzliche Hinweise und Texte zur Einführung in die Problematik; Bericht über das EU-/ESF-Projekt "Kleine und Mittlere Unternehmen und Wissenschaft im Dialog. Dekontamination von Kulturgütern. Frankfurt (Oder).

Zinke, Georg, Gottfried (1802): Kunst allerhand natürliche Körper zu sammeln, auf eine leichte Art für das Kabinett aufzubereiten und vor der Zerstörung feindlicher Insecten zu sichern. Jena: Göpferdt, J. C. G.

Zuska, Jan (1994): Haus- und Vorratsschädlinge. Hanau: Dausien.

Zweig, Stephan (2007): Die Welt von gestern. Erinnerungen eines Europäers. 5. Aufl. Frankfurt am Main: S. Fischer.

Appendix 1

Organic materials in museums and the most important insect pests

The most important pests in museums, insects, are listed in tabular form. The representation in relation to the materials is partly schematic because some insects are able to change their demands on the material and can therefore also infect materials with very different chemical compositions. This characteristic is well known, for example, in silverfish (*Lepisma saccharina*) and the brass beetle (*Niptus hololeucus*). On the other hand, the living conditions of some insect pests have not yet been fully researched. Also, some insects are dependent on climatic peculiarities, such as the silverfish on increased humidity, so that damage may not always occur. For the following table, extensive literature could also be drawn from the field of conservation sciences (Florian 1997; Sutter 2003; Weidner und Selllenschlo 2003; Zacher 1916, 1921, 1922a, 1922b, 1924a, 1924b, 1927; Zuska 1994).

Appendix 1 Table 1 Organic materials in museums and the main insect pests

Chemical group of substances of the materials	Organic materials	Insects German national nomenclature	Insects Scientific (Latin) nomenclature
Polysaccharides: Cellulose Hemicelluloses (Wood polyoses)	Softwood, lumber, roof beams, Wooden pallets, movable wooden objects	Common nail beetle	*Anobium punctatum*
	Softwood Lumber	House longhorn beetle	*Hylotrupes bajulus*
	Hardwood, also oak, but only sapwood	Lined sapwood beetle Common nail beetle Combed gnat beetle Southern nail beetle	*Lyctus linearis* *Anobium punctatum* *Ptilinus pectinicornis* *Oligomerus ptilinoides*
	Oakwood, also heartwood, if it is pre-damaged by fungi	Death watch beetle	*Xestobium Rufovillosum*

Appendix 1 Table 1 (Continued)

Chemical group of substances of the materials	Organic materials	Insects German national nomenclature	Insects Scientific (Latin) nomenclature
	Wood with tree bark in buildings	Brown sapwood beetle Soft nail beetle	*Lyctus brunneus* *Ernobium mollis*
	Outdoor wood, warm regions	Termites, several species	*Isoptera* (several species)
	Dry wood, warm regions	Drywood termites, some species	*Kalotermitidae* (some species)
	Exotic wood species	African sapwood beetle	*Lyctus africanus*
	Plant fibers, raffia, straw, and other materials	Brass beetle	*Niptus hololeucus*
Polysaccharides	Textiles: Linen Cotton Objects made of flour dough	Silverfish Bread beetle	*Lepisma saccharina* *Stegobium paniceum*
Proteins	Textiles: Silk Wool Fur, rawhide, leather	Silverfish Clothes moth Carpet beetle Fur beetle Common bacon beetle	*Lepisma saccharina* *Tineola bisselliella* *Anthrenus scrophulariae* *Attagenus pellio* *Dermestes lardarius*
	Insect preparations Gut, feathers, hair, mollusks, dried (mollusks)	Museum beetle (Synonyms: Cabinet beetle, woolly flower beetle)	*Anthrenus museorum* *Anthrenus verbasci*
Polysaccharides and proteins	Paper Possibly with gluing, with starch paste or with flour paste	Common nail beetle Silverfish Paper fish (new in Europe)	*Anobium punctatum* *Lepisma saccharina* *Ctenolepisma longicaudata*

References

Florian, Mary-Lou (1997): Heritage Eaters: Insects and Fungi in Heritage Collections. London: James and James.

Sutter, Hans-Peter (2003): Holzschädlinge an Kulturgütern erkennen und bekämpfen. Handbuch für Denkmalpfleger, Restauratoren, Konservatoren, Architekten und Holzfachleute. 4. überarb. und erw. Aufl. Bern: Paul Haupt.

Weidner, Herbert; Sellenschlo, Udo (2003): Vorratsschädlinge und Hausungeziefer. Bestimmungstabellen für Mitteleuropa. 6. Aufl. Heidelberg: Spektrum Akad. Verl.

Zacher, Friedrich (1916): Neue und wenig bekannte Schädlinge aus unseren Kolonien. II. Ein neuer Blattfloh als Gallenbildner an Kickxia. III. Einige Schädlinge des Tabaks an Kamerun. *Zeitschrift für angewandte Entomologie* (III), 418–425.

Zacher, Friedrich (1921): Eingeschleppte Vorratsschädlinge. Sitzungsberichte. Sitzung vom 28.11.1921. *Deutsche Entomologische Zeitschrift*, (4), 288–295.

Zacher, Friedrich (1922a): Eingeschleppte Vorratsschädlinge. In: *Verhandlungen der Deutschen Gesellschaft für angewandte Entomologie*, 55–58.

Zacher, Friedrich (1922b): Südamerikanische Kakaoschädlinge. *Tropenpflanzer*, (25), 119–121.

Zacher, Friedrich (1924a): Der Brotkäfer, ein schlimmer Haushaltsschädling. *Hof und Garten*, (46), 87–88.

Zacher, Friedrich (1924b): Pelz- und Kleidermotten. *Der Rauchwarenmarkt*, Jhrg. 12 (83), 2–3.

Zacher, Friedrich (1927): Die Vorrats-, Speicher- und Materialschädlinge und ihre Bekämpfung. Berlin: P. Parey.

Zuska, Jan (1994): Haus- und Vorratsschädlinge. Hanau: Dausien.

Appendix 2

Decontamination procedures for pesticide-contaminated cultural property and their assessment

In view of the sometimes uncontrolled entry of pesticides into museum collections, attempts to decontaminate art and cultural assets have become increasingly important in the conservation of art and cultural assets since the mid-1990s. Based on the minimization requirement, people who have contact with collection items contaminated in this way are to be protected. However, the objects themselves must also be protected from further deterioration caused by the former entry of old pesticides by means of suitable remediation or decontamination procedures (Tello and Unger 2010). At the same time, contaminated areas are to be made usable again, and secondary contamination is to be reduced (Zalewski 2014). A decision as to whether mechanical, thermal, preventive, or solvent processes can be used depends first of all on whether contaminated objects are transportable or fixed (Unger 2018). The following table describes in detail the methods and procedures with the exception of Soxhlet extraction. This solvent process represents the ideal principle of pesticide decontamination and is used, for example, to determine the concentration of polycyclic aromatic hydrocarbons (PAHs). Because this procedure is invasive and can only be performed in laboratories at a relatively high cost, it is not recorded in tabular form.

Appendix 2 Table 1 Decontamination procedure for pesticide-contaminated cultural property compiled according to Unger 2018.

Mechanical processes			
Procedures	*Materials/technology*	*Pros*	*Cons*
Dust removal	Dry cleaning of contaminated individual objects under a fume hood using compressed air; use of industrial vacuum cleaners type H or H+ in buildings	By removing the contaminated dust in buildings, significant reduction in pollutant load in room air	In the case of rough, uneven surfaces, effective dust removal is often difficult; loose color settings and gilding must be pre-strengthened

(Continued)

Appendix 2 Table 1 (Continued)

Mechanical processes			
Procedures	*Materials/technology*	*Pros*	*Cons*
Vacuum washing process	Procedure for removing dust containing pesticides from wooden surfaces with surfactant-containing water using spray extraction equipment	The decontamination rate for DDT and lindane on the surface lies between 50 % and 70 %	The method is only suitable for wood-faced surfaces; no depth effect; PCP can only be reduced to a small extent
Blasting techniques with dry ice	Removal of pesticide-contaminated surface layers or old layers of paint and wax using solid carbon dioxide ("dry ice") in the form of pellets or snow; combination of mechanical and thermal effects	Suitable for the treatment of fixed components, including cracks and joints	The process causes intense noise and dust pollution; risk of suffocation and damage to unprotected areas of skin caused by dry ice particles
Laser ablation	Application of short pulses of high energy, which are absorbed in particular by dark-colored dirt and dust particles; explosive erosion; combination of thermal and mechanical effects	Removal of pesticide-contaminated surface layers; minimal impairment of the carrier material	No depth effect; no thermal degradation of organochlorine biocides (DDT and lindane)

Thermal processes			
Procedures	*Materials/technology*	*Pros*	*Cons*
Microwave technology	Use of portable microwave generators in conjunction with cable-fed horn antennas	Accelerated outgassing of volatile pesticides from stationary components, such as wooden columns, floor timbers, and door frames	The process is not yet ready for application; metal parts and supports in/on the wood, as well as resin-rich woods, are problematic

Thermal processes			
Procedures	*Materials/technology*	*Pros*	*Cons*
Vacuum desorption	Desorption of volatile pesticides from transportable wooden objects by simultaneous application of a negative pressure and elevated temperature in an autoclave	Gentle reduction of highly volatile pollutants	Acceptable decontamination rates are only achieved after longer periods of treatment; the procedure is currently not practiced

Solvent processes			
Procedures	*Materials/technology*	*Pros*	*Cons*
Compress method	Use of solvent-moistened adsorbents (e.g., silica gel, cellulose fibers, and zeolites) on wooden surfaces with organochlorine pesticides	Chamber-free work in an open system is possible; applicable to fixed and variable objects; low cost	Low pesticide desorption with a long treatment time; risk of formation of new pesticide efflorescence on the wood surface as a result of solvent inclusion
Leaching process Solvent: Methyl *tert*-butyl ether (MTBE)	Storage of contaminated objects in a tight-fitting container with MTBE for approximately 4 weeks, followed by rinsing with fresh solvent	No complicated technical equipment and elaborate technologies necessary; old consolidation and wood preservatives can be greatly reduced	The solvent is flammable as a component of premium gasoline; the work must be carried out under an explosion-proof extraction system
Vapor phase extraction Solvent: 1,3-Dioxolan	Leaching of pollutants by conversion of the liquid solvent by negative pressure into the vapor phase in the nitrogen environment (explosion protection)	Gentle reduction of organo-soluble pesticides, oily wood preservatives, and wood consolidation agents	Depending on the wood preservative, its removal can take months or years; penetration depth is low with clogged cell lumens; unsuitable for oil-bound polychrome paints

(Continued)

Appendix 2 Table 1 (Continued)

Solvent processes			
Procedures	*Materials/technology*	*Pros*	*Cons*
Vapor phase extraction Solvent: Dichloromethane (DCM)	Use of a bone fat extraction system for anatomical and zoological specimens	Closed-loop process; high level of decontamination; advise for oily wood preservatives, such as Carbolineum	DCM is a chlorinated hydrocarbon and harmful to health; the solvent has a stripping effect; replacement of the solvent necessary when applying the technology to cultural property
Extraction with liquid CO_2	Use of a pilot plant for the degreasing of industrial goods; the cycle process is carried out in an autoclave at 15–20 °C and 50–60 bar	The objects, which are fixed in a rotating metal basket, are surrounded by CO_2; additional cleaning mechanisms result in a high pesticide reduction without damage to the object	Only smaller, transportable objects with a small cross section can be treated (see wood species collection); wood species with a high resin content are problematic
High-pressure extraction with supercritical CO_2	Extraction at 40°C and 250–350 bar according to the flow or batch principle in an autoclave; formation of a fluid	Closed-loop process with CO_2 recovery; semiporous materials are penetrated quickly and completely; as a result, very high decontamination rates for DDT and lindane (up to 90 %)	Difficult litigation; PCP and inorganic pesticides cannot be extracted in large quantities; at present, there is no facility available for the detoxification of cultural property

Shut-off processes			
Procedures	*Materials/technology*	*Pros*	*Cons*
Sealing with masking agents	Coating (painting, spraying, and rolling) of contaminated wooden components with film-forming preparations based on acrylates, urethanes, or natural resins; the alkaline primers are intended to cause a partial conversion of DDT, lindane, and PCP	Effective prevention of pesticide emissions after prior dust removal and crack treatment	Masking acts for a limited period of time, depending on climatic factors and the resistance to aging of the means are irreversible; they sometimes change the wood surface considerably in terms of color value and gloss

	Shut-off processes		
Procedures	*Materials/technology*	*Pros*	*Cons*
Wrapping with activated carbon fabric	Fabric with a large inner surface area	Adsorbs pesticides, among other things	No manufacturer's information on when saturation level is reached
Coating with pesticide-impermeable films	Aluminum composite foils; ceramic foils	Both foils are impermeable to vapor and water; can be welded together; ceramic foils are translucent and therefore viewable	Aluminum composite foils are opaque and therefore not viewable; relatively high costs
Storage in special showcases or rooms	Showcases with nitrogen generator that keep the oxygen content at <0.3 %	Suitable for individual objects	High costs

References

Tello, Helene; Unger, Achim (2010): Liquid and Supercritical Carbon Dioxide as a Cleaning and Decontamination Agent for Ethnographic Materials and Objects. *Smithsonian Contributions to Museum Conservation*, (1), 35–50.

Unger, Achim (2018): Vortrag. Dekontaminations-Verfahren für biozidbelastetes Kulturgut und ihre Bewertung. Tagung Wood Art Conservation vom 27–29.09.2018. Fachhochschule Bern, Folie 5.

Zalewski, Przemyslaw Paul (2014): Einführung in die Voraussetzungen, Ziele und Strategien des Projektes. Arbeitspolitik und Arbeitsmarkt als Ausgangsvoraussetzungen. In: Zalewski, Przemyslaw Paul; Tello, Helene; Meyer-Haake, Arne (eds.), Biozidbelastete Kulturgüter. Grundsätzliche Hinweise und Texte zur Einführung in die Problematik. Bericht über das EU-/ESF-Projekt "Kleine und Mittlere Unternehmen und Wissenschaft im Dialog. Dekontamination von Kulturgütern. Frankfurt (Oder), 7–10.

Appendix 3

Chronological overview of employees of the Generalverwaltung at the Königliche Museen and from 1918 at the Staatliche Museen zu Berlin in the middle of the 19th century to the beginning of the 20th century

In the following, only the employees of the Generalverwaltung der Königlichen/ Staatlichen Museen zu Berlin und des Chemischen Laboratoriums, who are mentioned by name in this book, are listed in chronological order. On the basis of the sources, not all personal data could be presented completely (Cf. Grabowski et al. 2010, 151; SMB-ZA. Internal employee database, unpublished). In this pyramid-like system, the directors were at the top of the list, headed by the director-general. He was followed by an administrative director, a legal advisor and a member of the board of directors, two chemists, an architect, a librarian, and a custodian for foreign ventures. Among them were the civil servants of the middle and lower ranks. The basis was formed by the museums of the former Königliche Museen and, from 1918, the Staatliche Museen zu Berlin (Anonymous 1924).

Appendix 3 Table 1 Employees of the Generalverwaltung of the former Königliche Museen and, from 1918, of the Staatliche Museen zu Berlin in chronological order.

Person	Time period and function
Ignaz Maria von Olfers (*1793–†1872)	1839–1869 Director-General of the Königlichen Museen zu Berlin
Max Junker (*?–†1925)	1892–1896 Office assistant in the Generalverwaltung der Königlichen Museen zu Berlin
	1896–1920 Secretary in the Generalverwaltung of the Königlichen- und ab 1918 der Staatlichen Museen zu Berlin
	1920–1924 Head of Office/Administrative Chief Secretary of the Staatlichen Museen zu Berlin
Wilhelm von Bode (*1845–†1929)	1905–1920 Director-General of the Königlichen- und ab 1918 der Staatlichen Museen zu Berlin
Kurt Stubenrauch (*? – †?)	1905–? Assessor in the Generalverwaltung of the Staatlichen Museen zu Berlin
Otto von Falke (*1862–†1942)	1920–1927 Director-General of the Staatlichen Museen zu Berlin
Wilhelm Waetzoldt (*1880–†1945)	1928–1934 Director-General of the Staatlichen Museen zu Berlin
Otto Kümmel (*1874–†1952)	1934–1945 Director-General of the Staatlichen Museen zu Berlin

Appendix 4

Chronological overview of the staff in
the Chemisches Laboratorium at the
Königliche Museen and from 1918 at the
Staatliche Museen zu Berlin in the middle
of the 19th century to the beginning of the
20th century

Appendix 4 Table 1 Employees of the Chemisches Laboratorium of the former Königliche
Museen and from 1918 of the Staatliche Museen zu Berlin in chronological order.

Person	Time period and function
Friedrich Rathgen (*1862–†1942)	1888–1927 Head of the Chemisches Laboratorium
Carl Brittner (*1883–†1958)	1907–1909 Auxiliary worker at the Chemisches Laboratorium
	1910 Assistant at the Chemisches Laboratorium
	1911 Directorial assistant at the Chemisches Laboratorium
	From 1920 Custodian at the Chemisches Laboratorium
	From April 1928 Professor and head of the Chemisches Laboratorium

Appendix 5

Chronological overview of employees at the Königliches Museum für Völkerkunde and from 1918 at the Staatliches Museum für Völkerkunde zu Berlin from the middle of the 19th to the beginning of the 20th century

In the following, in chronological order, the employees are mentioned by name and presented in tabular form. After the death of Adolf Bastian, the museum was divided into four departments with as many directorships. From 1934 onward, this structure was dissolved under Otto Kümmel and merged into a director's position. Analogous to the recording of the employees of the Generalverwaltung, not all personal data could be determined completely here either (Grabowski et al. 2010; SMB-ZA. Internal employee database, unpublished; Cf. Hermannstädter 2002; Westphal-Hellbusch 1969; Andree 1969; Cf. Peltz 2017).

Appendix 5 Table 1 Employees of the former Königliches Museum für Völkerkunde and from 1918 the Statatliches Museum für Völkerkunde zu Berlin in chronological order.

Person	Time period and function	
Leopold Freiherr von Ledebur (*1799–†1877)	1829–1873	Director of the Ethnographische Sammlung and Director of the Museum Vaterländischer Altertümer zu Berlin
Adolf Bastian (*1826–†1905)	1873–1905	Acting director of the Ethnographische Sammlung
	1876–1886	Director of the Sammlung Nordischer Altertümer at the Museum für Vor- und Frühgeschichte
	1876–1905	Director of the Königliches Museum für Völkerkunde zu Berlin
Albert Voß (*1837–†1906)	1874–1885	Research assistant in the collection of the Nordische Altertümer
	1886–1906	Director of the Department at the Vorgeschichtliche Altertümer in the Königliches Museum für Völkerkunde zu Berlin
Albert Grünwedel (*1856–†1935)	1882–1883	Directorial assistant, acting, in the Königliches Museum für Völkerkunde
	1883–1904	Deputy Director of the völkerkundliche Sammlungen
	1904–1921	Director of the Vorderasiatische and Indische Abteilung at the Königliches- and from 1918 at the Staatliches Museum für Völkerkunde zu Berlin
	1912–1915	Acting director of the Ozeanische Abteilung at the Königliches Museum für Völkerkunde zu Berlin

(Continued)

Person	Time period and function
Eduard Krause (*1847–†1917)	1884–1917 First conservator at the Königliches Museum für Völkerkunde zu Berlin
Felix von Luschan (*1854–†1924)	1885–1904 Directorial assistant
	1904–1911 Director of the Afrikanisch-Ozeanische Abteilung at the Königliches Museum für Völkerkunde zu Berlin
	1911–1924 Head of the Anthropologische Sammlung at the Königliche- and from 1918 at the Staatliche Museen zu Berlin
Karl von den Steinen (*1855–†1929)	1893–1900 Employee in the Königliches Museum für Völkerkunde zu Berlin
	1900–1903 Assistant director in the Königliches Museum für Völkerkunde zu Berlin
	1904–1906 Head of department of the Amerikanische Sammlungen at the Königliches Museum für Völkerkunde zu Berlin
Bernhard Ankermann (*1859–†1943)	1897–1902 Research assistant in the department (without further description) at the Museum für Völkerkunde zu Berlin
	1909–1916 Assistant in the Ethnologische Abteilung at the Königliches Museum für Völkerkunde zu Berlin
	1916–1921 Custodian with the title of a director in the Ethnologische Abteilung at the Königliches- and from 1918 at the Staatliches Museum für Völkerkunde zu Berlin
	Comment: In 1916 the division of the Afrikanisch-Ozeanische Abteilung was made
	1911–1921 Commissionar administration of the Afrikanische Sammlung at the Königliches- and from 1918 at the Staatliches Museum für Völkerkunde zu Berlin
	1921–1924 Director of the Afrikanische und Ozeanische Sammlung at the Staatliches Museum für Völkerkunde zu Berlin
Eduard Seler (*1849–†1922)	1901–1902 Volunteer in the Königliches Museum für Völkerkunde zu Berlin
	1902–1909 Research assistant in the Königliches Museum für Völkerkunde zu Berlin
	1903–1921 Head of the Amerikanische Abteilung at the Königliches- and from 1918 at the Staatliches Museum für Völkerkunde zu Berlin
	1904–? Director of the Amerikanische Abteilung at the Königliches Museum für Völkerkunde zu Berlin
	1919–? Director of the Nord- und Mittelamerikanische Abteilung at the Staatliches Museum für Völkerkunde zu Berlin (in 1919, the Amerikanische Abteilung was divided into a Nord- und Mittelamerikanische Abteilung)

(*Continued*)

Appendix 5 Table 1 (Continued)

Person	Time period and function	
Albert von Le Coq (*1860–†1930)	1904–?	Research assistant in the Königliches Museum für Völkerkunde zu Berlin
	1923–1925	Director of the Indisch-Asiatische Abteilung at the Staatliches Museum für Völkerkunde zu Berlin
Otto Kümmel (*1874–†1952)	1906–1911	Head of the Ostasiatische Sammlung at the Königliches Museum für Völkerkunde zu Berlin
	1912–1927	Director of the Ostasiatische Kunstsammlung at the Königliches- and from 1918 at the Staatliches Museum für Völkerkunde zu Berlin
	1928–1933	Director of the Asiatische Sammlungen at the Staatliches Museum für Völkerkunde zu Berlin
	1934–1945	Director of the Staatliches Museum für Völkerkunde zu Berlin
Friedrich Wilhelm Karl Müller (*1863–†1930)	1906–1928	Director of the Ostasiatische Abteilung at the Königliches- and from 1918 at the Staatliches Museum für Völkerkunde zu Berlin
	1921–1923	Acting Director of the Indisch-Asiatische Abteilung at the Staatliches Museum für Völkerkunde zu Berlin
Wilhelm Kissenberth (*1878–†1944)	1907–?	Volunteer at the Königliches Museum für Völkerkunde zu Berlin
	1922–1924	Employment in the entwicklungsgeschichtliche Abteilung at the Staatliches Museum für Völkerkunde zu Berlin
	Comment:	The Südamerikanische Sammlung and the Forschungs- und Lehrinstitut were merged to form the department of the Afrikanische, Ozeanische und Amerikanische Sammlung at the Staatliches Museum für Völkerkunde zu Berlin
Carl Schuchhardt (*1859–†1943)	1908–1925	Director of the Vorgeschichtliche Abteilung at the Königliches- and from 1918 at the Staatliches Museum für Völkerkunde zu Berlin
Alfred Schachtzabel (*1887–†1981)	1911–1913	Research assistant at the Königliches- and from 1918 at the Staatliches Museum für Völkerkunde zu Berlin
	1919–1921	
	1924–1945	Head of the Afrikanische Sammlung at the Staatliches Museum für Völkerkunde zu Berlin
	Documented:	Relocation commissionar for the relocation of the collections of the Staatliches Museums für Völkerkunde from Berlin-Mitte to Berlin-Dahlem in 1925
August Eichhorn (*1865–† after 1930)	1916–1929	Head of the Ozeanische Abteilung at the Königliches- and from 1918 at the Staatliches Museum für Völkerkunde zu Berlin
Konrad Theodor Preuß (*1869–†1938)	1921–1934	Director of the Amerikanische Abteilung and head of the Nord- und Mittelamerikanische Abteilung at the Staatliches Museum für Völkerkunde zu Berlin

Person	Time period and function	
Heinrich August Lösekrug (*1887–†?)	1922–1930	Technical assistant in the Abteilung der Afrikanischen, Ozeanischen und Amerikanischen Sammlungen at the Staatliches Museum für Völkerkunde zu Berlin
	1931–1935	Technical assistant in the Staatliches Museum für Völkerkunde zu Berlin
	1936–?	Assistant conservator at the Staatliches Museum für Völkerkunde zu Berlin
	and 1950–1952	
Erich Zorn (*1880–†?)	Juni 1925–?	Technical and scientific assistant in the Staatliches Museum für Völkerkunde zu Berlin
	April 1927–?	Assistant conservator at the Staatliches Museum für Völkerkunde zu Berlin
	1940–1943 and 1950–?	Conservator at the Staatliches Museum für Völkerkunde zu Berlin
Hermann Siebert (*1884–†?)	1926–?	Technical assistant in the Abteilung der Afrikanischen, Ozeanischen und Amerikanischen Sammlungen at the Staatliches Museum für Völkerkunde zu Berlin
	1934–?	Technical assistant at the Staatliches Museum für Völkerkunde zu Berlin
	1950–?	Operational worker at the Staatliches Museum für Völkerkunde zu Berlin
Walter Lehmann (*1878–†1939)	1927–1934	Director of the Afrikanischen, Ozeanischen und Amerikanischen Sammlungen at the Staatliches Museum für Völkerkunde zu Berlin
	1921–1933	Director of the Ethnologisches Forschungs- und Lehrinstitut at the Staatliches Museum für Völkerkunde zu Berlin
	Comment:	In 1927 the Afrikanisch-Ozeanische Abteilung, the Nord- und Mittelamerikanische Abteilung, the Südamerikanische Sammlung, and the Forschungs- und Lehrinstitut were merged to the Abteilung der Afrikanischen, Ozeanischen und Amerikanischen Sammlung at the Staatliches Museum für Völkerkunde zu Berlin
Heinrich Emil Snethlage (*1897–†1939)	1927–1928	Research assistant on contract at the Staatliches Museum für Völkerkunde zu Berlin
	1939–1940	
	1929–1934	Part-time employed at the Staatliches Museum für Völkerkunde zu Berlin
	1934–1935	Full-time employed in the Afrikanischen, Ozeanischen und Amerikanischen Sammlung at the Staatliches Museum für Völkerkunde zu Berlin

(Continued)

Appendix 5 Table 1 (Continued)

Person	Time period and function	
Walter Krickeberg (*1885–†1962)	1929–1946	Head of the Südamerikanische Sammlung at the Staatliches Museum für Völkerkunde zu Berlin
	1934–?	Head of the Amerikanische Abteilung at the Staatliches Museum für Völkerkunde zu Berlin (in 1934 the division of the Amerikanische Abteilung under the leadership of Walter Krickeberg was abolished again)

References

Andree, Christian (1969): Geschichte der Berliner Gesellschaft für Anthropologie, Ethnologie und Urgeschichte. In: Festschrift zum 100jährigen Bestehen der Berliner Gesellschaft für Anthropologie, Ethnologie und Urgeschichte 1869–1969, 9–139.

Anonymous (1924): Direktoralbeamte der Staatlichen Museen am 15. Oktober 1924. Berichte aus den Preussischen Kunstsammlungen. Beiblatt zum Jahrbuch der Preussischen Kunstsammlungen. *Berliner Museen*, XLV. Jhrg, Heft 4, 94.

Grabowski, Jörn; Winter, Petra; Ebelt, Beate; Pilgermann, Carolin (Hrsg.) (2010): Kunst recherchieren. 50 Jahre Zentralarchiv der Staatlichen Museen zu Berlin. Staatliche Museen zu Berlin-Preußischer Kulturbesitz, Ethnologisches Museum. Berlin: Deutscher Kunstverlag.

Hermannstädter, Anita (Hrsg.) (2002): Deutsche am Amazonas—Forscher oder Abenteurer? Expeditionen in Brasilien 1800 bis 1914; Begleitbuch zur Ausstellung im Ethnologischen Museum, Berlin-Dahlem in Zusammenarbeit mit dem Brasilianischen Kulturinstitut in Deutschland. Ethnologisches Museum Berlin; Ausstellung. 2., unveränd. Aufl. Münster: LIT (Veröffentlichungen des Ethnologischen Museums Berlin Fachreferat Amerikanische Ethnologie, Neue Folge 71; 9).

Peltz, Uwe (2017): Das Chemische Laboratorium bis zur Gründung als "Zwillingsinstitute" im geteilten Berlin. *Berliner Beiträge zur Archäometrie, Kunsttechnologie und Konservierungswissenschaft*, (25), 55–94.

Westphal-Hellbusch, Sigrid (1969): 100 Jahre Ethnologie unter besonderer Berücksichtigung ihrer Entwicklung an der Universität. *Festschrift zum 100jährigen Bestehen der Berliner Gesellschaft für Anthropologie, Ethnologie und Urgeschichte 1869–1969*, 166.

Substances Index

General Index